CONTENTS

FAST TRACK TO A 5

Preparing for the AP*
Statistics Examination

To Accompany
Statistics: Learning from Data
1st Edition
by Roxy Peck and Chris Olsen

Stephen R. Dartt
Shiloh High School, Snellville, Georgia

Vicki Greenberg
Kennesaw State University, Kennesaw, Georgia

Viva Hathaway
Norview High School, Norfolk, Virginia

Ed Moulton

Brendan Murphy
John Bapst Memorial High School, Bangor, Maine

BROOKS/COLE
CENGAGE Learning·

*AP and Advanced Placement Program are registered trademarks of the College Entrance Examination Board, which was not involved in the production of, and does not endorse, this product.

BROOKS/COLE
CENGAGE Learning·

Holt McDougal is pleased to provide Cengage Learning college-level materials to high schools for Advanced Placement*, honors, and college-prep courses. Our Advanced & Elective Programs department is dedicated to serving teachers and students in these courses. To contact your Advanced & Elective Programs representative, please call us toll-free at **1-800-479-9799** or visit us at **www.HoltMcDougal.com**

For permission to use material from this text or product, submit all requests online at **www.cengage.com/permissions** Further permissions questions can be emailed to **permissionrequest@cengage.com**

ISBN-13: 978-1-285-09464-9
ISBN-10: 1-285-09464-6

Brooks/Cole
20 Channel Center Street
Boston, MA 02210
USA

Cengage Learning is a leading provider of customized learning solutions with office locations around the globe, including Singapore, the United Kingdom, Australia, Mexico, Brazil, and Japan. Locate your local office at: **www.cengage.com/global**

Cengage Learning products are represented in Canada by Nelson Education, Ltd.

To learn more about Brooks/Cole, visit **www.cengage.com/brookscole**

Purchase any of our products at your local college store or at our preferred online store **www.cengagebrain.com**

*AP and Advanced Placement Program are registered trademarks of the College Entrance Examination Board, which was not involved in the production of, and does not endorse, this product.

Printed in the United States of America
1 2 3 4 5   20 19 18 17 16

ABOUT THE AUTHORS

STEPHEN R. DARTT has taught AP statistics for the last fifteen years at Shiloh High School in Snellville, Georgia. He has been involved in grading AP statistics exams as a reader for seven years and as a table leader for six years. Stephen's education includes a bachelor's degree in Chemical Engineering from Princeton University and a master's degree in the same field from Massachusetts Institute of Technology. After working as a chemical engineer for 19 years, Stephen has spent the last 24 years teaching mathematics and chemistry at Shiloh High School.

VICKI GREENBERG currently teaches mathematics at Kennesaw State University in Kennesaw, Georgia. Previously she taught AP Statistics at Woodward Academy, where she served as Assistant Director for the school's AP Summer Institutes. She holds a bachelor's degree in Mathematics from the University of North Carolina at Greensboro and a master's degree in Mathematics Education from Georgia State University. Vicki has been teaching mathematics for 12 years and AP Statistics for 4 years. She is a reader for the AP Statistics exam and a regular presenter at local and regional AP Summer Institutes. Her educational interests include creating hands-on and technology-infused activities to enhance her students' understanding of mathematics and statistics. In her spare time, Vicki enjoys photography, travel, baseball, football, and playing with her two cats and two dogs.

VIVA HATHAWAY has taught AP Statistics at Norview High School in Norfolk, Virginia for the past 10 years of her 20-year career teaching high school mathematics. She received both bachelor's and master's degrees in mathematics from Bowling Green State University in Ohio. Viva has served as a reader for the AP Statistics exam and routinely makes presentations on AP Statistics at the annual meeting of the National Council of Teachers of Mathematics. In addition, she teaches statistics as an adjunct professor for Saint Leo University. Viva is also a recipient of the *Presidential Award for Excellence in Mathematics and Science* for the state of Virginia. Besides teaching statistics and geometry at Norview High School, Viva serves on local and national committees regarding mathematics in secondary education.

ED MOULTON is presently pursuing a masters degree in statistics at Temple University in Philadelphia. Previously, he was a teacher of dual enrollment calculus, AP Calculus support, and Honors Algebra 2 at Norview High School in Norfolk, Virginia. Before going to Norview High School, he co-taught the AP Statistics Class at Dunbar High School in Baltimore, Maryland. Ed holds a bachelor's degree in Mathematics from Duke University and a master's degree in Arts and Teaching from Johns Hopkins University.

BRENDAN MURPHY, a teacher at John Bapst Memorial High School in Bangor, Maine, was the 2007 Awardee for the Presidential Award for Excellence in Mathematics and Science Teaching in Maine. In 2005, he was a finalist for the Maine Teacher of the Year and a Siemens Award Winner for Advanced Placement. Brendan has run weeklong summer institutes in AP Calculus, AP Statistics, and Vertical Teams/Pre-AP Math, and has presented numerous workshops around the country. He also ran a mentoring program for new AP Statistics teachers across Maine for the Maine Department of Education. Brendan is currently a Table Leader/Rubric Team for the AP Statistics exam and is also an approved reader for AP Calculus.

PREFACE

The field of statistics is relevant to nearly every career you will encounter. As such, most college majors will require at least a cursory knowledge of statistics. Statistics is an integral tool used in the fields of medicine, genetics, business, mathematics, engineering, psychology, sociology, education, political science, as well as a vast majority of other careers not mentioned. Because of this, we believe this course is critical for well-educated people to be successful in their careers.

Dealing with statistics in each field of study can vary depending on the subject matter. However, the basic statistical principles are the same in all fields and all majors. Thus the study of AP Statistics will allow you the unique opportunity to be well prepared for virtually any college major you choose. The statistician uses mathematics, but the science of statistics is found in what happens once a set of values is calculated. After these numerical values are obtained, the statistician tries to answer the question "what does this mean" by delving into connections between what one might expect to see in similar scenarios. Through the scientific use of experimental design, data analysis, probability theory, and inference testing, our decisions become critical to a vast number of careers in our world.

We hope this guide will help you successfully navigate the waters of an introductory statistics course as well as give you the ability to apply these concepts to your future studies. Hopefully, some of you will find such an interest in this field that you consider changing your career plans to become a statistician! We trust that you will find this guide to be a useful aid in your preparation for and success on the AP Statistics exam.

Stephen R. Dartt
Vicki Greenberg
Viva Hathaway
Edward Moulton
Brendan Murphy

ACKNOWLEDGEMENTS

I would like to thank Roxy Peck, who taught all of us how to grade AP statistics exams and who promised to dedicate her "next" book to me. I would also like to thank Christine Franklin and Allan Rossman for their outstanding leadership and for making it possible for me to be an ongoing part of the AP statistics program. Thanks also to Jeane Swaynos for her valuable input as a reviewer of the text.

Stephen Dartt

I would like to thank my wonderful husband for his support and encouragement during the writing of this book. Thank you also to my mentor, Paul Myers, for helping me to discover my love for statistics and continuously challenging me to grow as a teacher and person. In addition, I greatly appreciate the support, suggestions, patience, organizational skills, and occasional nudges of John Haley and Margaret Lannamann.

Vicki Greenberg

I would like to dedicate this to my four wonderful granddaughters who keep me moving with their excitement. I want to thank my husband, Bill, for always pushing me to go beyond my comfort zone. Without this encouragement I most likely would never have fallen in love with statistics in the first place. Thanks to my colleague Linda for keeping me real in analyzing data and my classes. Finally, without my mentor, Bill Speer, I would most likely not be as involved in mathematics. Thanks to you all! You're great!

Viva Hathaway

I would like to acknowledge my late father and thriving mother, both of whom dedicated their lives to teaching undergraduate math and computer science. My thanks also to the reviewers, Brendan Murphy and Chris Olsen, the editors, and the other professionals who contributed to the publication of this work.

Ed Moulton

Special thanks to my wife Jennifer for all her support, encouragement, and technical expertise over the years; without her none of this would be possible. Thanks as well to all the wonderful folks who I've met at the AP Statistics Readings and at my APSI's, thank you for sharing all of your expertise and helping me to understand the statistics on a deeper level. Also thanks to reviewer Jeane Swaynos for her insightful comments. Cheers!!

Brendan Murphy

Part I

Strategies for the AP Exam

PREPARING FOR THE AP* STATISTICS EXAMINATION

Advanced Placement Statistics is an exciting, practical course that is also very challenging academically. It is designed to help the student delve deeply into the methods and concepts of the appropriate use and analysis of research. Whether you are taking AP Statistics in a classroom or online, this year will prove to be both challenging and intellectually stimulating as you progress in your study of statistical analysis, inference, and experimental design.

The closer the school year gets to May, the more intimidating the looming Advanced Placement test may seem. Learning how to take all the information you have learned this year and apply it in a way that will demonstrate your expertise can be very overwhelming at times.

The best way to achieve success on any AP exam is to master it, rather than let it master you. If you think of these tests as a way to demonstrate how well your mind works, you will have an advantage—attitude does help.

This book is designed to put you on a fast track to a successful score. *Focused* review and practice time will help you master the examination so that you can walk in better prepared, more confident, and ready to do well on the test.

WHAT'S IN THIS BOOK

This book is keyed to *Statistics: Learning from Data,* AP edition, by Roxy Peck and Chris Olsen. This new text includes AP Tips that help students more easily remember and master key topics as well as AP Questions that give students practice in solving the types of problems they will encounter in the AP Statistics Exam. It also features expanded coverage of sampling, probability, and experimental design, and the addition of topics including cumulative relative frequency plots, non-linear models and transformations, and power and probability of type II error. The Technology sections in each chapter include the use of the graphing calculator. Because this book follows the College Board Topic Outline for AP Statistics, it is compatible with other statistics textbooks as well; we have designed this book to help students prepare for this exam regardless of the text being used. This book is divided into three sections. Part I offers suggestions for preparing for the exam, from signing up to take the test and having the correct calculators to writing a complete response for each free-response question. Part I ends with a diagnostic test that will help you determine which sections may need

---

more focus as you prepare for the exam. The diagnostic test is designed in the same format as the Advanced Placement Examination, with 40 multiple-choice questions and 6 free-response questions that follow the College Board Topic Outline.

The diagnostic test should help you identify which areas you need the most practice on; to make this easier for you, the content areas for each problem are noted. In reviewing the answers to the diagnostic test, you will be able to recognize gaps in your knowledge by noting when there are groups of questions from the same content area that gave you trouble. Page references with each answer will allow you to go to the textbook and review the content area where you need to focus your attention. You will also be able to analyze your responses to the free-response questions to identify weaknesses that may exist in conveying your written response to the AP Statistics test readers.

Part II consists of 12 review sections that encompass the 4 content areas covered by the AP Statistics examination. Each of these corresponds to a key topic in the College Board content outline. The percentages below indicate the approximate proportion of exam questions devoted to each of the 4 content areas.

Review Section 1	Exploring Data	Exploratory Analysis 20–30%
Review Section 2	Numerical Methods for Describing Data	
Review Section 3	Summarizing Bivariate Data	
Review Section 4	Collecting Data Sensibly	Planning & Conducting a Study 10–15%
Review Section 5	Probability	Probability 20–30%
Review Section 6	Random Variables and Probability Distributions	
Review Section 7	Sampling Distributions	
Review Section 8	Estimation Using a Single Sample	Statistical Inference 30–40%
Review Section 9	Hypothesis Testing Using a Single Sample	
Review Section 10	Comparing Two Populations or Treatments	
Review Section 11	Categorical Data and Goodness of Fit Tests; Chi-Square Tests	
Review Section 12	Inference for Linear Regression and Correlation	

The review sections are **not** intended to be a substitute for a textbook and class discussions; they offer review and help you prepare further for the exam. Each review section also has the textbook sections listed for further review if, as you prepare for the exam, you find a topic that needs more study. At the end of each review section, you will find 15 multiple-choice and 2 free-response questions based on the content of that section. Again, you will find page references with each answer directing you to the appropriate discussion on each point in *Statistics: Learning from Data*.

Part III offers two complete AP Statistics examinations. At the end of each test, you will find the answers, explanations, and text page references for the 40 multiple-choice and 6 free-response questions. Part IV contains formulas and tables useful for solving the problems you will encounter in the AP Statistics exam.

SETTING UP A REVIEW SCHEDULE

If you have been steadily doing your homework and keeping up with the coursework, you will be in good shape. The key to preparing for the examination is to begin as early as possible; don't wait until the exam is just a week or two away to begin your studying. But even if you've done all that—or if it's too late to do all that—there are other ways to pull it all together.

To begin, read Part I of this book. You will be much more comfortable going into the test if you understand how the test questions are designed and how best to approach them. Then take the diagnostic test and see where you stand.

Set up a schedule for yourself on a calendar. If you begin studying early, you can chip away at the review sections in Part II. Reviewing the AP Tips in the text will help you master key concepts more quickly and easily. You'll be surprised—and pleased—by how much material you can cover if you devote a half hour per day of study for a month or so before the test. Look carefully at the sections of the diagnostic test; if you missed a number of questions in one particular area, allow more time for the review sections that cover that area of the course. The practice tests in Part III will give you more experience with different kinds of multiple-choice questions and the wide range of free-response questions.

If time is short, reading the review sections may not be your best course of action. Instead, skim through the review sections to re-familiarize yourself with the main ideas. Spend the bulk of your time working on the multiple-choice and free-response questions at the end of each review. It also would be helpful to work through some of the AP Questions in the text. This will give you a good idea of your understanding of that particular topic. Then take the tests in Part III.

If time is really short, go straight from Part I to Part III. Taking practice tests repeatedly is one of the fastest, most practical ways to prepare.

BEFORE THE EXAM

By February, long before the exam, you should make sure that you are registered to take it. Many schools take care of the paperwork and handle the fees for their AP students, but check with your teacher or the AP coordinator to be certain that you are on the list. This is especially important if you have a documented disability and need test accommodations. If you are studying AP independently, call AP Services at the College Board for the name of an AP coordinator at a local school who will help you through the registration process.

The evening before the exam is not a great time for partying—nor is it a great time for cramming. If you like, look over class notes or skim through your textbook, concentrating on the broad outlines rather than the small details of the course. You might also skim over the AP Statistics approved formula sheets one last time to remind yourself of the formulas you can check during the test if you need to confirm your idea, and skim through the book and read the AP Tips. This is a great time to get your things together for the next day. Sharpen a fistful of number 2 pencils with good erasers; be sure your calculator is in good working order with either a spare or extra batteries in case you have a malfunction of some sort; check out your watch and make sure the alarm is off if it has one; package a healthy snack for the break like a piece of fruit, granola bar, and bottled water; make sure you have whatever identification is required as well as the admission ticket. Then relax, and get a good night's sleep.

On the day of the examination, it is wise to eat breakfast—studies show that students who eat a healthy breakfast before testing generally do better. Breakfast will give you the energy you need to power you through the test—and more. Remember, cell phones and other electronic devices, other than your approved calculator(s), are not allowed in the testing room. You will spend some time waiting while everyone is seated in the right room for the right test. That's before the test has even begun. With a short break between Section I and Section II, the statistics exam lasts well over 3 hours. Be prepared for a long test time. You don't want to be distracted by hunger pangs.

Be sure to wear comfortable clothes; take along a sweater in case the heating or air-conditioning is erratic.

You have been on the fast track. Now go get a 5!

TAKING THE
AP STATISTICS EXAM

The AP Statistics exam has two sections: Section I consists of 40 multiple-choice questions; Section II contains six free-response questions. You will have 90 minutes to complete the multiple-choice portion. The exam is then collected, and you will be given a short break. Then you have 90 minutes for the free-response questions. You should answer all six questions. Some AP exams allow you to choose among the free-response questions, but statistics is *not* one of them. Keep an eye on your watch and devote about 13 minutes to each of questions 1–5 and about 25 minutes on question number 6. Since question number six carries more weight on your overall score, you want to be sure to at least try the first couple of parts of this question. Please note that watch alarms are *not* allowed.

Below is a chart to help you visualize the breakdown of the exam:

Section	Multiple-Choice Questions	Six Free-Response Questions	
Weight	50% of the exam	50% of the exam (weighted as follows)	
Number of Questions	40	Questions 1–5 37.5% of exam	Question 6 12.5% of exam
Time Allowed	90 minutes	90 minutes	
Suggested Pace	2 minutes per question	13 minutes per question	25 minutes for this question

STRATEGIES FOR THE MULTIPLE-CHOICE SECTION

Here are some rules of thumb to help you work your way through the multiple-choice questions:

SCORING OF MULTIPLE CHOICE   There are five possible answers for each question. As of June 2010, there is no longer a penalty on the AP Statistics exam for wrong answers. You will simply not get credit for the wrong answer. Since there is no penalty for guessing, be sure to answer every question prior to the end of the 90 minutes allotted. This means, fill in a bubble for each of the 40 questions.

FIND QUESTIONS YOU KNOW FIRST   Find questions you are confident of and work those first. (The easier questions generally appear early in the exam.) Then return to the questions you skipped. Make a mark in the booklet on questions you are unsure of, then return to those questions later.

READ EACH QUESTION CAREFULLY Pressed for time, many students make the mistake of reading the questions too quickly or merely skimming them. By reading each question carefully, you may already have some idea about the correct answer. You can then look for that answer in the responses. Careful reading is especially important in EXCEPT questions (see the next section that describes the types of multiple-choice questions).

ELIMINATE ANY ANSWER YOU KNOW IS WRONG You can write on the multiple-choice questions in the test book. As you read through the responses, draw a line through every answer you know is wrong. This will help you in choosing correct solutions on questions you aren't sure about.

READ EACH RESPONSE, THEN CHOOSE THE MOST ACCURATE ONE AP examinations are written to test your precise knowledge of a subject. Sometimes more than one answer seems to be correct, but one of them is more specific and therefore the correct response.

AVOID ABSOLUTE RESPONSES These answers often include the words "always" or "never." For example, the statement "the data **are** normal" is an absolute answer. Instead, look for the phrase, "the data appear to be approximately normal."

MARK AND SKIP TOUGH QUESTIONS If you are hung up on a question, mark it in the margin of the question book. You can come back to it later if you have time. Make sure you skip that question on your answer sheet as well. In the end, be sure to answer ALL questions prior to the end of the 90 minutes.

TYPES OF MULTIPLE-CHOICE QUESTIONS

There are various kinds of multiple-choice questions. Here are some suggestions for how to approach each one.

CLASSIC/BEST-ANSWER QUESTIONS

This is the most common type of multiple-choice question. It simply requires you to read the question and select the correct answer. For example:

1. A firm claims that the percent of workers who express job satisfaction is higher than the 58% under the previous supervisors. Which of the following are the correct hypotheses to test this claim?
   (A) $H_0 : \mu = 0.58, H_a : \mu > 0.58$
   (B) $H_0 : p = 0.58, H_a : p > 0.58$
   (C) $H_0 : \mu = 0.58, H_a : \mu < 0.58$
   (D) $H_0 : p = 0.58, H_a : p < 0.58$
   (E) $H_0 : p > 0.58, H_a : p < 0.58$

Answer: B. This is the only correct answer as the data is categorical. The claim is about a population percentage, which is equivalent to a claim about a population proportion. The null is written as an = statement and the alternative is > because they claim the percentage has gone up since the change in supervisors.

EXCEPT QUESTIONS

In EXCEPT style questions, you will notice all of the answer choices but one are correct. The best way to approach these questions is to treat them as true/false questions. Mark a T or an F in the margin next to each possible answer. There should be only one false answer, and that is the answer you should select. For example:

1. A random sample selected from a population is helpful for all of the following reasons except
   (A) tends to produce a group that is representative of the population.
   (B) avoids selection bias.
   (C) ensures that all members of the population have an equal chance of selection.
   (D) allowing generalization from the sample to the population.
   (E) eliminates response bias.

Answer: E. Response bias is a type of bias that comes about because of the way a survey question is written or the way in which it is asked, leading the subjects to a certain response. Selecting a random sample will not eliminate this type of bias.

ANALYSIS/APPLICATION QUESTIONS

These questions will require you to apply your knowledge of statistics to a given situation or problem. For example:

1. A farmer wants to see which of two types of corn seed (seed A and seed B) will yield more corn. He has 52 plots marked off and he can plant either type of seed in any of the plots. Half of the plots (those on the west side) are bordered by a forest. The plots on the east side are not bordered by forest. Which of the following methods is the best way to decide how the two seed types should be assigned to the 52 plots?
   (A) Number the plots 01-52 and put these numbers in a hat. Pull out 26 numbers to determine the plots used for seed A. The remaining plots get seed B.
   (B) Number the plots 01-52 and go to a random digits table and locate two digit numbers. The plots corresponding to the first 26 numbers get seed A and the rest will get seed B.
   (C) Plant the 26 plots on the west side with seed A and the 26 east side plots with seed B.
   (D) Divide the 52 plots into two groups consisting of the 26 near the forest and the 26 not near the forest. Number the plots near forest 00-25 and put these numbers in a hat. Pull out 13 numbers to determine which plots would be planted with seed A; the remaining 13 plots would be planted with seed B. Repeat this process with those plots not near forest.
   (E) Pair up a plot near the forest with one not near the forest. Number them 1 and 2 and put numbers in a hat. First number

out gets assigned seed A and the other seed B. Repeat for all pairs.

Answer: D. This is the only correct solution because the forest may influence the results. So in this case, half of the plots near the forest will get seed A and half will get seed B. The same will occur for those away from the forest. Answer E might seem like a reasonable approach at first, but it is possible that most of the plots near the forest could have the same seed type with this method.

FREE-RESPONSE QUESTIONS

There are six free-response questions that should all be answered. These six questions account for 50% of the final score you receive on this exam and are weighted as follows. The first five free-response questions combine to make up 75% of your free-response score and the sixth question, the investigative task, will account for the remaining 25% of the grade. This means the first five make up 37.5% and the investigative task makes up 12.5% of the entire test. So it is important to answer all questions as completely as possible.

These questions can cover multiple content areas. Each of the four major content areas of the AP syllabus will be found in the free-response section. The following hints may help you prepare for this half of the exam.

You should start by quickly reading the six questions. As you are reading, you may want to make a quick note as to an idea you have on the subject of each question. Once you have done this, you should go to the easiest question you find in the first five questions. Answer it as completely as you can and then jump to question number six, the investigative task.

Now spend a little time, but no more than about 15 minutes, working on the investigative task. Since this carries ¼ of the weight for this portion of the test, it is very important that you take the time to at least attempt this task. While this task is intended to stretch you beyond the course, typically it will begin with one or two parts that should be familiar to an AP statistics student.

You should then go back and answer the remaining four questions. If time permits, you should reread and note any key phrases or points that are made in a given question. Once you have answered the questions, be sure you read the question again to ensure you have actually answered the question asked. Examination readers comment that sometimes students don't actually answer the question that is asked. While the information you give may be correct, if it doesn't answer the question, you won't get credit.

Don't assume any question is asking for a "cookie cutter" type response. The AP readers who will review your answer will be reading for a correct answer in the context of the question, and not just a memorized phrase or explanation. In that light, also be sure to use statistical vocabulary correctly. If you are not sure about the correct vocabulary, it is permissible to use a short phrase in place of the word you are unsure about. Correct notation is also important. It won't hurt to define your notation for clarity, but be sure you are using the notation correctly. For example, students may lose credit for using notation for a statistic when they should be using parameter notation and vice-versa.

You will have 90 minutes for this section of the exam, so watch your time. As a recommendation from the College Board, you should allow about 12 minutes for each of the first 5 questions and 25 minutes for the investigative task. Then use the remaining 5 minutes to read over your responses. However, earlier we recommended only about 15 minutes to begin with for the investigative task. This allows you to plan once you have begun answering the last question. In either case, it is important you take a watch and keep a close eye on your time. You want to be sure you have enough time to answer the questions you know and that you don't get bogged down on any one question.

KEY POINTS TO ANSWERING FREE-RESPONSE QUESTIONS

When answering the free-response questions, it is important to note the word choices used in the questions. Focus your writing for the AP Statistics Exam with these points in mind.

READ Carefully read the question and circle key phrases and comments you are to address. Watch for the specific wording and don't make assumptions about what you are being asked before completely reading the question.

CONTEXT Be sure to answer in the context of the problem. For a complete answer you are required to write the conclusions and interpretations in context.

DEFINE You are expected to define any non-standard symbols you use.

CORRECT NOTATION Be clear on parameter vs. statistic notation when answering. For example, when discussing a mean value, use the correct population and sample notation in your written responses. A sample mean notation $\bar{x}$ will be considered incorrect if you are supposed to use a population mean notation μ.

VOCABULARY Use statistical vocabulary correctly. If you are unsure of the meaning of a statistical term, write a couple sentences and either define the term or—better yet—let the sentences describe the concept and don't use the word.

CHECK ASSUMPTIONS Unless the question states that all assumptions are reasonable or have been checked, you should state and check assumptions whenever possible. Assumptions need to be checked or referred to as uncheckable when carrying out any hypothesis test or constructing any confidence interval.

ANSWER Did you actually answer the question? Be sure you have answered the question that was asked.

INTERPRET Be ready to justify your work in a variety of statistical ways. Know how to read printouts, graphical displays, and tables. Provide correct analysis and comparisons as required.

SCORING FREE-RESPONSE QUESTIONS

All six free-response questions will be scored using a holistic four-point scale. However, as stated earlier, question six will be weighted more heavily than questions one through five. The four-point score is holistic in nature. This means the readers will consider your entire solution in addition to each of the parts. Therefore, it is important that you answer the question and not contradict yourself. You should NOT attempt more than one solution; the reader will read each solution and your score will be that of your weakest solution.

Keep in mind that both statistical knowledge and verbal communication are essential on the AP Statistics exam. To score a four on a question, you must demonstrate that you are able to both correctly compute statistical values as well as properly interpret those values in the context of the question. Here are some general guidelines regarding the differences between the scores of zero to four on any free-response question. More specific guidelines can be found on the College Board website (http://www.collegeboard.com/student/testing/ap/sub_stats.html?stats) where you can download the PDF file of the course description.

Score	**Statistical Knowledge** (Be able to correctly use the statistical ideas and techniques.)	**Communication** (Demonstrate a clear and concise explanation of what work you performed and why.)
4	Demonstrates a complete understanding of the statistics and is able to perform correct statistical calculations. Even if you make minor arithmetic errors, the solution is plausible.	Your thoughts and explanations are correct, with all aspects of the explanation utilizing correct procedures, terminology, and appropriate conclusions, etc.
3	Able to demonstrate clear understanding of appropriate techniques but may have some minor gaps. Overall, this solution is substantially correct even if there are some mathematical errors.	Overall, the rationale in this solution is correct but there may be some missing assumptions, caveats, or even a bit unfinished in the final work. While there is a conclusion, it may be incomplete.
2	Has some parts correct but others have either misused or even possibly unreasonable answers. There is some understanding of the pieces yet little relationship demonstrated between them.	Vague or inappropriate explanations. May be missing components or diagrams needed for the explanation to be complete. Conclusions that are drawn are not complete.

	Statistical Knowledge	Communication
1	Misuses the statistical components and develops unreasonable solutions. Very limited knowledge of statistics given. *(A score of 1 may be awarded if the reader can identify something relevant to the question posed even if you have not answered the question.)*	Response is jumbled, messy, unclear, or difficult to follow. It may not match the statistical work provided by the student. Fails to provide a conclusion or answer the question posed. Incorrect diagrams or displays.
0	Very little, if any, understanding of the statistical pieces needed for the problem	No strategy that will lead to a solution is communicated.

FREE-RESPONSE SAMPLE QUESTION

Free response questions can differ in the number of parts as well as how much each part contributes to the overall question. Carefully analyze your final answer to make sure you have made a complete response by answering all of the parts of the question.

Be careful **not** to include parallel solutions. The AP readers are instructed to score the weakest solution, so try not to lose points you have already scored because you doubt yourself. Be confident in your choice and cross out any dual answers that you don't want considered.

Here is an example of a free response question that addresses hypothesis testing.

1. A party rental company rents out tents for large parties. These tents must be set-up prior to the party. The company usually allows 28 minutes for set up, but is now wondering if the mean time required to set up a tent is actually greater than 28 minutes. To answer this question, the company selects 34 past rentals at random from all its tent rentals and notes the time required to set up the tent for each of the selected rentals. The average set-up time for the sample of rentals was 35 minutes and the sample standard deviation was 6 minutes. Do these data provide convincing evidence that the mean tent set-up time is greater than the 28 minutes that the company allows?

Hypothesis test questions can usually be answered in four parts.

Part I—State the hypothesis.

$H_o : \mu = 28$ minutes, $\mu =$ average tent set-up time

$H_a : \mu > 28$ minutes

$\alpha = .05$ Since no alpha level given, we will use .05

Part II—Identify and test and any conditions or assumptions that must be met.

> Because the hypotheses are about a population mean and the population standard deviation is unknown, we consider a one-sample t test.
>
> Assumptions
> **Simple random sample**—The 34 rentals were selected at random from all of the company tent rentals.
> **Large sample or normal population distribution**—the sample size is large ($n > 30$).

Part III—Calculations

$$t = \frac{32 - 28}{\frac{8}{\sqrt{34}}} = 2.915, \text{ with } df = 33$$

$$p = .003$$

Part IV—Conclusion: Give your conclusion **both** in context and linked to alpha.

> Because .003 < .05, the null hypothesis is rejected. There is convincing evidence that the mean set-up time for a party tent is greater than the 28 minutes allowed by the company.

The usual scoring guidelines for a hypothesis test question classifies each of the four parts in the test as essentially correct, partially correct, or incorrect. A response that is judged to be essentially correct on all four parts would receive a score of 4.

More examples of free-response type questions and scoring guidelines are included in the sample exams and review sections that appear later in this book.

A DIAGNOSTIC TEST

This diagnostic test will give you some indication of how you might score on the AP Statistics Exam. Of course, the exam changes every year, so it is never possible to predict a student's score with certainty. This test will also pinpoint strengths and weaknesses on the key content areas covered by the exam.

AP Statistics Examination
Section I: Multiple-Choice Questions
Time: 90 minutes
Number of Questions: 40

Directions: Each of the following questions or incomplete statements is accompanied by five suggested answers or completions. Select the one that best answers the question or completes the statement.

1. If the average test score for one class of 30 students is 75.6 and the average test score for another class of 24 students is 68.4, what is the overall average for this test for all 54 students?
   (A) 71.5
   (B) 72
   (C) 72.4
   (D) 72.8
   (E) 74

2. The variable X has a normal distribution with a mean of 50. What standard deviation would be necessary to ensure that 80% of the values of X are within 12 of the mean?
   (A) 4.687
   (B) 9.375
   (C) 2.344
   (D) 7.129
   (E) 14.259

3. A simple random sample of 150 students at a local high school was taken to estimate the proportion who like strawberry-flavored licorice. Ninety-five of the students in the sample liked strawberry licorice. What would be an appropriate set-up for a 90% confidence interval?

   (A) $0.63 \pm 1.96\sqrt{\dfrac{0.63(0.37)}{150}}$

   (B) $0.63 \pm 1.96\sqrt{\dfrac{0.63(0.37)}{95}}$

   (C) $0.63 \pm 1.645\sqrt{\dfrac{0.63(0.37)}{150}}$

   (D) $0.63 \pm 1.645\sqrt{\dfrac{0.63(0.37)}{95}}$

   (E) $0.63 \pm 1.645\sqrt{\dfrac{0.63(0.37)}{149}}$

4. Amy is graduating in the spring from Howard University with a degree in accounting. AmeriGroup has offered her a job with a starting salary of $48,000 per year. How many standard deviations below or above the mean is her starting salary if the mean and standard deviation for beginning level accounting positions are $47,500 and 1470 respectively?
   (A) 0.17
   (B) 0.45
   (C) 0.34
   (D) 0.61
   (E) 32.65

5.
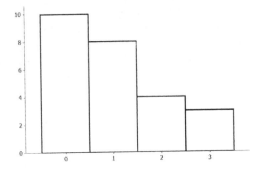

Data set A
Data set B

   Which of the following must be true for two boxplots shown? (Although the scale is not shown, the two boxplots are drawn using the same numerical scale.)
   (A) The IQR of data set A is smaller than the IQR of data set B.
   (B) The lower quartile of data set A is greater than the upper quartile of data set B.
   (C) The range of data set A and data set B are equal.
   (D) The median of data set B is greater than the median of data set A.
   (E) There are more data in data set A than in data set B.

6. The given histogram was constructed using data on X = number of pets for a sample of 25 students. Which of the following is closest to the value of the mean of this data set?

   (A) 0.5
   (B) 1.0
   (C) 1.5
   (D) 2.0
   (E) 3.0

7. To determine how students at a particular college feel about cigarette smoking in public places, all students at the college who chose to have their email address published in the college directory were sent an email with a link to an online survey. What is wrong with this sampling method?
   (A) Nothing, since students could go online to give their opinion.
   (B) Only smokers should have been surveyed.
   (C) They should have only sent the email to every 5th student in the directory.
   (D) This sampling method would result in a sample size that is too large.
   (E) Not all students would have an email listed in the student directory.

8. A cell phone company wants to know how happy its customers are with their service. The company sent a text message with a survey to every customer's phone. Out of the 100,000 customers, 42,000 replied. Based on the 42,000 responses, they concluded that their customers were highly satisfied. What type of bias should the company be worried about?
   (A) Selection bias
   (B) Response bias
   (C) Nonresponse bias
   (D) Measurement bias
   (E) There is no need to worry since all customers received the text message.

9. For employees of a large company, the correlation between X = years of experience and Y = pay increase is found to be $r = 0.74$. What would the correlation be if the Y variable was years of experience and pay increase was the X variable?
   (A) −0.74
   (B) −0.28
   (C) 0.28
   (D) 0.74
   (E) 1.41

10. The following computer output was taken from a data set where the adult height was used to predict the wrist circumference.

	Coefficients	Standard Error	t Stat	P-value
Intercept	−2.13	0.4836	−4.40	0.048
Height	0.13	0.0073	17.3	0.003

R Square	0.993
Standard Error	0.082

Which of the following is the equation of the least squares regression line for this data set?
(A) $\hat{y} = 0.13 - 2.13 \,(height)$
(B) $\hat{y} = -2.13 + 0.13 \,(height)$
(C) $\hat{y} = 0.08 - 2.13 \,(height)$
(D) $\hat{y} = 0.08 + 0.13 \,(height)$
(E) $\hat{y} = 0.99 + 0.13 \,(height)$

GO ON TO NEXT PAGE

11. A study found the correlation between the heights of men and the heights of their biological sons to be $r = 0.71$. What is the approximate value of r^2 and what does this value tell you?
    (A) $r^2 = 0.71$, it describes how well the least squares regression line fits the data.
    (B) $r^2 = 0.50$, it says that a son's height should be equal to 0.50 × (father's height).
    (C) $r^2 = 0.84$, it is the ration of father's height to son's height.
    (D) $r^2 = 0.50$, it is the proportion of the variability in sons' heights that can be explained by the approximate linear relationship between fathers' heights and sons' heights.
    (E) $r^2 = 0.84$, it is the proportion the son's height that is caused by father's height.

12. All but one of the following statements contain an error. Which of the following statements could be correct?
    (A) A social scientist found a correlation of $r = -1.23$ between the weight of monkeys and their height.
    (B) The correlation coefficient for X = political affiliation and Y = income was computed to be $r = 0.73$.
    (C) A psychology graduate student doing research found the correlation coefficient between gender and race to be $r = -0.81$.
    (D) The correlation coefficient for X = amount of fertilizer and Y = yield was found to be $r = 0.58$.
    (E) The correlation coefficient between X = the speed of a racing boat and Y = fuel consumption is found to be $r = 5.0$ mpg.

13. A new type of flame retardant created for clothing is ready for testing. There are two types of fabric that this retardant is to be tested on to see if it is better than the current substance used on these fabrics. There are 100 pieces of each type of fabric available for testing. Which of the following is the best method for assigning retardants to pieces of fabric?
    (A) Assign 100 pieces of each fabric to be tested with the new material.
    (B) Take one type of fabric first and number the 100 pieces 00–99. Using a random number table, find 50 different two-digit random numbers. The pieces of fabric corresponding to these numbers would get the new retardant and the remaining 50 get the old retardant. Repeat this process for the other fabric.
    (C) Take one type of fabric first and place the pieces of fabric in a stack. The first 50 pieces in the stack get the old retardant and the rest get the new retardant. Repeat this for the other fabric as well.
    (D) Put all 200 pieces of fabric in a stack with all of the pieces of the first type of fabric on the top of the stack. Select a piece and flip a coin. Heads get the old material and tails the new material. Do this until 100 pieces have been assigned to one of the retardants. The rest of the pieces would bet the other retardant.
    (E) Select one of the types of fabric (all 100 pieces) and flip a coin. If it comes up heads, then assign the old flame retardant to this fabric and the new to the other fabric.

14. Why is double-blinding used in a study?
    (A) To make sure the researcher does not know who is getting which treatment.
    (B) To keep the subjects from knowing which treatment they are receiving.
    (C) To keep the subjects from communicating with each other since they don't know which treatment they are getting.
    (D) To keep the subjects from telling the researcher which treatment they received.
    (E) To keep the subjects and person measuring the response from influencing the outcome.

15. A lottery ticket is sold for $2 with the promise that someone will win this particular jackpot of $500. Only 1,000 tickets are sold. What are the expected winnings for someone who purchases a single ticket?
    (A) $500.00
    (B) $1.00
    (C) $0.50
    (D) –$1.00
    (E) –$1.50

16. Every year, 50% of all children miss at least 3 school days due to illness. Which of the following is closest to the probability that 216 or more children out of a random sample of 400 would miss at least 3 school days this year?
    (A) 0.00
    (B) 0.03
    (C) 0.06
    (D) 0.50
    (E) 0.95

17. Scores on a standardized test have a normal distribution with a mean of 20 and a standard deviation of 6. What is the approximate interquartile range of the test scores?
    (A) 4
    (B) 6
    (C) 8
    (D) 10
    (E) 24

18. A 90% confidence interval has been computed to estimate the mean height in centimeters of newborn babies. Which of the following is a correct interpretation of the interval?
    (A) 90% of all babies' heights would be within the interval.
    (B) The sample mean will be in this interval 90% of the time.
    (C) 90% of the time, the method used to construct the interval results in an interval that includes the true mean height of newborns.
    (D) Any baby with a height not contained within the interval is an outlier.
    (E) The probability that the true mean height of newborns is contained in this interval is 90%.

19. Each person in a random sample of 300 elementary school children in New York City was asked if he or she preferred outside recess in the morning. A 98% confidence interval for the proportion of New York elementary school children who prefer outside recess is (0.32, 0.46). Which of the following would be the 95% confidence interval computed from this sample data?
    (A) (0.30, 0.48)
    (B) (0.32, 0.42)
    (C) (0.32, 0.96)
    (D) (0.32, 0.50)
    (E) (0.33, 0.45)

GO ON TO NEXT PAGE

20. A city official in a storm-damaged region of the United States believes that the percentage of homes with contaminated water is higher than the recently reported value of 9%. The city inspected 168 homes and found 23 homes with contaminated water. Is there enough evidence to conclude that the initial report of 9% is too low at the 0.05 significance level?
    (A) No, because the P-value = 0.017, which is smaller than α.
    (B) No, because the P-value = 0.034, which is bigger than α.
    (C) Yes, because the P-value = 0.017, which is smaller than α.
    (D) Yes, because the P-value = 0.034, which is bigger than α.
    (E) No, because p = 0.14, which is larger than α.

21. In a study of a random sample of 100 popular songs, the average length was found to be 3.2 minutes and the sample standard deviation was 0.04 minutes. Which of the following would be a true statement?
    (A) If we took another random sample of 100 popular songs, the average length for this sample would be 3.2 minutes and the standard deviation would be 0.04 minutes.
    (B) The sampling distribution of the sample mean song length would be approximately normal with a larger variance than that of the population.
    (C) The population standard deviation is likely to be larger than 0.04 minutes and the mean for the population will probably be smaller than 3.2 minutes.
    (D) The population standard deviation is likely to be smaller than 0.04 minutes and the population mean will probably be greater than 3.2 minutes.
    (E) The sampling distribution of sample mean song length is approximately normal, with a standard deviation that is likely to be smaller than 0.04.

22. A corporation must pay for a service person to work on the phone lines within the building at a rate of $50 per visit. In any given month, the probability distribution of x = number of phone service visits is shown in the accompanying table.

x	0	1	2	3	4
P(x)	0.38	0.19	0.34	0.06	0.03

On average, about how much should the company expect to pay per month for phone service calls?
    (A) $50
    (B) $59
    (C) $75
    (D) $100
    (E) $158

Questions 23–24 refer to the following table, which is the probability distribution of X = the number of times a copy machine needs repair in a given month.

x	0	1	2	3	4	5
P(x)	0.12	0.18	0.23	0.20	0.14	0.13

23. What's the probability of fewer than two repairs in a month?
    (A) 0.12
    (B) 0.18
    (C) 0.23
    (D) 0.30
    (E) 0.53

24. What is the probability the machine will need repair at least twice but not more than four times in a given month?
    (A) 0.23
    (B) 0.24
    (C) 0.43
    (D) 0.57
    (E) 0.87

25. Which of the following is not a property of a binomial experiment?
   (A) All trials are independent.
   (B) The outcome of each trial can be classified as success or failure.
   (C) The probability of success is the same for each trial.
   (D) There are a fixed number of trials.
   (E) Trials continue until a success occurs.

26. Jerome decides to take a simple random sample from a population that is normally distributed. However, σ is not known. If his sample size is 20, and Jerome wishes to compute a 95% confidence interval for the population mean, which formula would be correct?

   (A) $\bar{x} \pm 2.093 \dfrac{s}{\sqrt{20}}$

   (B) $\bar{x} \pm 2.093 \dfrac{s}{\sqrt{19}}$

   (C) $\bar{x} \pm 1.960 \dfrac{s}{\sqrt{20}}$

   (D) $\bar{x} \pm 1.960 \dfrac{s}{\sqrt{19}}$

   (E) $\bar{x} \pm 1.725 \dfrac{s}{\sqrt{20}}$

27. Marvin is planning to use a simple random sample of 30 to estimate a population mean. What would happen to the margin of error if he were to increase the sample size to 150?
   (A) It will increase.
   (B) It will decrease.
   (C) It will not be affected.
   (D) It will be tripled.
   (E) There is not enough information to determine what will happen to the margin of error.

28. A video game company is interested in the proportion of people who currently play one of their online games who would be willing to pay for an enhanced version of the game. They plan to market the enhanced game if they are convinced that more than 40% of the current users would pay for the enhanced version. They plan to survey a random sample of 200 players of the current version of the game. What is an appropriate null hypothesis in this situation?
   (A) $\mu = 80$
   (B) $\mu > 80$
   (C) $p = 0.40$
   (D) $p > 0.40$
   (E) $p < 0.40$

29. A newspaper claims that the proportion of people using public transportation is 0.39. An environmental group disagrees, stating the proportion is smaller than what the newspaper claims. The environmental group plans to conduct a survey to support their position. What hypotheses should the environmental group test?
   (A) $H_0 : p = 0.50, H_a : p > 0.50$
   (B) $H_0 : p = 0.39, H_a : p \neq 0.39$
   (C) $H_0 : p = 0.39, H_a : p > 0.39$
   (D) $H_0 : p = 0.39, H_a : p < 0.39$
   (E) $H_0 : p = 0.50, H_a : p \neq 0.50$

30. A scientist wants to use data from a survey of 1,000 randomly selected registered voters to determine whether there is a relationship between gender and whether or not a person voted in the last election. What is an appropriate test to use in this situation?
   (A) Chi-square goodness of fit test
   (B) Chi-square test of independence
   (C) Two-sample t test
   (D) Two-sample z test
   (E) One proportion test

GO ON TO NEXT PAGE

31. A researcher wants to study the effectiveness of two different medications for ADHD in children. Fifty boys and 50 girls who are diagnosed with ADHD have volunteered to participate in the study. Which of the following could cause the introduction of a confounding variable?
    (A) The subjects were not randomly selected.
    (B) For each child, a coin flip will determine which medication the child will receive.
    (C) The 50 girls will be given medication 1 and the 50 boys will be given medication 2.
    (D) Double blinding is used in the study.
    (E) No control group is used in the study.

32. The population {4, 6, 8, 10} has mean $\mu = 7$ and standard deviation $\sigma = 2.24$. If sampling is done with replacement, there are 16 different possible samples of size 2 that can be selected from this population. Each of these samples has a sample mean. For example, one of the samples is (8, 10), which has a mean of 9. The mean of the sampling distribution of the sample mean is denoted by $\mu_{\bar{x}}$ and the standard deviation is $\sigma_{\bar{x}}$. Which of the following statements is true?

    (A) $\mu_{\bar{x}} = 7$ and $\sigma_{\bar{x}} = 2.24$
    (B) $\mu_{\bar{x}} = 7$ and $\sigma_{\bar{x}} > 2.24$
    (C) $\mu_{\bar{x}} = 7$ and $\sigma_{\bar{x}} < 2.24$
    (D) $\mu_{\bar{x}} > 7$ and $\sigma_{\bar{x}} > 2.24$
    (E) $\mu_{\bar{x}} < 7$ and $\sigma_{\bar{x}} < 2.24$

33. A local store owner wants to know if a new line of souvenir glass figurines is a better sales product than a line he has carried for years. To compare sales of the two products, he decides to keep track of sales of each item for a period of one month. He randomly assigns one item to the checkout lane and the other to an aisle in the store. The new item is placed to the left of the checkout lane and the old item is randomly assigned to the second aisle. Is this a good study design?
    (A) Yes, because he has assigned the location at random.
    (B) Yes, but the new item should be placed in the aisle and the old item by the register.
    (C) No, because a confounding variable of location within the store has been introduced.
    (D) Yes, because the new item will be seen more because it is by the register.
    (E) No, because one month is too short of a time to compare sales.

34. A weight loss group wants to see if a new program seems to be effective in helping individuals who want to lose weight. Nine individuals have been weighed before the program and then after two months on the program. The results are in the table below.

Subject	1	2	3	4	5	6	7	8	9
Before	150	182	118	197	208	215	134	107	168
After	151	180	118	203	206	216	133	105	170

What would be an appropriate set of hypotheses to use if the group wants to see if the program was effective for weight loss?

(A) $H_0 : \mu_d = 0, H_a : \mu_d < 0$,

where μ_d = mean change in weight (*before – after*), with df = 8

(B) $H_0 : \mu_d > 0, H_a : \mu_d = 0$,

where μ_d = mean change in weight (*before – after*), with df = 8

(C) $H_0 : \mu_d = 0, H_a : \mu_d < 0$,

where μ_d = mean change in weight (*before – after*), with df = 9

(D) $H_0 : \mu_d = 0, H_a : \mu_d > 0$,

where μ_d = mean change in weight (before-after), with df = 8

(E) $H_0 : \mu_d > 0, H_a : \mu_d \neq 0$,

where μ_d = mean change in weight (*before – after*), with df = 17

35. Which of the following actions will increase the power of a test?
   I.   Increasing the sample size
   II. Decreasing the sample size
   III. Increasing the significance level
   IV. Decreasing the significance level

   (A) I only
   (B) III only
   (C) IV only
   (D) I and III only
   (E) II and IV only

36. In a particular town, 40% of the registered voters are registered as Democrats, 35% are registered as Republicans, and 25% are registered as Independents. What is the expected number of Independents in a random sample of 200 registered voters from this town?
   (A) 20
   (B) 25
   (C) 40
   (D) 50
   (E) 75

37. A newspaper reported that based on a recent survey of reality show watchers, 58% of the people preferred the reality show *The Bachelor* to the reality show *Survivor*. The margin of error reported was ± 4%. What does this ± 4% mean?
   (A) 4% of the reality TV watchers don't know what show they prefer.
   (B) There is a 4% chance of error in these results from this particular sample.
   (C) The results are based on 4% of the reality TV show watchers.
   (D) The actual percentage who prefer The Bachelor to Survivor is probably somewhere between 54% and 62%.
   (E) 54% to 62% of the population watch *The Bachelor*.

38. Each person in a random sample of high school students was asked whether he or she takes Vitamin C on a regular basis and whether he or she had a cold in the last 6 months. A chi-square test of independence resulted in a P – value of 0.37. What conclusion is reasonable based on the P –value?
    (A) There is an association between taking Vitamin C and whether or not a person gets a cold.
    (B) The data prove that taking Vitamin C and whether or not a person gets a cold are independent.
    (C) There is convincing evidence that there is an association between taking Vitamin C and whether or not a person gets a cold.
    (D) There is convincing evidence that taking Vitamin C and whether or not a person gets a cold are independent.
    (E) There is not convincing evidence that there is an association between taking Vitamin C and whether or not a person gets a cold.

39. If a 95% confidence interval for the slope of a regression line were computed, what would it mean if the interval contained 0?
    (A) There is sufficient evidence to conclude that the slope of the true regression line is 0.
    (B) There is sufficient evidence to conclude that the slope of the true regression line is not 0.
    (C) There is insufficient evidence to conclude that the slope of the regression line is 0.
    (D) There is insufficient evidence to conclude that the slope of the regression line is not 0.
    (E) The slope of the population regression line is equal to 0.

40. A recent poll of 1,000 adults included 540 people who said they would be willing to give up television for the summer if they could be sure that doing so would result in weight loss. Assuming it is reasonable to regard this sample of 1,000 as a random sample of adult Americans, is there convincing evidence that a majority of adult Americans would be willing to do this?
    (A) Yes, P – value < 0.01
    (B) Yes. P – value > 0.01
    (C) Yes. P – value > 0.05
    (D) No. P – value = 0.51
    (E) No, because 0.54 is not enough larger than 0.50 to be convinced.

STOP

END OF SECTION I
IF YOU FINISH BEFORE TIME IS CALLED, YOU MAY CHECK YOUR WORK ON THIS SECTION. DO NOT GO ON TO SECTION II UNTIL YOU ARE TOLD TO DO SO.

Section II: Free-Response Questions
Part A
Questions 1–5
Spend about 65 minutes on this part of the exam
Percent of Section II grade—75

Directions: Show all your work. Clearly indicate the methods you use, as you will be graded on the correctness of the methods as well as the accuracy of your results and explanation.

1. Strep throat is a common infection among teenagers. It must be treated quickly to avoid serious complications. The Rapid Strep Test (RST) is a diagnostic test that may be performed by a doctor in his or her office. RST gives a false positive in about 4 percent of the healthy students and gives a negative result in about 20 percent of students who do have strep throat. Suppose there is a lot of strep throat going around, and 5% of the student body has been infected.
   (a) What is the probability that a student selected at random would test positive for strep throat if the RST is used? Show your work.
   (b) What is the probability that a child selected at random who tests positive for strep throat using the RST does not have the strep throat? Show your work.

2. *Panda's Pearls* grows cultured pearls in a section of the Chesapeake Bay area. Cultured pearls are formed by placing a bit of sand in an oyster and allowing the oyster to sit for a time on the ocean bed "growth area." The growth area of the bay has a rocky region at one end and a soft sandy region at the other end, and the region type might affect pearl growth. The growth area can accommodate 600 oysters, with 200 placed in the rocky section and 400 placed in the sandy section. Black pearls are considered more valuable than pink pearls. Two grades of sand are being considered for use in producing pearls, and it is not known if one type of sand might produce more black pearls than the other type of sand.
   (a) You are asked to design an experiment to decide if there is a difference in the proportion of black pearls produced for the two types of sand. You can use 600 oysters that have been placed in the growth area described. Explain why it would be a good idea to incorporate blocking and describe the blocks you would use.
   (b) Describe the method you would use to assign a type of sand to each of the 600 oysters in the growth area.

3. Scores on a standardized exam are known to be normally distributed with a mean of 500 and a standard deviation of 100.
   (a) What is the probability that a randomly selected student taking this exam will score 650 or higher?
   (b) What is the probability that a random sample of 16 students taking this exam will have a mean score of 650 or higher?

4. A cookie manufacturer sells two types of chocolate chip cookies, regular and fat free. The manufacturer claims that the fat-free cookies have, on average, the same number of chocolate chips as the regular cookies. A group of students wanted to test this claim. They counted the number of chips in a sample of 20 regular cookies and in a sample of 20 fat-free cookies. The data are summarized in the following dotplots.

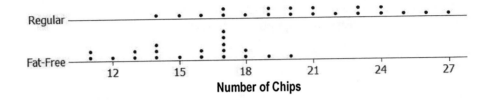

(a) In a few sentences, compare the distributions of number of chips for regular and fat-free cookies.
(b) What hypotheses should the students test if they want to determine if there is convincing evidence against the manufacturer's claim?
(c) Assuming it is reasonable to regard these two samples as independent random samples from the populations of regular and fat-free cookies, would it be reasonable to use a two-sample *t* test to test the hypotheses in part (a)? Explain why or why not.

5. LeRhonda and Nick want to consider the weights of *Snickers* fun-size candy bars to see if the advertised weight of 1 ounce is accurate. Eight candy bars are weighed. The weights (in ounces) were as follows: 0.99   0.97   0.98   0.96   1.01   1.03   0.99   1.02

(a) Assume that it is reasonable to consider this sample of eight bars as a random sample of all Snickers fun-size bars. Estimate the mean weight for Snickers fun-size bars using a 90% confidence interval.
(b) Based on your confidence interval from part (a), do you think that the manufacturer's advertised weight of 1 ounce is reasonable? Why or why not?

Section II
Part B
Question 6
Spend about 25 minutes on this part of the exam
Percent of Section II grade—25

Directions: Show all your work. Clearly indicate the methods you use as you will be graded on the correctness of the methods as well as the accuracy of your results and explanation.

6. The Federal Bureau of Investigation handles requests by law enforcement officers to gain top-secret clearance status. In years past, the requests were rarely granted. However, in recent years, more requests have been granted. The data below show the number of top-secret clearances granted by year.

Year	2001	2002	2003	2004	2005	2006	2007	2008	2009
# Top-Secret Clearances	8	57	102	136	258	317	489	743	1,823

Computer software was used to fit the least-squares regression line to these data, resulting in the output below.

```
Predictor      Coef   SE Coef      T      P
Constant    -342852     89460   -3.83  0.006
Year         171.22     44.62    3.84  0.006

     S = 345.612   R - Sq = 67.8%   R - Sq(adj) = 63.2%
```

(a) What is the equation of the least-squares regression line?

(b) What is the residual value for the year 2006? Did the line over- or under-predict this value?

(c) Based on the regression line, the predicted number of requests granted in the year 2020 is 2972. Explain why this is or is not a good estimate to use.

(d) Below are a scatterplot and a residual plot from the linear regression for these data. Based on these plots, an exponential model has been recommended as a better way to describe the relationship between number of requests granted and year. What aspect of these plots supports this recommendation?

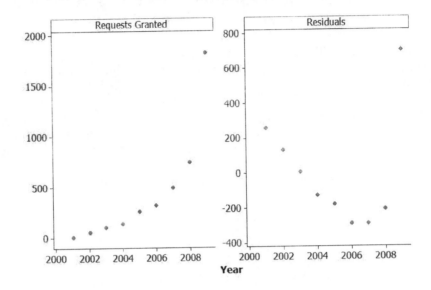

$$\hat{y} = 11.82(1.74)^{year^*}$$

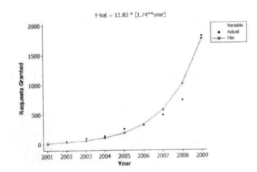

END OF EXAMINATION

ANSWERS TO DIAGNOSTIC TEST

MULTIPLE-CHOICE ANSWERS

Using the table below, score your test.

Determine how many questions you answered correctly and how many you answered incorrectly. You will find explanations of the answers on the following pages.

1. C	9. D	17. C	25. E	33. C
2. B	10. B	18. C	26. A	34. D
3. C	11. D	19. E	27. B	35. D
4. C	12. D	20. C	28. C	36. D
5. A	13. B	21. E	29. D	37. D
6. B	14. E	22. B	30. B	38. E
7. E	15. E	23. D	31. C	39. D
8. C	16. C	24. D	32. C	40. A

ANSWERS AND EXPLANATIONS

MULTIPLE-CHOICE ANSWERS

1. **C.** The combined average is calculated by $\dfrac{30(75.6)+24(68.4)}{54}=72.4$
    (*Statistics: Learning from Data,* 1st ed. pages 148–158/4th ed. pages 375–380).

2. **B.** The z score for a central area of 80% (10% in each tail) is about 1.28. Values of x that are within 12 of the mean are those between 38 and 62. Using the upper value of 62, we get $z=\dfrac{62-50}{\sigma}$ equal to 1.28 and solve for the value of σ. This results in $1.28=\dfrac{62-50}{\sigma}$; $\sigma=\dfrac{62-50}{1.28}=9.375$ Using the lower value of 38 would give σ = -9.375, which is also correct, because the sign in front of σ signifies direction from the mean (negative means left and positive means right). (*Statistics: Learning from Data,* 1st ed. pages 375–380).

3. **C.** The z* for a 90% interval is 1.645, 95/150 = 0.63 and the sample size of 150 goes the denominator for the standard error calculation (*Statistics: Learning from Data,* 1st ed. pages 375–380).

4. **C.** Using $z = \frac{x - \mu}{\sigma}$, Amy can solve for her z score, which is the answer. This would give $z = \frac{48000 - 47500}{1470} = 0.34$ (*Statistics: Learning from Data,* 1st ed. pages 396–413/4th ed. pages 375–380).

5. **A.** The IQR is equal to the width of the box in the boxplot (Q3-Q1), so the IQR for data set A is smaller than the IQR for data set B (*Statistics: Learning from Data,* 1st ed. pages 144–147).

6. **B.** Since 15 values are ≥1 and 18 values are ≤ 1. You could also compute the mean by adding the 10 zeros, 8 ones, 4 twos, and 3 threes and then dividing by 25. (*Statistics: Learning from Data,* 1st ed. pages 144–147).

7. **E.** The use of the emails included in the directory would mean that only those who chose to be in the directory would be able to respond. This plan excludes students who are not in the directory and is not an SRS. (*Statistics: Learning from Data,* 1st ed. pages 13–15).

8. **C.** They should worry about nonresponse bias because the survey was returned by only 42% of the customers. The majority of those who responded might be those who are highly satisfied with their service. (*Statistics: Learning from Data,* 1st ed. pages 19–20).

9. **D.** The value of the correlation coefficient does not depend on which variable is x and which is y. (*Statistics: Learning from Data,* 1st ed. pages 185–187).

10. **B.** The slope of the line is the coefficient for explanatory variable, height. The other coefficient value is the *y*-intercept (*Statistics: Learning from Data,* 1st ed. pages 198–211).

11. **D.** r^2 is calculated by squaring the value of the correlation coefficient. It is the coefficient of determination defined exactly as stated in choice D (*Statistics: Learning from Data,* 1st ed. pages 217–220).

12. **D.** Correlation has no units so this will remove choice E. Also, the correlation must be between -1 and +1, which eliminates choice A. B and C both have at least one categorical variable and the correlation coefficient is for two numerical values (*Statistics: Learning from Data,* 1st ed. pages 185–187).

13. **B.** For the method in A, only the new fire retardant will be used. For the method described in C, there is no random assignment and if the first 50 pieces in the stack differ from the last 50 in the stack in some important way, this would be a problem. The method described in D will tend to produce an uneven distribution of fabric types for the two retardants. The method in part E results in retardant being confounded with fabric type. The method described in B is the best choice (*Statistics: Learning from Data,* 1st ed. pages 24–31).

14. **E.** The subject as well as the person measuring the response would not know which treatment the subject is getting in order to keep this knowledge from influencing the study results (*Statistics: Learning from Data,* 1st ed. page 35).

15. **E.** The probability of winning with a single ticket is 1/1000 (0.001). Because the ticket costs $2, the possible amounts won are $498 (with probability 0.001) and –$2 (with probability 0.999). The expected amount won is then (498)(0.001) + (–2)(0.999) = –1.50 (*Statistics: Learning from Data,* 1st ed. pages 356–360).

16. **C.** This is a binomial probability B(400, 0.5) with n = 400 and p = 0.5. This means we want $1 - P(X \le 215)$. You can evaluate this using the following calculator command: 1 – binomcdf(400, 0.5, 215) = 0.06. (*Statistics: Learning from Data,* 1st ed. page 367).

17. **C.** The z-scores for a central area of 0.50 are $z = \pm 0.67$. Since this distribution is normal, the upper quartile is approximately 0.67 standard deviations above the mean and the lower quartile is about 0.67 standard deviations below the mean. The IQR is then about 2(0.67) – 1.34 standard deviations or (1.34)(6) = 8.04 (*Statistics: Learning from Data,* 1st ed. pages 144–147).

18. **C.** A confidence interval uses a point estimate to generate an interval of plausible values for the true height of babies. Since it was a 90% interval, a method was used that will result in an interval that includes the true mean height about 90% of the time in the various intervals (*Statistics: Learning from Data,* 1st ed. pages 467–468).

19. **E.** The 95% confidence interval will be narrower than the 98% confidence interval, since it is less likely that the mean will be included in a narrower interval. It will also be centered in the same place as the 98% confidence interval, so the correct answer must be E (*Statistics: Learning from Data,* 1st ed. pages 467–468).

20. **C.** This will give a z statistic of 2.12 and *P*-value = 0.017. Since 0.017 < 0.05 we would have strong evidence that the earlier claim of only 9% is too low and would reject the null hypothesis that the population proportion is 0.09 (*Statistics: Learning from Data,* 1st ed. pages 376–380).

21. **E.** For large samples, the sampling distribution is approximately normal with standard deviation = population standard deviation divided by the square root of the sample size (*Statistics: Learning from Data,* 1st ed. pages 441–442).

22. **B.** The expected number of service calls is 1.17, $\sum p(x) \cdot x$. This would mean that, on average, they should expect to pay (1.17)(50) = $58.50 per month (*Statistics: Learning from Data,* 1st ed. pages 356–360).

23. **D.** The $P(X < 2) = P(X = 0) + P(X = 1) = 0.12 + 0.18 = 0.30$ (*Statistics: Learning from Data,* 1st ed. pages 339–344).

24. **D.** $P(X = 2) + P(X = 3) + P(X = 4) = 0.57$ (*Statistics: Learning from Data,* 1st ed. pages 339–344).

25. **E.** Continuing trials until a success occurs is a property of a geometric experiment (*Statistics: Learning from Data,* 1st ed. pages 369–371).

26. **A.** Because the population standard deviation is not known, a t interval should be used. The appropriate t^* value of 2.093 is based on 19 *df* and a central area of 0.95. (*Statistics: Learning from Data,* 1st ed. pages 580–581).

27. **B.** Increasing sample size decreases margin of error, since you are dividing by the square root of n (*Statistics: Learning from Data,* 1st ed. pages 480–482).

28. **C.** The null hypothesis must include the = case and the question of interest is about a population proportion (*Statistics: Learning from Data,* 1st ed. pages 507–511).

29. **D.** The alternative is less than 0.39, as the group wants to determine if the actual proportion is less than the claimed proportion of 0.39 (*Statistics: Learning from Data,* 1st ed. pages 507–511).

30. **B.** The chi-square test for independence is designed to decide whether or not there is an association between two categorical variables (*Statistics: Learning from Data,* 1st ed. pages 715–716).

31. **C.** If all the girls received medication 1 and all the boys received medication 2, we would not be able to distinguish if the change in behavior was the result of the medication or of gender medication on each gender (*Statistics: Learning from Data,* 1st ed. pages 25 and 29).

32. **C.** The mean of the sampling distribution will be the same as the population mean but the standard deviation of the sampling distribution will be smaller than the population standard deviation because the standard deviation of the sampling distribution is the population standard deviation divided by the square root of the sample size (*Statistics: Learning from Data,* 1st ed. pages 441–442).

33. **C.** Since each is in a different location, the study has now introduced a confounding variable. A person may be more prone to last-minute shopping, for example, and thus choose the ones by the register more often (*Statistics: Learning from Data,* 1st ed. page 25).

34. **D.** If the program is effective, the after weight should be less than the before weight. This would mean the before – after differences should tend to be greater than 0. Also, since this is a matched pair, the degrees of freedom would be $n - 1 = 8$ (*Statistics: Learning from Data,* 1st ed. pages 635–639).

35. **D.** Increasing sample size is one way to increase power. Increasing the significance level decreases the probability of a type II error, β, and increases power, which is $1-\beta$. So, both I and III increase power (*Statistics: Learning from Data*, 1st ed. pages 527–530).

36. **D.** The expected number = $200(0.25) = 50$ (*Statistics: Learning from Data*, 1st ed. pages 353–363).

37. **D.** The margin of error of $\pm4\%$ tells us that the sample percentage is expected to be within 4% of the actual population percentage. Therefore, the population percentage is probably between 52% and 62% (*Statistics: Learning from Data*, 1st ed. pages 443–444).

38. **E.** With a *P*–value as large as 0.37, we would not reject the null hypothesis of independence at any reasonable level of significance, α. There is not evidence convincing of an association (*Statistics: Learning from Data*, 1st ed. pages 715–716).

39. **D.** Since the interval contains 0, it is plausible that the slope of the population least-squares regression line could be 0 (*Statistics: Learning from Data*, 1st ed. pages 644–649).

40. **A.** A one-proportion z test is appropriate for testing $H_0 : p = 0.50$ versus $H_a : p > 0.50$. Since the value of $z = 2.530$, the *P*–value for this test is 0.0057, which is very small and indicates that the null hypothesis should be rejected (*Statistics: Learning from Data*, 1st ed. pages 514–526).

FREE-RESPONSE PROBLEMS

1. (a) Those who test positive will consist of those who have strep and test positive and those who do not have strep but still test positive.

   $P(\text{have strep} \cap \text{test positive}) + P(\text{don't have strep} \cap \text{test positive})$

   $= P(\text{strep})P(+ \,|\, \text{strep}) + P(\text{no strep})P(+ \,|\, \text{no strep})$

   $= 0.05(0.80) + 0.95(0.04)$

   $= 0.078$

   This means that 7.8% of the students would test positive.

   It is helpful to diagram the probability as shown on the following page. This will also demonstrate that you are familiar with all the correct probabilities.

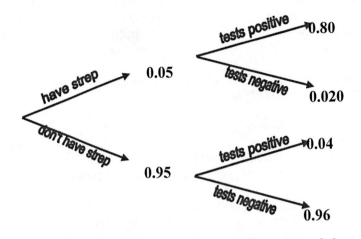

(b) This is the conditional probability that a student doesn't have strep even though the student tested positive. The solution can be found as follows

$P(A\,|\,B)$

$$= \frac{P(A \cap B)}{P(B)} = \frac{P(A)P(B)}{0.078} = \frac{0.95(0.04)}{0.078}$$

$$= 0.487$$

This means that 48.7% of those who test positive don't actually have strep.

(*Statistics: Learning from Data,* 1st ed. pages 290–306).

Scoring Question 1:

Each part of this problem can be scored as essentially correct (E), partially correct (P), or incorrect (I).

Part (a) is scored
  E if the probability is correctly computed and supporting work is shown.

  P if a correct probability is given but no supporting work is shown OR if the probability computed only considers one of the two possible cases (testing positive and has strep or testing positive and does not have strep).

Part (b) is scored
  E if the probability is correctly computed and supporting work is shown.

  P if a correct probability is given but no supporting work is shown OR if the numerator in the conditional probability calculation is not correct but the computed probability is still reasonable (a number between 0 and 1).

Question 1 is scored a
  4 if both parts are E
  3 if one part is E and the other is P
  2 if one part is E and one part is I or if both parts are P
  1 if one part is P and one part is I
  0 if both parts are I

2. (a) Whether oysters are grown in a rocky region or a sandy region might affect the proportion of black pearls produced. We want the blocks to be homogeneous within each other and heterogeneous between each other. Therefore, we should use region to create blocks, since the growth area consists of a rocky region and a sandy region. One block would consist of the 200 oysters placed in the rocky region and the other block would consist of the 400 oysters placed in the sandy region.

(b) Assign sand type to each of the 200 oysters in the rocky region. Number the oysters from 1 to 200. Write the numbers from 1 to 200 on slips of paper, mix the slips well, and then select 100 slips of paper. The oysters corresponding to these 100 numbers would get sand type 1. The other 100 oysters would get sand type 2. Repeat this process (using numbers 1 to 400) to assign a sand type to the 400 oysters in the sandy region. Compare the proportion of black pearls produced (*Statistics: Learning from Data,* 1st ed. pages 25–33).

Scoring Question 2:

Each part of this problem can be scored as essentially correct (E), partially correct (P), or incorrect (I).

Part (a) is scored
  E if the regions are chosen as the blocks and the response states that the region may affect the proportion of black pearls produced.

  P if regions are chosen as blocks but the response is not clear in stating that the region (rocky or sandy) might be related to the proportion of black pearls produced.

Part (b) is scored
  E if a correct method of assigning sand type at random to oysters is described and it is clear that this is done separately for each block (rocky region and sandy region). Note: Other random assignment schemes such as using a random number table or flipping a coin are also correct.

  P if a correct method of assigning sand type at random to oysters is described but it is not clear that this is done separately for each block.

Question 2 is scored
4 if both parts are E
3 if one part is E and the other is P
2 if one part is E and one part is I or if both parts are P
1 if one part is P and one part is I
0 if both parts are I

3. (a) The desired probability is the area to the right of 650 under the normal curve with mean 500 and standard deviation 100. The z-score is $z = \dfrac{650 - 500}{100} = 1.50$. The area to the right of 1.50 under the standard normal curve is 0.0668.

(b) This is a question about the sample mean. Because the population distribution is normal, the sampling distribution of $\bar{x}$ will also be normal. The mean of the sampling distribution is equal to the population mean, 500, and the standard deviation is $\dfrac{\sigma}{\sqrt{n}} = \dfrac{100}{\sqrt{16}} = 25$. Then $z = \dfrac{650 - 500}{25} = 6.00$ and

$P(\bar{x} \geq 650) = P(z \geq 6.00) \approx 0$.

(*Statistics: Learning from Data,* 1st ed. pages 591–604).

Scoring Question 3:

Each part of this problem can be scored as essentially correct (E), partially correct (P), or incorrect (I).

Part (a) is scored
   E if the normal distribution is used to compute a correct probability and supporting work is shown. Note: supporting work can consist of showing the computation of the z – score or of a sketch of the normal distribution with the appropriate area shaded. Also, since exam score can be viewed as discrete, a response that uses 649.5 as the boundary for the area computed should also be scored as correct as long as supporting work is shown. The calculator command normalcdf(650,E99,500,100) is correct if a sketch of the normal distribution is shown and correctly labeled or if the equation for calculating z is shown and the values of the symbols in the equation are correctly labeled.

   P if the normal distribution is used to compute a correct probability but no supporting work is shown, OR if a two-sided probability is computed with supporting work.

Part (b) is scored
   E if a correct probability is computed using the standard deviation of the sampling distribution, supporting work is shown, and there is an explanation of why the sampling distribution is normal. The calculator command normalcdf(650,E99,500,25) is correct if a sketch of the normal distribution is shown and correctly labeled or if the equation for calculating z is shown and the values of the symbols in the equation are correctly labeled.

   P if a correct probability is computed using the standard deviation of the sampling distribution, supporting work is shown, but there is no explanation of why the sampling distribution is normal OR a correct probability is computed, an explanation of why the normal distribution is appropriate is given, but no supporting work is shown.

   Note: for part (b), if the standard deviation of the sampling distribution is not used in the probability calculation and the response duplicates the work in part (a), part (b) is scored as incorrect.

Question 3 is scored a
   4 if both parts are E
   3 if one part is E and the other is P
   2 if one part is E and one part is I or if both parts are P
   1 if one part is P and one part is I
   0 if both parts are I

4. (a) Both of the distributions are approximately symmetric. The variability appears to be a little greater for regular cookies than for fat-free cookies (the range for regular cookies is 13 and the range for fat-free cookies is 11). The centers of the two distributions appear to be different, with number of chips for regular cookies centered at a higher value than number of chips for fat-free cookies.

(b) The claim is that the mean number of chips is the same for both types of cookies. Since no direction is implied, this would be a two-sided test with hypotheses

$H_0 : \mu_R = \mu_F$

$H_a : \mu_R \neq \mu_F$

where μ_R = mean number of chips for regular cookies

and μ_F = mean number of chips for fat-free cookies

(c) We are told that we can regard the samples as independent random samples. The other condition required for the t test is that the samples sizes are large or that the population distributions are approximately normal. Because both sample sizes are less than 30, we need to look at the distributions. Both of the dotplots show distributions that are approximately symmetric and there are no outliers, so this condition is reasonable and it would be appropriate to carry out a t test.

(*Statistics: Learning from Data,* 1st ed. pages 618–633).

Scoring Question 4:

Each part of this problem can be scored as essentially correct (E), partially correct (P), or incorrect (I).

Part (a) is scored

   E if the two distributions are compared on the basis of all three of shape, center, and spread.

   P if the distributions are compared on only 2 of shape, center, and spread

   I if shape, center, and spread are described for both distributions, but no comparative statements are made.

Part (b) is scored

   E if correct hypotheses are given and if symbols other than μ_R and μ_F are used, they are correctly defined. Note: it is also correct to write the hypotheses in terms of the difference in means: $H_0 : \mu_R - \mu_F = 0$ and $H_a : \mu_R - \mu_F \neq 0$.

   P if the null hypothesis includes the = case, but the alternative hypothesis is one sided.

Part (c) is scored

   E if the response says use of the t test is appropriate and mentions that independent random samples is given and judges the normality conditions to be reasonable based on the given dotplot or on some other appropriate graphical display that is constructed and included in the response.

   P if the response says use of the t test is appropriate and mentions that independent random samples is given and judges the normality conditions to be reasonable but does not link this judgment to a graphical display.

Question 4 is scored a
  4 if all three parts are E
  3 if two parts are E and the one is P
  2 if two parts are E and no parts are P, or one part is E and two parts are P, or if three parts are P
  1 if two parts are P and one part is I or if one part is E and no parts are P
  0 if one or no parts are P

5. (a) Because we need to estimate a population mean and the population standard deviation is not known, we will use a one-sample t confidence interval.

Check Assumptions
1. We are told that it is reasonable to regard the sample as a random sample, so this condition is met.
2. The sample size is not large, so we need to check to see if it is reasonable to assume the population distribution is approximately normal. A dotplot of the sample is

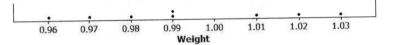

The dotplot is approximately symmetric and there are no outliers, so it is reasonable to use the t confidence interval.

Calculations:

$\overline{x}$ =0.994, s=0.024, n=8, df=8-1=7, t*=1.895

$$\overline{x} \pm t \cdot \frac{s}{\sqrt{n}} = (0.98, 1.01)$$

Interpretation:
We can be 90% confident that the true mean weight of Snickers fun-size bars is between 0.98 ounces and 1.01 ounces.

(b) Because 1 ounce is in the confidence interval, it is a plausible value for the population mean. There is no reason to think that the manufacturer's claim is not reasonable.

(*Statistics: Learning from Data,* 1st ed. pages 578–591).

Scoring Question 5:

Each part of this problem can be scored as essentially correct (E), partially correct (P), or incorrect (I).

Part (a) is scored
   E if the *t* confidence interval is identified by name or by formula, computations are correct, and a correct interpretation of the interval is given. If the calculator command is used, the formula must be shown and the statistics in the formula must be correctly identified.

   P if the *t* confidence interval is identified by name or by formula, computations are correct, but an interpretation is not given or the interpretation is incorrect.

Part (b) is scored
   E if the response says that the manufacturer's claim is reasonable and justifies this based on the fact that 1 is in the computed interval.

   P if the response says the claim is reasonable but does not explicitly state that this is because 1 is in the interval.

   Note: if the interval in part (a) is incorrectly computed, the response in part (b) must be consistent with the interval in part (a). Part (b) should be scored based on the interval in (a).

Question 5 is scored a
   4 if both parts are E
   3 if one part is E and the other is P
   2 if one part is E and one part is I or if both parts are P
   1 if one part is P and one part is I
   0 if both parts are I

6. (a) $\hat{y} = -342{,}852 + 171.2x$

   where $\hat{y}$ = predicted number of requests granted and x = year.
   (b) Predicted value for year 2006:

   $\hat{y} = -342{,}852 + 171.2(2006) = 575$

   residual=y $- \hat{y}$ =317 $-$ 575= $-$258
   The least-squares regression line overpredicted for 2006. The predicted value was quite a bit larger than the actual value.
   (c) The year 2020 is far outside the range of the data (which only goes to 2009). It would be unwise to use the line to predict number of requests granted for 2020.
   (d) Both the scatter plot and the residual plot show strong curved patterns, indicating that a linear model is not the best way to describe the relationship between number of requests granted and year. The pattern in the scatterplot looks like an exponential curve, so it is reasonable to consider an exponential model.

   (*Statistics: Learning from Data*, 1st ed. pages 205–206 and 229).

Scoring Question 6:

Each part of this problem can is scored as correct (E) or incorrect (I).

Part (a) is scored as correct if a correct equation for the least squares regression line is given and $\hat{y}$ and x are correctly defined.

Part (b) is scored as correct if the predicted value and residual are correctly computed using the line from part (a) and the prediction is identified as an over-prediction.

Part (c) is scored as correct if the response says it is not a good idea to use the least squares regression line to make a prediction for 2020 and identifies extrapolation as the reason.

Part (d) is scored as correct if the response comments on the curved pattern in both plots.

Question 6 is scored a
    4 if all four parts are scored E
    3 if three parts are scored E
    2 if two parts are scored E
    1 if one part are scored E
    0 if no parts are scored E

CALCULATING YOUR SCORE

SECTION I: MULTIPLE-CHOICE QUESTIONS

[_____] × 1.25 = _____
Number Correct   Weighted Section I Score
(out of 40)   (do not round)

SECTION II: FREE-RESPONSE PROBLEMS

Question 1 _____ × (1.875) = _____
     (out of 4)   (Do not round)
Question 2 _____ × (1.875) = _____
     (out of 4)   (Do not round)
Question 3 _____ × (1.875) = _____
     (out of 4)   (Do not round)
Question 4 _____ × (1.875) = _____
     (out of 4)   (Do not round)
Question 5 _____ × (1.875) = _____
     (out of 4)   (Do not round)
Question 6 _____ × (3.1255) = _____
     (out of 4)   (Do not round)
                 Sum = _____
                    Weighted Section II Score
                    (Do not round)

COMPOSITE SCORE

_____ + _____ = _____
Weighted   Weighted   Composite Score
Section I Score   Section II Score (Round to nearest
                             whole number)

Composite Score Range	Approximate AP Grade
70–100	5
57–69	4
44–56	3
33–43	2
0–32	1

Part II

A Review of AP Statistics

1

GRAPHICAL METHODS FOR DESCRIBING DATA

Data are everywhere. You might be interested in data concerning the SAT scores needed to get into specific colleges. When confronted with this or any other set of data, you might have questions about typical values, or whether or not the data are consistent with currently held beliefs. You might ask if this set of data is similar to a second set of data, or if particular observations are unusual in the context of the data. In this review section, you will review techniques for organizing and describing data using graphs and tables to better address these questions.

OBJECTIVES

- Use an appropriate graph to display categorical data.
- Use an appropriate graph to display numerical data.

Before we begin, we need to be able to distinguish between two different kinds of data: categorical data and numerical data. Categorical data typically is not numerical and usually is the product of categorizing observations and counting the number of times each category occurs. Examples of categorical data include hair color, phone number, make or model of a car, and gender. Numerical data usually arises from a measurement. Examples of numerical data include height, the age of a car, number of players on a sports team, and a person's salary.

EXAMPLE Decide if the following data sets are categorical or numerical.

(a) heights of mothers

(b) numbers on football player uniforms

(c) number of years teachers have taught statistics

(d) colors of statistics books

SOLUTION

(a) A mother's height is numerical.

(b) The number on a football player's uniform has no meaning as a number and so it is categorical. There is nothing in the real world that would have any meaning corresponding to adding or subtracting or finding an average of the numbers.

(c) The number of years a teacher has taught statistics is numerical.

(d) The color of a statistics book is categorical.

CATEGORICAL DATA: BAR CHARTS, PIE CHARTS, AND SEGMENTED BAR CHARTS

(*Statistics: Learning From Data*, 1st ed. pages 692–733)

Categorical data are typically presented in a data table. Such is the case for the data given below on the state of residence for a sample of people attending a large music festival. Although we can glean some information from looking at the numbers, it is often easier to get a quick overall idea of the data from a picture.

One appropriate picture for categorical data is a **bar chart**. A bar chart is shown below the data table. A correct bar chart has both axes labeled and an appropriate scale on the vertical axis. The numeric values above each bar are not necessary but provide a more precise value than could otherwise be estimated from the graph. In this example, the order of the states on the horizontal axis is arbitrary.

State	Relative Frequency
Georgia	0.163
Tennessee	0.131
South Carolina	0.146
Florida	0.154
North Carolina	0.132
Mississippi	0.132
Other States	0.20

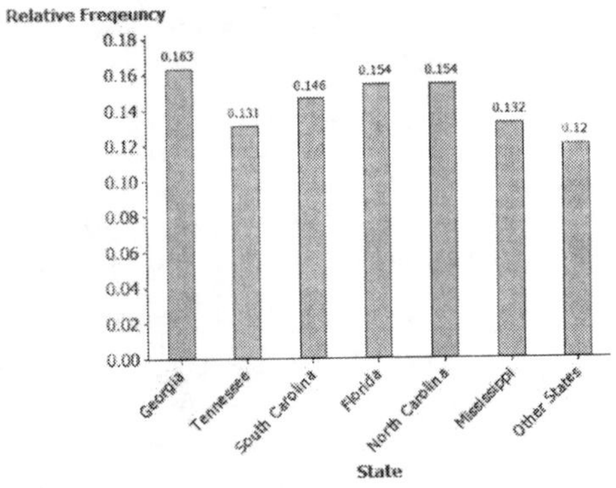

SAMPLE PROBLEM 1 Three large bags of M&M candies were purchased: one bag of regular candies, one bag of peanut candies, and one bag of peanut butter candies. The color of each candy was recorded. The accompanying frequency distribution summarizes the data. Draw a comparative bar chart to represent the data and describe what the bar chart tells you about the color distributions.

Color	Regular	Peanut	Peanut Butter
Orange	165	120	112
Blue	135	106	88
Green	145	111	75
Yellow	103	98	64
Brown	92	71	51
Red	65	53	43

SOLUTION TO PROBLEM 1 Because the bags have different total numbers, we will convert the counts into percents to make our comparisons easier (this is equivalent to working using relative frequencies). Below are the percentages for each type of candy and a comparative bar chart constructed using these percentages.

Color	Regular Percent	Peanut Percent	Peanut Butter Percent
Orange	23	21	26
Blue	19	19	20
Green	21	20	17
Yellow	15	18	15
Brown	13	13	12
Red	9	9	10

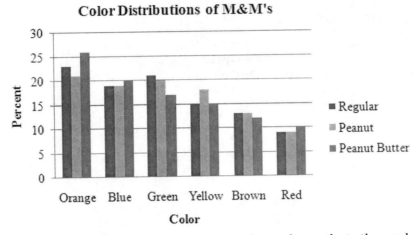

We can see from the comparative bar chart that the color distribution is similar in each of the three different types of M&M candies.

Another way to visually represent numerical data is to use a **pie chart**. A pie chart shows how a whole is divided into parts.

EXAMPLE Using the large bag of regular M&M candies referenced in the previous example, construct a pie chart to show the color distribution of the regular M&M's.

SOLUTION

Color	Regular	Percent	Central Angle Measure
Brown	92	13%	47°
Yellow	103	15%	53°
Red	65	9%	33°
Blue	135	19%	69°
Orange	165	23%	84°
Green	145	21%	74°

To find the percent column, you must first find the total number of M&M's that are in the bag: 92 + 103 + 65 + 135 + 165 + 145 = 705. Then find the percent for each color by dividing the color count by the total count and then multiplying by 100. For example, the percent of brown regular M&M's is $\left(\dfrac{92}{705}\right)(100) = 13\%$. To find the measure of the central angle that forms the piece of the pie chart, we multiply the percent by 360°. For example, the brown sector has a central angle that measures $0.13 \times 360° \approx 47°$. Drawing pie charts is much easier with software, and the use of such technology is strongly encouraged.

Regular M&M's Color Distribution

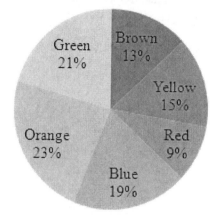

SAMPLE PROBLEM 2 Consider the following data. Does it make sense to draw a pie chart for these data on the percent of people 25 years old and over with a bachelor's degree or more in 2005 in the southeastern states? Explain your reasoning.

State	Percent
Georgia	26.9
Tennessee	21.6
South Carolina	24.2
Florida	25.5
North Carolina	25.4
Alabama	19.8
Mississippi	21.9

SOLUTION TO PROBLEM 2 No, it does not make sense to construct a pie chart for this data. A pie chart shows how a whole group is divided into parts based on one characteristic. This table does not describe the distribution of a categorical variable. (One clue that a pie chart would be problematic is that the percents add up to something different from 100%.)

An additional way to display categorical data is a **segmented bar graph**. The segmented bar graph is similar to the pie chart from the previous example, but here each of the bars shows 100%. Within each bar you can see how the whole is divided into parts just as a pie chart.

EXAMPLE Each student in a random sample of 70 students at Westside High School is asked what his or her favorite social networking site is. Below is a summary of the data. Construct a segmented bar graph to display the data.

Favorite Site	Frequency
Facebook	35
MySpace	20
LinkedIn	7
Other Site	3

SOLUTION First, we convert the counts into percents or relative frequencies. Here are the percentages for each category. The columns do not add up exactly to 100% because of rounding in computing the percentages.

Favorite Site	Percent
Facebook	54
MySpace	31
LinkedIn	11
Other Site	5

SAMPLE PROBLEM 3 The table below summarizes data on opinion regarding a change in public education in New York City under the present administration for a sample of New Yorkers who make $50,000 or less per year and a sample of New Yorkers who make over $50,000 a year. Construct segmented bar graphs to compare the opinions of New Yorkers with income $50,000 or less and adults with income over $50,000.

Opinion	$50,000 or Less	Over $50,000
Satisfied	33	25
Not satisfied	52	62
Don't know	15	13

SOLUTION TO PROBLEM 3 Because both sample sizes are equal to 100, the percentages are equal to the frequencies in the given table so there is no need to compute them.

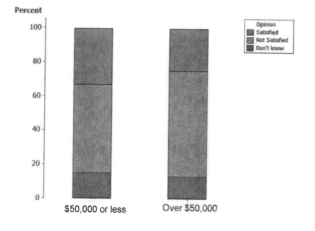

NUMERICAL DATA: STEM–AND–LEAF DISPLAYS, FREQUENCY DISTRIBUTIONS, AND HISTOGRAMS

(*Statistics: Learning From Data*, 1st ed. pages 56–127)

There are many graphs that are useful in the display of numerical data. Each graph has benefits and disadvantages. We will discuss the most common graphs for summarizing numerical data here.

A **stem–and–leaf plot** is an informative graph that shows each value in a set of data. Each data value is divided into a stem and leaf. Often (but not always), the leaf is the last digit of the data points (assuming they are rounded to the same decimal place) and the stem consists of the other digits. For example, if a data point were 3.07, 30 would be the stem and 7 the leaf.

EXAMPLE The grade point averages for 40 students are listed below. Construct a stem–and–leaf plot to display the data.

```
2.0  3.2  2.8  1.8  0.9  4.0  3.4  2.9  3.6  0.8  3.1  2.1
2.4  2.3  1.6  1.6  4.0  3.1  3.2  1.8  2.2  2.2  1.7  0.5
3.6  3.4  1.9  2.0  3.0  1.1  3.0  4.0  4.0  2.1  1.9  1.1
0.5  3.2  3.0  2.2
```

SOLUTION The ones digit is the stem and the tenths digit is the leaf. So, the stems are 0, 1, 2, 3, and 4.

The first GPA given is 2.0; go to the stem of 2 and place 0 next to it. The next GPA is 3.2; go to the stem of 3 and place 2 next to it. The next GPA is 2.8; go to the stem of 2 and place 8 next to the 0 that is already there. Keep repeating this process until you have put every data value in the stem–and–leaf plot. Be sure to include a key when you are done.

Student GPAs

```
0 | 9855
1 | 866879191
2 | 08914322012        Key:
3 | 246112640020       Stem = ones
4 | 0000               Leaf = tenths
```

The last step is to put the leaves in order from smallest to largest. (Note: This step is optional.)

Student GPAs

```
0 | 5589
1 | 116678899
2 | 00112223489        Key:
3 | 000112224466       Stem = ones
4 | 0000               Leaf = tenths
```

You may have noticed a drawback to stem–and–leaf plots: it is only reasonable to use with small data sets.

SAMPLE PROBLEM 4 The numbers of home runs that Sammy Sosa hit in each of his 18 years in major league baseball are listed below. Also listed are the numbers of home runs that Barry Bonds hit in the first 18 years of his major league baseball career. Construct a comparative (or back-to-back) stem–and–leaf plot and comment on the similarities and differences in the distribution of home runs for the two players.

Sosa: 4 15 10 8 33 25 36 40 36 66 63 50
       64 49 40 35 14 21

Bonds: 16 25 24 19 33 25 34 46 37 33 42 40
       37 34 49 73 46 45

SOLUTION TO PROBLEM 4 The stems are the tens digit and the leaves are the ones digit.

Sosa's Homeruns   **Bonds' Homeruns**

84	0	
540	1	69
51	2	455
6653	3	334477
900	4	025669
0	5	
643	6	
	7	3

The distributions of homeruns are similar, with the centers in the 30s. The Sosa values are more variable than Barry Bonds'. Sammy Sosa did not seem to have any unusual seasons whereas Barry Bonds had an unusual season (the one season where he hit 73 homeruns), which could be considered as an outlier.

A **histogram** is another compact way to display larger data sets with ease.

EXAMPLE The grade point averages for 40 students are listed below. Construct a histogram to display the data. Then, describe the data set.

```
2.0  3.2  2.8  1.8  0.9  4.0  3.4  2.9  3.6  0.8  3.1  2.1
2.4  2.3  1.6  1.6  4.0  3.1  3.2  1.8  2.2  2.2  1.7  0.5
3.6  3.4  1.9  2.0  3.0  1.1  3.0  4.0  4.0  2.1  1.9  1.1
0.5  3.2  3.0  2.2
```

SOLUTION The first step is to decide on a scale. We might choose to look at GPAs grouped in intervals of 0.5 (for instance, 3.0–3.5) or in intervals of 1 (for instance, 3.0–4.0). Let's use 1 as the interval width. Now, construct a frequency (or relative frequency) table. Make sure that whenever a GPA falls on the border of two different intervals, you make the same decision as to which bin it goes into. From here on, we will put those GPAs in the bin to the right (that is, 3.0 would not go in the 2.0–3.0 bin, it would be counted in the 3.0–4.0 bin).

GPA	Frequency	Relative Frequency
0.0 ≤ GPA < 1.0	4	0.1
1.0 ≤ GPA < 2.0	9	0.225
2.0 ≤ GPA < 3.0	11	0.275
3.0 ≤ GPA < 4.0	12	0.3
4.0 ≤ GPA < 5.0	4	0.1

You can decide which you would like to show on the vertical axis—frequency or relative frequency (or equivalently, percent). The graphs will convey the same information.

Note: You lose information about the counts and data size when you change to relative frequency.

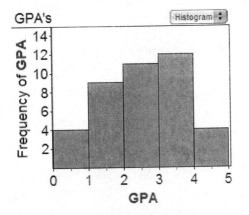

The distribution of GPAs is unimodal and slightly skewed left. The typical value appears to be around 3.0. The GPAs appear to have a range of approximately 4. (We could find the exact range by looking at the raw data.)

AP Tip

When commenting on what you see in a graphical display of numerical data, you should address the following:

1. center or typical value (mean or median)

2. spread or variability (examples: standard deviation, IQR, range, spread)

3. general shape, including number of peaks (examples: skewed left, skewed right, uniform, bimodal, symmetric, mound shape [if you use normal, state APPROXIMATELY normal])

4. unusual features (which include outliers, gaps, and clusters)

If you are using the mean for the center, then you should use the standard deviation as the spread. It is the standard deviation of a data set with mean___.

If you are giving the median as the center, then the IQR most times should be used to describe the spread.

When you are comparing data, be careful if the IQR for both sets are the same but the range may be different. Sometimes you need two different measures to accurately describe the spread, especially for boxplot comparisons.

(Remember GSOCS: Gaps/clusters, shape, outliers, center, spread.)

Notice the similarities between the histogram and the stem–and–leaf plot constructed earlier. They are almost identical in the information they show. The main difference is that in a histogram we can no longer see the individual data points.

SAMPLE PROBLEM 5 Suppose the following data are the average SAT math score for public high schools in Los Angeles. Construct a histogram of the average SAT math scores and describe the distribution.

509	493	501	516
506	494	503	519
505	497	504	518
498	500	506	520
497	500	508	518
496	501	511	515
494	501	512	515
493	502	511	515
492	501	514	
492	500	514	

SOLUTION TO PROBLEM 5

Average Score	Frequency	Relative Frequency
490 ≤ Average Score < 494	4	0.105
494 ≤ Average Score < 498	5	0.132
498 ≤ Average Score < 502	8	0.211
502 ≤ Average Score < 506	4	0.105
506 ≤ Average Score < 510	4	0.105
510 ≤ Average Score < 514	3	0.079
514 ≤ Average Score < 518	6	0.158
518 ≤ Average Score < 522	4	0.105

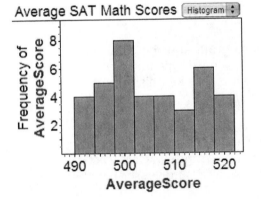

Make sure you do not have too many or too few bins for your data. The size of the bins can completely change the shape of the distribution.

The distribution of the average math SAT scores appears to be bimodal with scores concentrated around 500 and again at 515. The center is around 505 and the range is about 30. There do not appear to be any unusual average SAT math values.

A **cumulative relative frequency plot** (or ogive) shows the cumulative proportions of the data plotted against the interval endpoints.

EXAMPLE Use the data on the average SAT math score to construct a cumulative relative frequency plot.

Average Score	Frequency	Relative Frequency
490 ≤ Average Score < 494	4	0.105
494 ≤ Average Score < 498	5	0.132
498 ≤ Average Score < 502	8	0.211
502 ≤ Average Score < 506	4	0.105
506 ≤ Average Score < 510	4	0.105
510 ≤ Average Score < 514	3	0.079
514 ≤ Average Score < 518	6	0.158
518 ≤ Average Score < 522	4	0.105

SOLUTION The cumulative relative frequencies are shown below. To find a cumulative relative frequency for data value for an interval, you add the relative frequencies up to and including the relative frequency for that interval. For example, the cumulative relative frequency associated with 494 (the endpoint of the first interval) is 0.105. The cumulative relative frequency associated with 498 (the endpoint of the second interval) is 0.105 + 0.132 = 0.237.

Average Score	Relative Frequency	Cumulative Relative Frequency
490 ≤ Average Score < 494	0.105	0.105
494 ≤ Average Score < 498	0.132	0.237
498 ≤ Average Score < 502	0.211	0.448
502 ≤ Average Score < 506	0.105	0.553
506 ≤ Average Score < 510	0.105	0.658
510 ≤ Average Score < 514	0.079	0.737
514 ≤ Average Score < 518	0.158	0.895
518 ≤ Average Score < 522	0.105	1

To construct the cumulative relative frequency plot, we graph the cumulative relative frequency against the endpoints of the intervals and then connect these points with line segments.

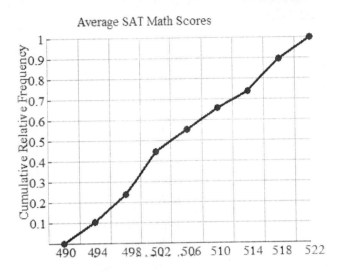

Average SAT Math Scores

BIVARIATE NUMERICAL DATA: SCATTERPLOTS

(*Statistics: Learning From Data,* 1st ed. pages 96–100)

Bivariate data are data with two variables that are paired in some logical way. Typically one variable is termed the explanatory variable and the other the response variable. The explanatory variable is placed on the *x* axis and the response variable is placed on the *y* axis. Unless there is the intention to use one variable as the basis for predicting the other variable (as in a regression setting), either variable can be designated as the response variable.

EXAMPLE It is well known that the frequency (chirps per minute) of crickets changes with the temperature. The following data shows the temperatures (in degrees Fahrenheit) at 15 times when cricket chirps per 15 seconds were observed. Draw a scatterplot of the data.

Temperature	89	72	93	84	81	75	70	82	69	83	80	83	81	84	76
Chirps	20	16	20	18	17	16	15	17	15	16	15	17	16	17	14

SOLUTION Since it seems more reasonable that temperature might affect the number of chirps, rather than the other way around, temperature is selected as the independent or explanatory variable and the number of chirps per 15 seconds is selected as the dependent variable.

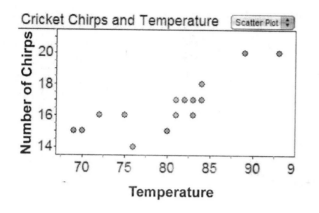

GRAPHICAL METHODS FOR DESCRIBING DATA: STUDENT OBJECTIVES FOR THE AP EXAM

- You should be able to construct and interpret a frequency table.
- You should be able to construct and interpret a bar chart and a comparative bar chart.
- You should be able to construct and interpret a pie chart.
- You should be able to construct and interpret a stem-and-leaf plot and a back-to-back stem-and-leaf plot.
- You should be able to construct and interpret a histogram.
- You should be able to construct and interpret a frequency plot and a cumulative frequency plot (ogive).
- You should be able to construct and interpret a scatterplot.

MULTIPLE-CHOICE QUESTIONS

1. Which of the following is a categorical variable?
   (A) Number of dogs in a house
   (B) TV channel number
   (C) Shoe size
   (D) Number of songs played on a radio station in one hour
   (E) Number of songs stored on an iPod

2. Which variable is most likely skewed to the right?
   (A) Student's grades on an easy statistics test
   (B) Heights of male high school students
   (C) Number of goals scored in a soccer game
   (D) Ages of male high school students
   (E) Color of a student's car

Questions 3–6 refer to the following histograms:

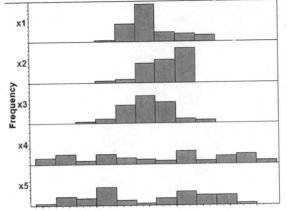

3. Which of the variables has a skewed left distribution?
   (A) *x1*
   (B) *x2*
   (C) *x3*
   (D) *x4*
   (E) *x5*

4. Which of the variables would be described as mound-shaped and symmetric?
   (A) *x1*
   (B) *x2*
   (C) *x3*
   (D) *x4*
   (E) *x5*

5. Which of the variables is most likely the distribution hair length of high school students?
   (A) *x1*
   (B) *x2*
   (C) *x3*
   (D) *x4*
   (E) *x5*

6. Which of the distributions is the most representative of the costs of properties on a Monopoly board? (The distribution of costs is fairly uniform and there is a lot of variability.)
   (A) *x1*
   (B) *x2*
   (C) *x3*
   (D) *x4*
   (E) *x5*

7. Which of the following are false statements about stem–and–leaf plots?
   I. They can be used to display both categorical and numerical variables.
   II. They are useful for both small and large data sets.
   III. One can easily see the shape of the distribution and unusual data values.
   (A) I only
   (B) III only
   (C) I and III
   (D) I and II
   (E) II and III

8. Consider the bar chart on average yearly tuition for public four-year universities. Which of the following is true?
   I. The graph is a segmented bar chart.
   II. Tuition increased less from 2006 to 2007 than in 2008 to 2009.
   III. Starting the vertical axis at $13,000 would result in a misleading picture.
   (A) II only
   (B) III only
   (C) II and III
   (D) I and II
   (E) I, II, and III

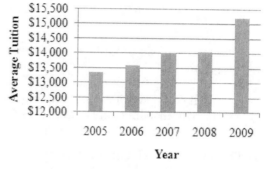

9. Which of the following are true statements about histograms?
   I. The vertical axis can indicate either frequencies or relative frequencies.
   II. They are particularly useful for large datasets.
   III. One can easily see the shape of the distribution and unusual data values.
   (A) I and II
   (B) I and III
   (C) II and III
   (D) I, II, and III
   (E) I only

10. Which of the following are true statements?
   I. For a given dataset, histograms can appear to be different if interval widths used in the two histograms are different.
   II. For every bar chart, you can construct a pie chart with the same information.
   III. Every symmetric distribution is unimodal.
   (A) I and II
   (B) I and III
   (C) I, II, and III
   (D) I only
   (E) III only

11. Given the following stem–and–leaf plot, which of the following is true?

Student Test Scores

5	13779
6	2
7	35789
8	2347778
9	1112455579

Key:

Stem = tens

Leaf = ones

(A) The mean would be a good description of a typical value for the data.
(B) The distribution is positively skewed.
(C) If 60 is passing, most students did not pass.
(D) The distribution is mound shaped.
(E) The range of the scores is about 50.

12. Given the following comparative bar chart, which of the following is true?
(A) The percentage of women who prefer cats is higher than the corresponding percentage for men.
(B) The number of male dog lovers in the data set is greater than the number of female dog lovers.
(C) A high percentage of women prefer a pet other than cats or dogs.
(D) About 30% of the people in the sample who prefer cats are male.
(E) Almost half of the dog lovers are female.

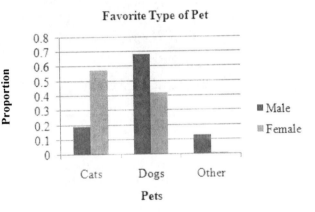

13. Given the following pie chart for data on type of car driven for a sample of students, which of the following is true?

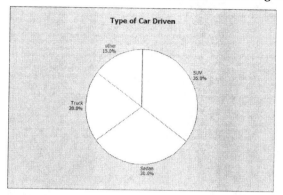

Type of Car Driven

other
15.0%

SUV
35.0%

Truck
20.0%

Sedan
30.0%

(A) Students tend to drive cars made in the USA.
(B) Most teachers do not drive trucks.
(C) More students in the sample drove SUVs than all other types of vehicles combined.
(D) A higher percentage of students drive SUVs than any other type of vehicle.
(E) More students drive jeeps than trucks.

14. The most appropriate graph to display data collected to show the relationship between the number of TVs and the number of people in a group of randomly selected households is
(A) a bar chart.
(B) a scatterplot.
(C) a back-to-back stem-and-leaf plot.
(D) a histogram.
(E) a cumulative frequency plot.

15. The most appropriate graph to display data on favorite food for the students in your statistics class is
(A) a bar chart.
(B) a scatterplot.
(C) a back-to-back stem-and-leaf plot.
(D) a histogram.
(E) a cumulative frequency plot.

FREE-RESPONSE PROBLEMS

1. The following cumulative frequency plot was constructed using data on number of siblings for each of the students in an AP Statistics class. Construct a histogram to display the class data and describe the distribution.

Number of Siblings	Cumulative Percent
0	0.33
1	0.83
2	0.96
3	0.96
4	1

Number of Siblings for Students in an AP Statistics Class

2. There are 200 students in the graduating class at a local high school. Construct a bar chart based on the data used to construct the pie chart below.

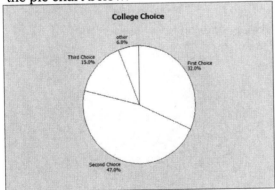

College Choice

other 6.0%

Third Choice 15.0%

First Choice 32.0%

Second Choice 47.0%

Answers

MULTIPLE-CHOICE QUESTIONS

1. **B.** (*Statistics: Learning From Data,* 1st ed. pages 62–68)

2. **C.** (*Statistics: Learning From Data,* 1st ed. pages 142–149)

3. **B.** (*Statistics: Learning From Data,* 1st ed. pages 142–149)

4. **C.** (*Statistics: Learning From Data,* 1st ed. pages 133–142)

5. **E.** We expect hair lengths to be bimodal—one cluster for the females and another cluster for the males (*Statistics: Learning From Data,* 1st ed. page 86).

6. **D.** (*Statistics: Learning From Data,* 1st ed. pages 133–142)

7. **D.** (*Statistics: Learning From Data,* 1st ed. pages 72–76)

8. **C.** (*Statistics: Learning From Data,* 1st ed. pages 63–67)

9. **D.** (*Statistics: Learning From Data,* 1st ed. pages 76–87)

10. **D.** (*Statistics: Learning From Data,* 1st ed. pages 63-67 and 76-87)

11. **E.** (*Statistics: Learning From Data,* 1st ed. pages 72–76)

12. **A.** (*Statistics: Learning From Data,* 1st ed. pages 63–67)

13. **D.** (*Statistics: Learning From Data,* 1st ed. pages 101–104)

14. **B.** (*Statistics: Learning From Data,* 1st ed. pages 63–67, 76–87 and 72–76)

15. **A.** (*Statistics: Learning From Data,* 1st ed. pages 63–67, 76–87 and 72–76)

FREE-RESPONSE PROBLEMS

1. Find the percent associated with each interval. For the interval 0 to 1, the percent is 0.33. For the interval 1 to 2, the percent is 0.83 – 0.33 = 0.5, since the percentages are cumulative. For the interval 2 to 3, the percent is 0.96 – 0.83 = 0.13. For the interval 3 to 4, the percent is 0.96 – 0.96 = 0. For the interval 4 to 5, the percent is 1 – 0.96 = 0.04. Here is the fully labeled histogram.

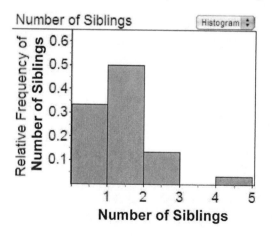

The distribution is slightly skewed right and unimodal. The center appears to be between 1 and 2 siblings. The range is 4 siblings. There do not appear to be any unusual features (*Statistics: Learning From Data*, 1st ed. pages 63–67, 76–87 and 72–76, 76–87).

2. Use the percentages given in the pie chart to determine the bar heights.

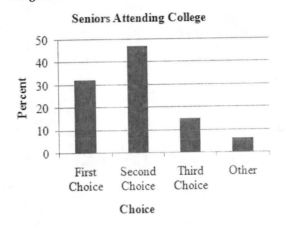

(*Statistics: Learning From Data*, 1st ed. pages 63–67, 76–87 and 72–76, 63–67).

2

NUMERICAL METHODS FOR DESCRIBING DATA

Now we will focus on summarizing data by describing the overall shape, center, and spread of a set of data. This will allow us to compare distributions of data as well as make predictions based on the data we have.

OBJECTIVES

- Compute and interpret measures of center.
- Calculate and interpret measures of spread within a data set.
- Use measures of center and spread to compare two or more sets of data.

DESCRIBING THE CENTER OF A DATA SET

(Statistics: Learning From Data, 1st ed. pages 128–133)

Two key measures of center are the mean and median. Measures of center give a single numerical value that is representative of the data set. This gives us a "typical" value for the data set. Figuring out what is "typical" within a data set helps us in making decisions about the population as a whole. Both mean and median provide key information about the data set (and are called measures of central tendency).

MEAN

The mean is the arithmetic average of any numerical data set. This measure gives a value that represents the data by adding up all the numbers and dividing by the total number of observations in the data set. For example, if there were 20 observations, you would divide the sum by twenty.

$$\overline{x} = \frac{\Sigma x}{n}, \quad \text{where}$$

$\overline{x}$ = sample mean

Σx = the sum of all the values in the data set

n = number of values in the data set

To use this formula, add all of the values in the data set. Then divide this sum by the total number of observations in the data set.

EXAMPLE

Let's take a set of ages collected from a class of nontraditional night students in an Introductory Statistics course at a local university. The 18 ages were as follows:

46, 29, 44, 30, 47, 30, 31, 31, 39, 27, 25, 45, 40, 38, 48, 29, 42, 27

To calculate the mean, add all the values and divide by 18.

$$\frac{\left(\begin{array}{c}46 + 29 + 44 + 30 + 47 + 30 + 31 + 31 + 39 + \\ 27 + 25 + 45 + 40 + 38 + 48 + 29 + 42 + 27\end{array}\right)}{18} = 36$$

For these data, the average age is 36.

The sample mean is commonly used to make inferential decisions in the later units of AP Statistics.

An important consideration regarding the mean is its sensitivity to **any** change in the data entries. No matter how small the change, the mean will also be affected. In the class age data from the example above, assume that the age of 25 was supposed to be a 35. In this case, the mean would now be 36.56 years of age.

$$\frac{\left(\begin{array}{c}46 + 29 + 44 + 30 + 47 + 30 + 31 + 31 + 39 + \\ 27 + 35 + 45 + 40 + 38 + 48 + 29 + 42 + 27\end{array}\right)}{18} = 36.56$$

The mean does not need to be one of the data points.

Even a change of one year would affect the mean. Because the mean is so sensitive, it is said to be **non-resistant** to change. This implies that an outlier or extreme value will pull the mean even further from the rest of the data set. This sensitivity to change makes the mean non-resistant.

The mean is a single number that represents a typical value in the data set. This allows us to use the mean to represent the data. For example, if 3 people with an average age of 41 are in a room and a 22-year-old walks in we can still calculate the new mean. Since the average age of the first three is 41, then the sum of the first three folks would be 123. (3(41) = 123). Now, add 22 to this sum and divide by four individuals. The new mean would be 36.25.

MEDIAN

The median is another measure of the center of a data set. Here, the center is measured by placing all the data entries in numerical order (from smallest to largest or largest to smallest) and then identifying the middle value. Should there be an even number of entries, then the median is the average of the middle two values.

EXAMPLE Again, utilizing the university age data given in the prior example, the median calculation is computed by placing all the numbers in order and then finding the middle value. In the example, since there were 18 entries, the 9th and 10th values would be averaged.

$$25, 27, 27, 29, 29, 30, 30, 31, \boxed{31, 38}, 39, 40, 42, 44, 45, 46, 47, 48$$

$$\Downarrow$$

$$\text{median} = \frac{31 + 38}{2} = 34.5$$

Unlike the mean, the median is **resistant** to changes (or outliers) within the data set. Because the median is based on the ranked ordering of the numbers, it isn't affected as much by an outlier in the same way as the mean. Instead, the median maintains its place in the middle of the numerical set no matter how extreme an outlier value may be.

AP Tip

A key strategy when analyzing data sets using the mean and median values is to compare two data sets using these values. It is important to state not just the numerical facts for each data set but to also stress what these values tell us. Questions on the AP exam may ask you to state what the difference is between distributions of data. For example, consider two data sets A and B where both are roughly normal and have about the same spread. However, mean A is 6 and mean B is 4. It is not sufficient to just list these facts. You need to make a comparison. Here, you might say,

"Distribution A is centered at 6 whereas distribution B is centered near 4. Since the mean in set B is lower than A, set B would tend to have lower overall values, on average than set A."

Comparison statements may include choices such as larger, smaller, wider, less skewed.

SAMPLE PROBLEM 1   A student recorded the time (in seconds) required to travel from the 1st to the 3rd floor in a school building on 10 separate occasions. Calculate the mean and median time and explain why these values are different.

39	32	28	36	40
37	34	36	29	37

Another student had found the mean of 6 trips to be 41 seconds. However, she forgot to include a 7th trip, which took her 55 seconds. What is the new mean for all seven of her trips?

SOLUTION TO PROBLEM 1 While the median is 36, the mean is 34.8 seconds. The difference in the two centers is based on their calculation. The median is the middle value, which in this case is the average of the 5th and 6th values of 36. However, the mean is being pulled to a lower value because 28 is smaller than the other values and far to the left of the rest of the values in the data set. The mean is sensitive to any numerical change making the mean value smaller. Because the data is skewed, the median is not the same as the mean. For left skewed data where the tail is on the left side, the mean tends to be less than the median. For right skewed data where the tail is on the right side, the mean tends to be greater than the median.

Since the mean is the numerical average, this value can be multiplied by the total number of entries and this will give a total number of seconds for all 6 trips. Then add 55 seconds to this total and divide by 7.

$$41 \times 6 = 246$$

$$\frac{246 + 55}{7} = \frac{301}{7} = 43 \text{ seconds}$$

CATEGORICAL DATA

We typically summarize categorical data by the category relative frequencies. This calculation is based on the number of observations in a given category out of the total number of observations. If there are only two possible categories, one is often denoted as a "success" and the other as "failure." To find the sample proportion of successes, denoted by $\hat{p}$, compute

$$\hat{p} = \frac{\text{number of successes in the data set}}{\text{total number of observations}}$$

The notation $\hat{p}$ is used for the sample proportion of successes and the population proportions of successes is denoted by p. The p-hat is a statistic and will have variation. The "p" is a parameter and will not have variation; we just do not know its value.

EXAMPLE If 12 students in an AP classroom are female and there are a total of 20 students in all, then we say that the proportion of females is 0.60 or that 60% of the class is female. The meaning of "success" is arbitrary; but here we say that "success" is being female. The result would be written as a simple fraction.

$$\hat{p} = \frac{\text{number of successed}}{n} = \frac{12}{20} = 0.60$$

In this scenario, the proportion of successes in the sample is 0.60. This sample proportion is often used to make inferences about the corresponding population proportion (p). The Greek symbol pi is sometimes used to denote the population proportion.

DESCRIBING VARIABILITY IN A DATA SET

(*Statistics: Learning From Data*, 1st ed. pages 133–142)

While the center of a data set is important for describing what a typical value is. It does not give a complete picture of the data set. Two sets of data could have the same mean, for example, and yet look quite differently in terms of the overall spread of the values. Consider the following two sets of data.

<u>Set A</u>
10, 10, 11, 12, 12

<u>Set B</u>
2, 4, 11, 18, 20

In both cases the mean is 11, however, a quick look at the dotplots reveals that data set B is more variable than data set A.

Set A _____

Set B _____

A description of variability is also important information to have when looking at a set of data. The most common way to measure variability is to use one of two statistics: the **variance** and **standard deviation**. Before considering these statistics, however, we will look quickly at another measure of variability, the **range**.

RANGE

The range is calculated by subtracting the minimum value from the maximum value in a data set, and is just a single value or number. This describes variability in a quick and simplistic form. It is incomplete in giving information about the rest of the data set as it is calculated using only the two extreme observations of the entire data set. Consider the following two sets of data and their corresponding dotplots.

<u>Set C</u>
2, 10, 11, 12, 20

<u>Set D</u>
2, 4, 11, 18, 20

Set C _____

Set D _____

While both sets have the same range (20 − 2 = 18), data set C appears to have less overall variation than data set D. This is because data set C has more numbers tightly packed around the mean than those in data set D. These data sets demonstrate why it is important to look at more than just mean or median values. By taking variability into account, we have a better overall description of each data set.

VARIANCE AND STANDARD DEVIATION

Both the sample variance and the sample standard deviation take into account all the values within a data set and their individual distance from the sample mean. The standard deviation is calculated by taking the square root of the variance. So understanding variance will help us calculate standard deviation.

The variance for any set of sample data is calculated by subtracting the mean from each value in the data set to find its directed distance (positive and negative values) from the mean value. If these directed distances are all added together, they sum to 0 because some values are below the mean, yielding negative values and some are above the mean, yielding positive values. To get around this problem, the deviations are squared before adding. Then, the sum of these values is divided by n – 1 (called degrees of freedom). The formula is for the sample variance, s^2 is shown below.

$$s^2 = \frac{\Sigma(x - \bar{x})^2}{n-1}$$

Notice how squaring the $(x - \bar{x})$ values eliminates the negative quantities that lead to the deviations having a sum of 0. However, the squaring leads to a process of averaging "square units." To fix this, and get a more interpretable result, the square root of this value is taken and it returns the original units of measure (similar to the distance formula from Algebra I and Geometry). This value is known as the sample standard deviation.

$$s = \sqrt{\frac{\Sigma(x - \bar{x})^2}{n-1}}$$

In a very general sense, the standard deviation calculation allows the researcher to see how far a typical observation is from the mean. While some values will be closer to the mean and others farther away, this computation will be used to describe the typical distance from the mean value and allow the data analyst to compare data sets that have smaller versus larger deviation values.

By using both the measures of center and variability, different data sets can be compared and differences and similarities can be described. Unfortunately, the standard deviation, like the mean, is sensitive to the presence of extreme outlier values.

You will also notice that the formula for the variance uses a denominator of $n – 1$ instead of n. While dividing by n will give you the average squared deviation for the data set, it tends to underestimate the population variance.. The graphing calculator gives us both values; we typically want to use S_x.

AP Tip

When asked to **DESCRIBE** a data set, you should always mention the *shape*, *center*, and *spread* of the data set and interpret this information in the context of the problem. When using the mean as a measure of center, use the S_x as a measure of spread. When using the median, use the IQR as a measure of spread.

INTERQUARTILE RANGE

The **interquartile range (IQR)** is a measure of variability that is helpful when there are extreme values present in the data set. The IQR is the range of the middle half of the data set. The IQR is resistant to outliers since any extreme values would be found below the lower quartile or above the upper quartile of the data set rather than in the middle of the data set. As with the median, to calculate the IQR, place the data in ranked order. The median of the lower half of the data is the lower quartile, Q_1, and the median of the upper half of the data is the upper quartile, Q_3. Subtract Q_1 from Q_3 and this is the IQR. The formula is shown below.

$$IQR = Q_3 - Q_1$$

When using your graphing calculator to analyze a list of data, notice that simple 1-variable summary statistics will include the values in a 5-number summary (min, Q_1, median, Q_3, max). This will allow you to quickly make a boxplot as well as calculate the value of the IQR.

NOTE: In calculating Q_3: if there is an odd number of observations in the upper half of the data, the middle value will be Q_3; if there is an even number of observations in the upper half of the data, the average of the two middle values will be Q_3. The same is true for calculating Q_1.

EXAMPLE Recall the 18 ages from the section on median.

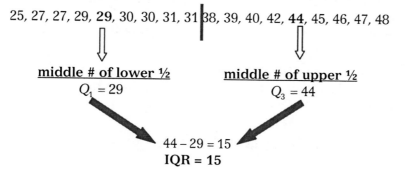

SAMPLE PROBLEM 2 The following stem-and-leaf plot summarizes data on the minutes it takes to get to work from home for 20 teachers at a regional rural high school in Virginia.

1|3 = 13 minutes to drive to work from home

0	1 2 5 6
1	0 1 3 5
2	1 2 4 7 8
3	2 4 9
4	3 6
5	8
6	
7	4

(a) Calculate the mean and median and comment on the difference between these two values.

(b) Which measure would better represent a typical travel time for teachers at this school? Explain.

(c) Calculate the IQR for this data set.

Solution to Problem 2

(a) The mean is 25.55 minutes and the median is only 23 minutes. The reason the mean is almost 3 minutes higher is that there is a very large observation (74 minutes) that is quite different from the other values in the data set. This large value is pulling the mean up even though more than half the teachers take less than 30 minutes to get to work.

(b) The median, which is 23, is the better choice since the data is skewed to the right. The mean is being pulled up quite a bit due to the large value of 74 minutes as well as the 58 minutes that make the value of the mean further from what seems typical for this data set.

(c) The IQR = 26. It was found by taking $Q_3 - Q_1$ (36.5 – 10.5). Since the upper half of the data has 10 values in it, Q_3 is the average of the middle two numbers.

$$Q_1 = \frac{10+11}{2} = 10.5 \quad \text{and} \quad Q_3 = \frac{34+39}{2} = 36.5$$

therefore

$$IQR = 36.5 - 10.5 = 26$$

SUMMARIZING A DATA SET: BOXPLOTS

(*Statistics: Learning From Data,* 1st ed. pages 149–159)

A nice way to present the data and get a quick idea of the data set's center, spread, and even the overall symmetry or skewness is with a boxplot. Boxplots are constructed using the five number summary rather than all the data, and presents a quick summary of a distribution that shows center, spread, and outliers.

A boxplot is easy to draw and only requires 5 values to make. So even in a large set of data, there will only be 5 numerical values that we can see in the actual plot. These numbers are usually referred to as the "**5-number summary**" and consist of the following values.

Minimum, Q_1, Median, Q_3, Maximum

These values can be marked on a number line and a quick sketch then made to look like a box with two whiskers as shown in the next diagram. Consider the mathematics and science departments at a local school and the ages of the teachers within these departments. These boxplots were drawn using the same numerical scale.

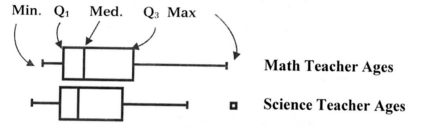

Math Teacher Ages

▫ **Science Teacher Ages**

The two boxplots make it easy to see a few key things. First, while both graphs are modified boxplots (which means they will show outliers if any exist), in the Science Teacher plot, notice that the maximum value is displayed as a single dot, which marks it as an outlier. This will be discussed further in the next section. The modified plot allows us to see the outliers clearly and to quickly calculate if any such outliers may exist.

Next, we can look at the youngest and oldest teachers. The Science Department has the youngest as well as oldest teacher in the two departments. Now, focusing on center, the median ages (middle line of each box) seem to be relatively close.

We can also detect skew. Each section of the boxplot represents 25% of the observations in the data set. With this in mind, you can think of each section containing 25% of the data and note if the data in any section is more tightly grouped than the other. For example, the smallest 50% of the ages in both departments seem to be pretty similar while the largest 50% of the ages are more spread out. In this case, the ages of the older 50% of the teachers are more variable than the ages of the younger 50%.

The mean age, while not something that can be determined by just looking at the boxplot, will be pulled to the right of the median since the data in both departments appears to be skewed to the right. We can not tell how many data points are included by looking at a boxplot; a "longer" boxplot may or may not have more data points than a "shorter" boxplot.

OUTLIERS

An outlier is a value that is "far enough" away from other numbers in the data set. The determination of how far away is "far enough" is left to a very common rule of thumb, accepted on the AP Exam. We will consider any observation that is beyond 1.5(IQR) from either Q_1 or Q_3 to be an outlier. (See the example below.) A nice way to think of this calculation is to look at the width of the box in a boxplot. Visualize a string of that length. Then make the string 1.5 times this box length. Now, place one end of the string at Q_1 and extend it to the left. Any observations that are beyond this 1.5(box length) will be considered outliers. The same process is repeated at Q_3 to determine outliers at the upper end of the data set. We will now perform the calculations that implement this visualization.

EXAMPLE Let's consider the length of time it takes Tiffany to apply color or highlight treatments to her clients at a local hair salon. The following 20 observations are recorded in minutes.

11	12	13	14	16	16	17	18	19	21
21	22	22	23	24	26	28	33	36	49

To determine if there are outliers, we perform the following steps.

1. Find Q_1 and Q_3.

2. Calculate the $IQR = Q_3 - Q_1$.

3. Compute 1.5(IQR).

4. Add this value to Q_3 and subtract this value from Q_1.

5. Any observations beyond these calculated values are considered outliers.

$$Q_1 = 16$$
$$Q_3 = 25$$
$$IQR = 25 - 16 = 9$$
$$1.5(IQR) = 1.5(9) = 13.5$$
$$Q_1 - 1.5(IQR) = 16 - 13.5 = 2.5$$
$$Q_3 + 1.5(IQR) = 25 + 13.5 = 38.5$$

Since there are no values below 2.5 minutes, there are no outliers on the lower end of Tiffany's coloring times. However, the maximum time of 49 minutes is beyond 38.5 minutes. This makes 49 minutes an outlier using this rule. This is also evident in the boxplot. Also notice the whisker on the right side stops at the last data point prior to 38.5, which is 36 minutes.

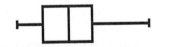

Tiffany's time to color or highlight hair

AP Tip

Since we have the data we can see that 49 is an outlier and there are no others. If we are given the boxplot without the data there may be more than one outlier and we should state there is "at least" one outlier at the upper end.

INTERPRETING CENTER AND VARIABILITY: THE EMPIRICAL RULE AND z-SCORES

(IStatistics: Learning From Data, 1st ed. pages 160–165)

The mean and standard deviation are both useful in describing a data set. However, using these values together can give additional information in terms of relative position of any particular observation within the data set. These strategies allow for an interpretation that not only takes into account the mean value but also the standard deviation by describing any individual observation in terms of the number of standard deviations it falls above or below the mean of the data set.

EMPIRICAL RULE

When a data set is roughly normal, its distribution has a mounded, "bell shaped" appearance. Both of the data sets portrayed below are roughly normal as can be seen by the smooth curve over the top of each set.

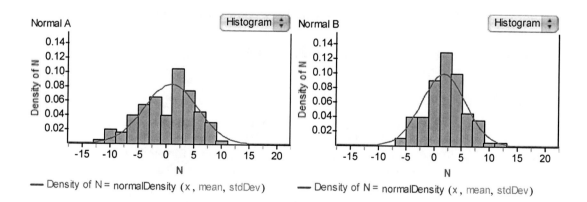

While both examples are of different sets of data and Normal A looks a little skewed left, you can easily see that both would be considered roughly normal.

The characteristics common to all normal distributions allow us to say that on average, a standard percentage of the data will fall within 1, 2, or 3 standard deviations of the mean. This rule, known as the

Empirical Rule, states that for distributions that are approximately normal,

About 68% of the data lies within ±1 standard deviation of the mean,

About 95% of the data lies within ±2 standard deviations of the mean, and

About 99.7% of all the data lies within ±3 standard deviations of the mean.

The displays below are shaded to visually represent these areas.

68% within
±1 standard deviation

95% within
±2 standard deviations

99.7% within
±3 standard deviations

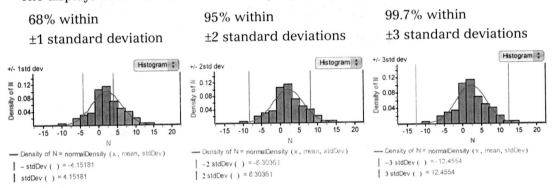

SAMPLE PROBLEM 3 The commute times during rush hour traffic on a local interstate have a mean of 22 minutes and a standard deviation of 2 minutes. The data from repeated studies of this section of interstate suggest it is reasonable to consider the drive times to be approximately normally distributed.

(a) A local businessman claims his drive takes him 26 minutes. Draw this distribution and label it. Include the businessman's commute time.

(b) What percent of commute times are between 20–24 minutes to work daily?

(c) What percentage of commute times are shorter than the time for the businessman in (a)?

SOLUTION TO PROBLEM 3

(a) Sketch a normal curve and locate a mean of 22 min. at the center. Then mark off 3 standard deviation units above and below the mean. Add the standard deviation (s.d.) to the mean for each mark above and subtract the s.d. for each mark below the mean as shown.

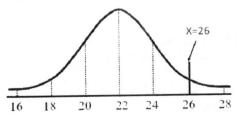

(b) The empirical rule says that approximately 68% of the observations are between 20–24 minutes. In other words, approximately 68% of the commutes on this section of interstate take 20–24 minutes.

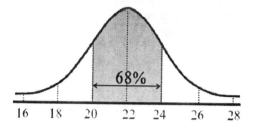

(c) 97.5% of the commutes are at or shorter than 26 minutes. Since 95% of the observations are within 2 s.d. of the mean (18–26 minutes), then 5% would be outside this range with 2.5% in each tail. So adding 0.025 + 0.95 = 0.975 = 97.5%.

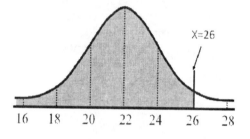

CALCULATING Z-SCORES

Many times in data settings it is important to compare values that have been derived using different scales. One problem with such comparisons is the difficulty in comparing data from distributions that have different means and standard deviations. We can use the mean and standard deviation to convert any observation in a data set into standard deviation units. These scores are called z-scores. They are simply the directed distance away from the mean in standard deviation units. Converting an observation into standardized units from the mean allows for comparison between two sets of data with different means and/or standard deviations. The calculation is made using the following equation.

$$z = \frac{x - \text{mean}}{\text{standard deviation}}$$

Once the z-score is found, the measures in different data sets can be compared by considering their relative standing.

A z-score can be used for any distribution—not just for normal distributions only.

SAMPLE PROBLEM 4 Consider once again the daily commute to work for a section of interstate that has a mean of 22 minutes and a standard deviation of 2 minutes (see Sample Problem 3). In the same community, the mean commute time on a non-interstate route is also normally distributed with a mean of 19.5 minutes and a standard deviation of 4.1 minutes.

(a) Joey uses the interstate section to get to work and his commute takes 25 minutes. Calculate Joey's z-score.

(b) Toyanda uses the main stretch of town to get to work and her commute takes 25.5 minutes. With this in mind, whose route is longer relative to the distribution of commute times for the respective routes?

SOLUTION TO PROBLEM 4

(a) $z_{Joey} = \dfrac{25 - 22}{2} = 1.5$

(b) $z_{Toyanda} = \dfrac{25.5 - 19.5}{4.1} = 1.46$

Since Toyanda's z-score is slightly smaller than Joey's, this means she is not as far above the mean as Joey. So here, Joey takes longer on his route than Toyanda relative to the other drivers on the same route. In other words, there was a higher percent of commuters on Joey's route that had commute times that are less than his time. Joey's commute is 1.5 standard deviations above the mean. Toyanda's commute is 1.46 standard deviations above the mean relative to their respective groups.

NUMERICAL METHODS FOR DESCRIBING DATA: STUDENT OBJECTIVES FOR THE AP EXAM

▪ You should be able to compare measures of center and discuss differences in two or more data sets.
▪ You should be able to calculate sample variance and standard deviation.
▪ You should be able to discuss the similarities and differences in data sets based on variability.
▪ You should be able to calculate and use a five-number summary to draw a boxplot.
▪ You should be able to describe how to find the IQR from Q_1 and Q_3 values.
▪ You should be able to describe features of a data set using either a boxplot or a five-number summary.
▪ You should be able to identify outliers using the 1.5(IQR) calculation.
▪ You should be able to draw and use/interpret a modified boxplot.
▪ You should be able to use the Empirical Rule as it relates to a data set that is approximately normal.
▪ You should be able to use a z-score to compare relative positions in data sets.

MULTIPLE-CHOICE QUESTIONS

1. Each of the boxplots shown summarizes 24 student scores on a recent test in two different classes. The boxplots are drawn using a common numerical scale. Which of the following must be true?

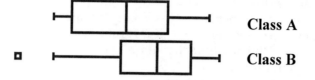

        Class A

        Class B

(A) Class A had more people between the median and the lower quartile than Class B.
(B) More people scored below the lower quartile for Class A than Class B.
(C) The distribution for Class A is somewhat symmetric, while the distribution for Class B is skewed right with a noticeable outlier as well.
(D) The distribution for Class B is skewed left and has a noticeable outlier as well, while the distribution for Class A is somewhat skewed right.
(E) The lower 25% of Class B are much more spread out than the lower 25% of Class A.

Use the following scenario to answer Questions 2–3.
Heights of 250 men and 200 women were measured. The distribution of men's heights had a mean of 68″ and a standard deviation of 3″ while the distribution of women's heights had a mean of 65″ and a standard deviation of 2.5″. Both distributions were approximately normal.

2. Which of the following is true?
(A) The percentage of women shorter than 62.5″ is smaller than the percentage of men shorter than 65″.
(B) There are a higher percentage of women between the heights of 62.5″–67.5″ than men between the heights of 65″ and 71″.
(C) More than ½ of the men are taller than 68″ while only ¼ of the women are taller than 67.5″.
(D) Approximately 95% of the women have heights between 60″ and 70″ and approximately 95% of the men have heights between 62″ and 74″.
(E) The percentage of men with heights between 68″ and 71″ and the percentage of women with heights between 65″ and 67.5″ is approximately 68%.

3. Which of the following is **not** a correct statement?
   (A) The men's heights have more variation than the women's heights.
   (B) The men's height distribution is more skewed than the women's height distribution because the standard deviation for the men is greater than the standard deviation for the women.
   (C) On average the men are taller than the women.
   (D) Since the standard deviation of the men's heights is larger than the standard deviation of the women's heights, you would expect the graphical display of the height distribution to be more spread out for men than for women.
   (E) Although more men were sampled, you can still expect about 68% of the women to be within one standard deviation of the mean.

4. A teacher found that the mean number of hours her 12 students spent studying for an AP Statistics exam per week was 6.7. A new student from across town transferred in and reported studying 11 hours per week. What is the new mean?
   (A) 7.6 hours
   (B) 7.0 hours
   (C) 6.7 hours
   (D) 8.9 hours
   (E) 8.3 hours

5. A small department at a university is seeking faculty raises. The university claims the pay for faculty in the department is comparable to other departments on campus, but the department claims that they are paid less. The university and the department both based their statements on a measure of "typical" salary using data that included salary information for all faculty and administrators at the university. Administrator salaries tend to be much higher than faculty salaries. Which of the following best explains their differing interpretations of the same data?
   (A) The university compared the mean salary for the department to the mean salary of all faculty and administrators and the department also used the means for its comparison but made an error in the calculation.
   (B) The university compared the median salary for the department to the median salary of all faculty and administrators while the department used the means for its comparison.
   (C) The university compared the mean salary for the department to the mean salary of all faculty and administrators while the department used the medians for its comparison.
   (D) The university compared the median salary for the department to the median salary of all faculty and administrators while the department used the modes for its comparison.
   (E) The university compared the mode salary for the department to the mode salary of all faculty and administrators while the department used the means for its comparison.

6. Two data sets have different variances. A constant of k units is added to all the numbers in both data sets. What is the affect on their new variances?
   (A) Both variances will stay the same.
   (B) Both variances will increase by k units.
   (C) The smaller variance will now be closer to the larger variance.
   (D) Both variances will increase by a factor of k units.
   (E) The smaller variance will now be farther away from the larger variance.

7. Which measures are considered to be not resistant to outliers in the data set?
   (A) mean, standard deviation, and interquartile range
   (B) median, interquartile range, and variance
   (C) mean, variance, interquartile range, and median
   (D) median, range, and standard deviation
   (E) mean, variance, and standard deviation

8. Every observation in a data set is multiplied by a constant k. What best describes the affect on summary measures for this data set?
   (A) The mean will be unchanged, and the standard deviation will increase by a factor of k.
   (B) The mean and standard deviation will both change by a factor of k.
   (C) The standard deviation will be unchanged, but the mean will change by a factor of k.
   (D) The mean and standard deviation will be unchanged because k is a constant.
   (E) Both the mean and standard deviation will change, but it is not possible to determine how they would change.

Use the following boxplots to answer Questions 9–10.

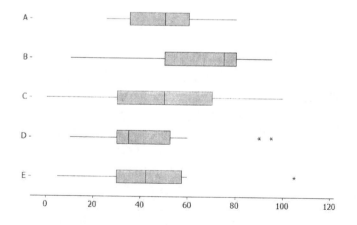

9. Which boxplot summarizes a data set that is skewed left?
   (A) A
   (B) B
   (C) C
   (D) D
   (E) E

10. Which boxplot corresponds to the following histogram?

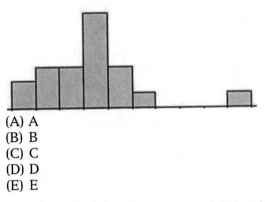

   (A) A
   (B) B
   (C) C
   (D) D
   (E) E

11. Which of the following are true statements?
   I. Variance is smaller when extreme outliers are present.
   II. The interquartile range (IQR) describes spread in the middle 50% of the data.
   III. The standard deviation gives relative position of a value from the mean.
   (A) I only
   (B) II only
   (C) I and II
   (D) II and III
   (E) I, II, and III

12. A group of 15 workers on an assembly line have a mean completion time of 15.2 minutes per part. A second group of 21 workers on a second assembly line have a mean completion time of 16.8 minutes. What is the mean time for the combined group of workers?
   (A) 0.89 minutes
   (B) 1.81 minutes
   (C) 16.0 minutes
   (D) 16.13 minutes
   (E) 32.0 minutes

13. The distribution of heights of 150 students is approximately normally distributed with a mean of 65 inches and a standard deviation of 2.5 inches. Which of the following intervals would contain approximately 102 student heights?
   (A) 57.5"–62.5"
   (B) 57.5"–72.5"
   (C) 60"–70"
   (D) 62.5"–67.5"
   (E) 65"–72.5"

14. A two-year study of high school students in the 1980s measured the times to complete a one-mile run. The mean for boys was 460 seconds and the standard deviation was 55 seconds. The mean for girls was 591 seconds and the standard deviation was 70 seconds. Both distributions are approximately normal. Darnell ran the mile in 394 seconds and Christina ran the mile in 500 seconds. Assuming that the distributions of run times have not changed, which student actually performed better (ran faster) when compared to high school students of the same sex?
   (A) Darnell performed better with a z-score of −1.2 being further to the right of Christina's z-score of −1.3 on a normal distribution.
   (B) Christina performed better with a z-score of −1.3 being further to the left of Darnell's z-score of −1.2 on a normal distribution.
   (C) Darnell performed better with a z-score of −1.3 being better than a z-score of −1.2 for a run time.
   (D) They both performed equally well within their groups respectively.
   (E) These scores can't be compared to see who performed better relative to students of the same sex because the means and standard deviations are different for males and females.

15. A study was conducted to investigate the length of time (in seconds) it took for a chocolate chip to melt in the mouth. Melt time was measured for each of 80 students who participated in the study. The five number summary for the resulting data set was:

12	29	35	38	41

   Approximately how many chips had a melt time between 12–29 seconds?
   (A) 6
   (B) 12
   (C) 17
   (D) 20
   (E) 40

FREE-RESPONSE PROBLEMS

1. Two AP teachers recorded test scores on the district exam that was taken at the end of the first semester of the course. Both classes covered the same material prior to the first semester exam. The scores for each of the two classes are shown below.

   Teacher A   59, 86, 92, 42, 71, 73, 84, 78, 73, 78, 80, 75
   Teacher B   67, 68, 70, 55, 60, 95, 86, 72, 85, 80, 74, 59

   (a) Draw comparative boxplots that would allow you to compare the two score distributions.
   (b) Are there any outliers? Explain how you arrived at this answer.
   (c) Which class appears to have higher scores? Justify your response.

2. Shavonna will only eat the red M&M's from any bag of the candies. The dotplot below shows the count of red M&M's in 26 randomly selected bags of plain M&M's.

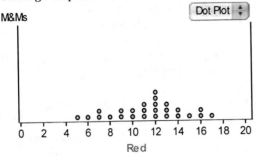

Describe the distribution and comment on what Shavonna might expect in a typical bag based on these results. What is the approximate percentile rank for 15 red candies in a bag? How did you arrive at this percentile?

Answers

MULTIPLE-CHOICE QUESTIONS

1. **E.** The outlier seen in Class B is farther from the median than the minimum value in Class A and the lower whisker is longer (*Statistics: Learning From Data,* 1st ed. pages 151–155).

2. **D.** Approximately 95% of the data in a normally distributed data set is within 2 standard deviations of the mean (*Statistics: Learning From Data,* 1st ed. page 161).

3. **B.** The standard deviation tells us nothing about the shape of the distribution (*Statistics: Learning From Data,* 1st ed. pages 137–139).

4. **B.** The mean is the average of all 12 students. 12(6.7) = 80.4, which is the total number of hours. Then add the new student's hours to this value (80.4 + 11 = 91.4). Finally, divide by 13 (*Statistics: Learning From Data,* 1st ed. pages 134–135).

5. **B.** The mean would be pulled up by the higher administrator salaries, whereas the median would not be sensitive to these actual salaries. So, if the department compared the mean salaries, the department mean would appear lower than the university mean (*Statistics: Learning From Data,* 1st ed. pages 134–135).

6. **A.** Adding a constant to each number in a data set shifts the entire set, but the spread within the data set remains the same (*Statistics: Learning From Data,* 1st ed. pages 355–357).

7. **E.** These three measures are affected most by outliers in the data sets. The median is the most resistant. Also, the IQR resists change due to outliers and extreme values so it is resistant to

outliers as well (*Statistics: Learning From Data,* 1st ed. pages 415–416).

8.   **B.** Multiplying by a constant changes both the mean and standard deviation (*Statistics: Learning From Data,* 1st ed. pages 352–357).

9.   **B.** The lower part of the distribution spreads out more than the upper part (*Statistics: Learning From Data,* 1st ed. pages 151–155).

10.   **E.** This boxplot is the one with a single outlier on the high side (*Statistics: Learning From Data,* 1st ed. pages 151–155).

11.   **B.** The relative position is given by a z-score and is calculated using the standard deviation, but this is not the definition of standard deviation. Outliers tend to increase the variance. (*Statistics: Learning From Data,* 1st ed. pages 137–139 and 144–147).

12.   **D.** $\dfrac{15.2 \times 15 + 16.8 \times 21}{36}$ (*Statistics: Learning From Data,* 1st ed. pages 134–135)

13.   **D.** 102/150 = 68%. The Empirical Rule says about 68% of the data is within one standard deviation of the mean so 65 ± 2.5 is the correct answer (*Statistics: Learning From Data,* 1st ed. page 161).

14.   **B.** Christina's z-score is –1.3 and Darnell's is –1.2. In running, the faster time would be further to the left of the mean. In this case, Christina has the better run time when compared to high school girls (*Statistics: Learning From Data,* 1st ed. pages 161–164).

15.   **D.** The five-number summary divides the data into quarters, with 25% of the data in each section. So 25% of the 80 (0.25 x 80) observations would be between the minimum (12) and the lower quartile (29) (*Statistics: Learning From Data,* 1st ed. pages 151–155).

FREE-RESPONSE PROBLEMS

1.   (a) The two boxplots are shown here. The scale should be labeled and the plots each marked with the correct teacher.

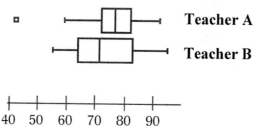

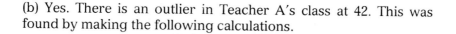

(b) Yes. There is an outlier in Teacher A's class at 42. This was found by making the following calculations.

$$IQR = Q_3 - Q_1 = 82 - 72 = 10$$
$$1.5(IQR) = 1.5(10) = 15$$
$$Q_1 - 1.5(IQR) = 72 - 15 = 57$$

Any number smaller than 57 will be considered an outlier. So, 42 is an outlier.

(c) While Teacher B has both a higher Q_3 and the highest score of 95, this teacher's scores are also more spread out. However, 75% of Teacher A's scores are above the value of 71 and only 50% of Teacher B's students are at or above this value. Teacher A appears to have better scores overall. Also, it should be noted that teacher A's students in the top 75% are tightly packed so this teacher's student scores are more consistent (*Statistics: Learning From Data,* 1st ed. pages 151–155).

2. A complete response to part (a) would include an appropriate response about the shape, center, and spread of the data in the context of red M&M's. A complete response to part (b) would show a simple calculation.

    (a) Based on these 26 bags, the distribution of red M&M's appears to be skewed left with a mean of about 11 and a median of 12. The range of counts of reds is 12, indicating that the number of red candies in a bag is quite variable. On average, it would appear that one should expect about 11–12 reds in a bag.

    (b) Since the count of 15 is the 23rd data entry of 26, the percentile rank is about 23/26 = 0.88. This means that 15 is about the 88th percentile (*Statistics: Learning From Data,* 1st ed. pages 68–72).

3

SUMMARIZING BIVARIATE DATA

In this section, we explore methods for describing relationships between two numerical variables and assessing the strength of such relationships. These methods allow us to use existing data to make predictions.

OBJECTIVES

▦ Discuss the strength of a linear relationship.
▦ Fit a line to bivariate data.
▦ Discuss the appropriateness of a linear model.
▦ Use transformations in describing nonlinear relationships.

THE CORRELATION COEFFICIENT

(*Statistics: Learning From Data*, 1st ed. pages 185–193)

The correlation coefficient (r) is a numerical assessment of the strength and direction of a linear relationship between two numerical variables. The correlation coefficient is given by

$$r = \frac{\sum z_x z_y}{n-1}$$

This numerical value is always between –1 and 1, inclusive ($-1 \le r \le 1$). A value of 1 indicates a perfect linear relationship with a positive slope whereas a value of –1 indicates a perfect negative linear relationship (which almost never happens in the real world). A value close to 1 (or –1) describes a moderate to strong positive (negative) linear relationship. A value close to zero denotes a weak linear relationship or no linear relationship. The scatterplots below illustrate these relationships.

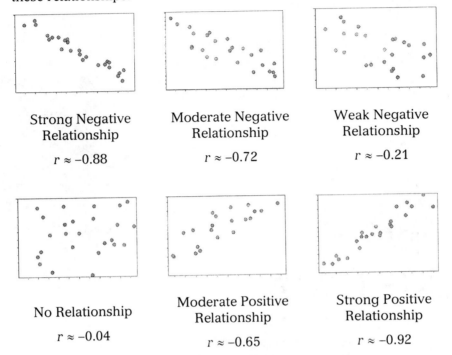

Strong Negative Relationship	Moderate Negative Relationship	Weak Negative Relationship
$r \approx -0.88$	$r \approx -0.72$	$r \approx -0.21$

No Relationship	Moderate Positive Relationship	Strong Positive Relationship
$r \approx -0.04$	$r \approx -0.65$	$r \approx -0.92$

Notice in a positive relationship that large values of y tend to be paired with large values of x, while in a negative relationship, large values of y tend to be paired with small values of x.

EXAMPLE Manatees are large marine mammals found along the coast of Florida. They are noted for their friendly nature, large size, and paddle-like flippers. Many manatees, unfortunately, are injured or killed by collisions with powerboats each year. The table below shows data on the number of powerboat registrations (in thousands) and number of manatees killed by powerboats for the years 1977 to 2007.

Year	Powerboat Registrations	Manatees Killed	Year	Powerboat Registrations	Manatees Killed
1977	447	13	1993	678	35
1978	460	21	1994	696	49
1979	481	24	1995	713	42
1980	498	16	1996	732	60
1981	513	24	1997	755	54
1982	512	20	1998	809	66
1983	526	15	1999	830	82
1984	559	34	2000	880	78
1985	585	33	2001	944	81
1986	614	33	2002	962	95
1987	645	39	2003	978	73
1988	675	43	2004	983	69
1989	711	50	2005	1010	79
1990	719	47	2006	1024	92
1991	681	53	2007	1027	73
1992	679	38			

Describe the relationship between the number of powerboat registrations (in thousands) and the number of manatees killed by powerboats.

SOLUTION The linear relationship between the number of powerboat registrations and the number of manatees killed is relatively strong and positive. If you are not getting r and r^2 on your calculator when doing a linear regression, go to catalog and scroll down to Diagnostic On. Hit enter one or two times to turn this feature on and you will get these values when doing regressions.

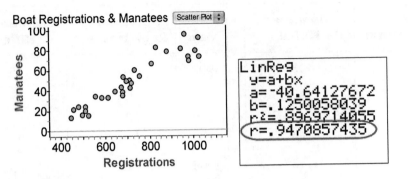

Boat Registrations & Manatees [Scatter Plot ♦]

Be Careful!

You must look at a scatterplot before interpreting the correlation coefficient. You can fit a straight line to any bivariate numerical dataset regardless of the pattern shown in the scatterplot. Do not assume that a correlation coefficient value near one or negative one means a strong linear relationship. A value of $r = 0.99$ doesn't necessarily mean the relationship *is* linear as there may be a strong curved pattern in the scatterplot (and that does not mean a slope of near one either). Similarly a value of $r = 0$ doesn't mean there is *no* relationship. You need to look at the scatterplot before interpreting the value of r.

EXAMPLE The scatterplot below displays a relationship where the correlation between x and y is 0.92. Clearly, this relationship is not linear.

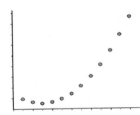

AP Tip

When describing a scatterplot, give form, direction, and strength in context. There is no need, unless asked, to speculate on a formula to describe the nonlinear model.

The description of the form of the relationship may include words such as curved, more than one cluster, quadratic, nonlinear, and so on.

Association vs. Causation

A strong association does not mean that x *causes* y. Correlation only measures the strength of the relationship between variables that have been assumed to be linearly related. For example, there is a strong positive linear relationship between ice cream sales and the number of swimming injuries at the beach. Increasing ice cream sales will not cause an increase in number of injuries. A more realistic theory is that another variable (in this case outside temperature) may cause both ice cream sales and swimming injuries to increase at the beach. When it is hot outside, people tend to buy more ice cream and more people are swimming at the beach (which unfortunately increases the number of swimming injuries).

SAMPLE PROBLEM 1 A college professor recorded the outside temperature and number of students absent from his statistics seminar for nine consecutive class meetings. Construct a scatter plot of the data and describe the relationship between temperature and number of students absent.

Outside Temperature (°F)	60	62	63	65	67	71	72	74	75
Number of Students Absent	1	0	2	5	6	5	8	7	8

SOLUTION TO PROBLEM 1 The scatterplot shows a strong, positive, and linear relationship between outside temperature and the number of absent students. The correlation coefficient, $r = 0.9026$, confirms that the relationship is strong and positive for this range of temperatures. A properly displayed scatterplot will include labels on both axes.

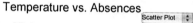
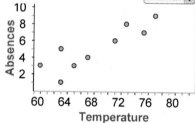

Temperature vs. Absences

LINEAR REGRESSION

(*Statistics: Learning From Data*, 1st ed. pages 198–211)

One potential goal in analyzing bivariate numerical data is to be able to use one variable, *x*, to predict the other variable, *y*. It is common to call the *x* variable the *explanatory variable* and the *y* variable the *response variable*. The linear relationship between *x* and *y* is expressed as

$$\hat{y} = a + bx$$

where *a* is the y-intercept, *b* is the slope and $\hat{y}$ is a predicted *y* value.

There are many lines that could be drawn through a set of data. However, we are most interested in the line that in some sense minimizes the sum of vertical distances. That is, we want a line that gets as close to each of the points as possible. Any line that contains the mean point $\left(\overline{x}, \overline{y}\right)$ will have a sum of the vertical distances (residuals) equal to zero. There is only one line that minimizes the sum of the squared vertical distances: the *least squares regression line* (often abbreviated as least squares regression line). An illustration of the squared vertical distances is shown below.

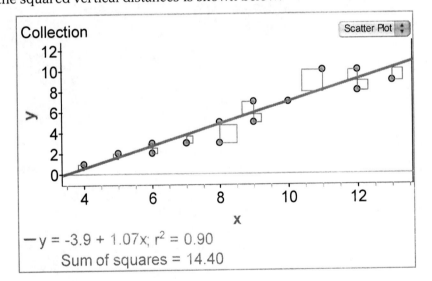

We write the least squares regression line as

$$\hat{y} = a + bx$$

The y has a "hat" to denote that the y value is a predicted value based on the value of x. There is variation in the least squares regression line and not every data set will give the same slope and y-intercept. The y-hat designates this as a "statistic" that has variation and is based upon our small data set).

The slope of the least squares regression line is interpreted as "for a one-unit increase in x, the y increases (or decreases if the slope is negative) *on average b* units."

The intercept of the least squares regression line is interpreted as "when x equals zero, the y <u>is predicted to be</u> a." However, in many contexts it does not make sense to interpret the value of a because an x value of 0 is not reasonable in that context or because 0 is far removed from the x values in the data set.

AP Tip

Be sure to distinguish between actual data values and predicted values. In the above statements, the words "on average" and "predicted" accomplish this goal. And make sure to always identify your variables, that is, what do x and y represent in the specific problem.

There is a clear relationship between the correlation between x and y and the slope of the least squares regression line:

$$b = r\left(\frac{s_y}{s_x}\right)$$

where r is the correlation coefficient, b is the slope of the least squares regression line, and s_x and s_y are the standard deviations of the x's and y's, respectively. This relationship shows that slope and correlation have the same sign: either both are positive or both are negative since standard deviations are always positive.

Once we have found the least squares regression line, we can use it to make predictions. When making predictions using the least squares regression line, we need to stay within the given range of the x values (interpolation). In the manatees example, we would only want to make predictions for x values between 447,000 and 1,027,000 boat registrations. Predictions based on values of x outside this range should be regarded as suspect and are less and less reliable the further away an x value is from the x values in the data set. Predicting beyond the range of the data values of x is called *extrapolation, and this is not appropriate.*

EXAMPLE Use the Minitab regression output below for the manatee data to answer the following questions.

a. What is the equation of the least squares regression line?

b. What is the interpretation of the slope? Does it make sense to interpret the intercept in this context?

c. What would you predict for number of manatees killed when 700,000 powerboats are registered?

Regression Analysis: Manatees versus Registrations

The regression equation is
Manatees = - 40.6 + 0.125 Registrations

Predictor	Coef	SE Coef	T	P
Constant	-40.641	5.840	-6.96	0.000
Registrations	0.125006	0.007867	15.89	0.000

S = 7.87985   R-Sq = 89.7%   R-Sq(adj) = 89.3%

SOLUTION

a. Notice in the printout that the equation of the regression line is given. We can also get the equation of the regression line from the numbers under the "Coef" (coefficient) column. The equation is $\hat{y} = -40.6 + 0.125x$ where x is the number of powerboat registrations in thousands and y is the predicted number of manatees killed.

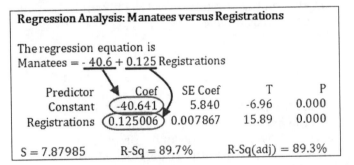

Regression Analysis: Manatees versus Registrations

The regression equation is
Manatees = - 40.6 + 0.125 Registrations

Predictor	Coef	SE Coef	T	P
Constant	-40.641	5.840	-6.96	0.000
Registrations	0.125006	0.007867	15.89	0.000

S = 7.87985   R-Sq = 89.7%   R-Sq(adj) = 89.3%

b. The slope of 0.125 means "for an increase of 1,000 powerboat registrations, the number of manatees killed increases on average by 0.125 manatees."

The intercept means "when zero powerboats are registered, the regression line predicts negative 40.6 manatee deaths." This has no real-world meaning. Since zero powerboats is far away from the range of the number of registered powerboats, our prediction of manatees killed when the number of powerboat registrations is 0 is extrapolation beyond the data. Furthermore, the intercept in this case does not have any practical meaning; we cannot have a negative number of manatee deaths.

c. To predict the number of manatee deaths, we substitute 700 into the equation of the regression line for x. (Remember that the data are in thousands so we substitute 700 instead of 700,000.)

$$\hat{y} = -40.641 + 0.125006(700) = -40.641 + 87.5042 = 46.8642$$

If there are 700,000 powerboat registrations, we predict approximately 46.8632 manatees to be killed.

AP Tip

Be sure to know how to read computer output. Know where you can find the statistics to answer the questions you might be asked. Some of the information on the computer output may be extraneous and you will not need this information nor are you expected to know what it means.

SAMPLE PROBLEM 2 Find the equation of the least squares regression line and interpret the slope in context. Then predict the number of students absent when the outside temperature is 70°F.

Outside Temperature (°F)	60	62	63	65	67	71	72	74	75
Number of Students Absent	1	0	2	5	6	5	8	7	8

The computer output is given here:

```
Predictor      Coef  SE Coef      T      P
Constant    -28.612    5.917  -4.84  0.002
Temp        0.49180  0.08719   5.64  0.001


S = 1.36191   R-Sq = 82.0%   R-Sq(adj) = 79.4%
```

SOLUTION TO PROBLEM 2 From the computer output, the equation of the least squares regression line is

NumAbsent = - 28.612 + 0.49180 * Temp

The interpretation of the slope is that for a one-degree increase in outside temperature, the number of students absent is expected to increase by approximately 0.49 students. Note that the value of the intercept has no real meaning in this context.

Predicting y when $x = 70$ gives

$$\hat{y} = -28.6120 + 0.4918(70) = -28.6120 + 34.426 = 5.814$$

When the outside temperature is 70°F, we predict that 5.814 students will be absent.

ASSESSING THE FIT OF A LINE

(*Statistics: Learning From Data*, 1st ed. pages 211–228)

Once you have found the equation of the least squares regression line, you will want to assess how effectively this line summarizes the relationship between the two variables.

RESIDUALS

A residual is the difference between an actual data value and the corresponding predicted or "expected" value. A residual indicates the distance to the least squares regression line and the direction; positive if the data value is above the regression line or negative if the data value is below the regression line.

EXAMPLE A positive residual and a negative residual are illustrated in the scatterplot below.

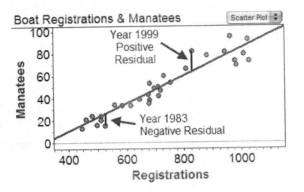

To calculate the residual for 1999, we first need to find the actual number of manatees killed in 1999 (look in the data table given earlier in this review section). We find the predicted number of manatees killed in 1999 using the least squares regression line.

Actual number of manatees killed = 82

Predicted number of manatees killed = −40.641 + 0.125(830) = 63.109

Residual = 82 − 63.109 = 18.891

This means that for the year 1999, our model *under*estimated the number of manatees killed by almost 19.

Likewise, for the residual in 1983,

Residual = 15 − (40.641 + 0.125(526)) = 15 − 25.109 = −10.109.

This means that for the year 1983, our model *over*estimated the number of manatees killed by approximately 10 manatees.

RESIDUAL PLOTS

To assess the overall fit of a line, one should look at a bigger picture—*all* of the residuals. A residual plot gives the "big" picture of how well our line fits the data.

A desirable residual plot is one that displays a random scatter of points. A residual plot that shows a distinct nonlinear pattern suggests that a different model would be a better choice for summarizing the data.

To create a residual plot, make a scatterplot of the residuals versus the corresponding *x* values. When looking at a scatterplot and a residual plot, take note of data points that are unusual compared to the rest of the data. An *influential point* is a data point that significantly changes the slope of the regression line.

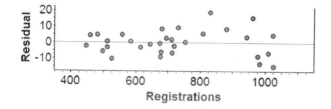

For our manatee example (residual plot shown above), we see what appears to be a random scatter of points until about 800,000 powerboat registrations (800 registrations on the graph) when we begin to see a curved pattern. The overall plot does not show an obvious curved pattern, although there is a hint of curvature for large registrations. Thus, we judge the least squares regression line to be an appropriate summary of the data. Patterns are not just curves; whenever the residuals are not random this is considered a pattern.

THE COEFFICIENT OF DETERMINATION

The coefficient of determination, r^2, gives the proportion of variability in y that can be explained by the linear association with x.

$$r^2 = 1 - \frac{SSResid}{SSTo}$$

where $SSResid$ is the sum of the squared residuals $\sum (y - \hat{y})^2$ and

$SSTo$ is the total sum of squares $\sum (y - \overline{y})^2$.

EXAMPLE Suppose that your teacher asks you to predict the number of years she has been teaching. She tells you that you are allowed to collect data on one variable from a sample of 10 other teachers at your school. Here are the data you collect:

Teacher	1	2	3	4	5	6	7	8	9	10
Number of Years Teaching	8	8	11	18	15	10	8	9	17	14

You plot the data and calculate the mean number of years teaching is 11.8 years. Based on this data, you estimate your teacher has taught for 11.8 years.

While you don't know how far off this prediction is, you could use the same method of prediction (using the sample mean $\overline{y}$ as the prediction) to make predictions for the 10 teachers in the sample. Because you know the actual y values for these teachers, you can compute the sum of the squared prediction errors to get a sense of how good this method of prediction is overall.

$$\sum\left(y - \overline{y}\right)^2 = 135.6$$

This is the SSTo, or the sum of the squares, from the mean.

Your teacher now tells you that you should collect data on a second variable for the same 10 teachers and see if you can improve your predictions. You decide that age might be related to number of years teaching so you also collect data on age. You now have the following data:

Teacher	1	2	3	4	5	6	7	8	9	10
Age	30	36	35	41	37	33	31	34	39	39
Number of Years Teaching	8	8	11	18	15	10	8	9	17	14

For the new data is where x is the teacher's age and y is the number of years teaching, you calculate the equation of the least squares regression line. In order to estimate the number of years your teacher has taught, you need to know her age. She tells you she is 33 years old. Using the regression line, you predict that your statistics teacher has taught for 9.44 years.

$$\widehat{y} = -21.719 + .944x$$

Let's again calculate the sum of squared prediction errors by using the regression line to make predictions for the 10 teachers in the sample. The sum of the squared difference between actual number of years teaching and predicted number of years teaching is

$$\sum\left(y - \widehat{y}\right)^2 = 31.7373$$

How much better did the second method for making predictions do? Let's find the proportion reduction in the sum of squared prediction error achieved by adding the second variable. Remember that proportion change is found by dividing the change by the original value.

$$\frac{\text{change}}{\text{original value}} = \frac{135.6 - 31.7373}{135.6} = 0.7659$$

This is r^2! By using the regression line to make predictions, you have achieved a 76.59% reduction in the sum of squared prediction errors. This can also be interpreted as meaning that 76.59% of the variability in number of years teaching is explained by the linear relationship created with age.

EXAMPLE We can also find the value of r^2 in computer regression output.

```
Regression Analysis: Manatees versus Registrations

The regression equation is
Manatees = - 40.6 + 0.125 Registrations

     Predictor       Coef     SE Coef        T        P
     Constant      -40.641      5.840     -6.96    0.000
  Registrations   0.125006    0.007867    15.89    0.000

S = 7.87985       R-Sq = 89.7%      R-Sq(adj) = 89.3%

Analysis of Variance

     Source     DF      SS       MS        F        P
   Regression    1    15677    15677    252.48    0.000
 Residual Error  29    1801      62
     Total       30   17477
```
←SSResid
←SSTo

AP Tip

Notice that in the computer output, r^2 is given but not r. To find r, take the square root of r^2 and then look at the sign of the slope to determine the sign of r.

THE STANDARD DEVIATION ABOUT THE LEAST SQUARES REGRESSION LINE

The standard deviation about the least squares regression line is given by

$$s_e = \sqrt{\frac{SSResid}{n-2}}$$

Roughly speaking, s_e is a typical amount by which an observation deviates from the least squares regression line. Remember that the standard deviation for one variable data is a typical distance from the mean. Now for two variable data, we calculate typical vertical distance from the least squares regression line.

EXAMPLE From the printout, which indicates that a typical difference between the actual number of manatee deaths and the predicted number of manatee deaths is 7.87985 deaths. This number is quite small, indicating that our model is a good predictor of manatee deaths.

Regression Analysis: Manatees versus Registrations

The regression equation is
Manatees = - 40.6 + 0.125 Registrations

Predictor	Coef	SE Coef	T	P
Constant	-40.641	5.840	-6.96	0.000
Registrations	0.125006	0.007867	15.89	0.000

S = 7.87985   R-Sq = 89.7%   R-Sq(adj) = 89.3%

SAMPLE PROBLEM 3 Is the least squares regression line you found in Sample Problem 2 an appropriate model for the data?

Outside Temperature (°F)	60	62	63	65	67	71	72	74	75
Number of Students Absent	1	0	2	5	6	5	8	7	8

SOLUTION TO PROBLEM 3 The residual plot (shown below) shows a random scatter with no clear pattern and a high r^2 value (). Thus, we believe that a linear model is an appropriate way to summarize the data.

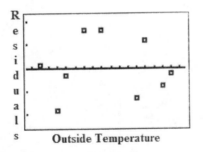

NONLINEAR RELATIONSHIPS AND TRANSFORMATIONS

(*Statistics: Learning From Data*, 1st ed. pages 231–248)

Unfortunately, the relationship between two numerical variables is not always linear. When the residual plot shows a curved pattern, we should fit a curve to the data. A common method for fitting a curve to a nonlinear bivariate data is through the use of transformations.

EXAMPLE The study of prehistoric birds depends on fossil information, which typically consists of imprints in stone of a prehistoric creature's remains. To study ancient ecosystems effectively, it would be useful know the actual mass of individual birds, but this information is not preserved in the fossil record. It seems reasonable that the biomechanics of birds is much the same today as in the past. For example, the relationship between the wing length and total weight of a bird today should be very similar to what it was in the distant past. The wing lengths of ancient birds can be obtained from the fossil record, but the weight cannot.

Data are available for some species of modern birds of prey and are given below. We will attempt to "predict" the total weight of a bird using the wing length.

Wing length and total weight of modern species of birds of prey

Bird species	Wing length (cm)	Total weight (kilograms)
Gyps fulvus	69.8	7.27
Gypaetus barbatus grandis	71.7	5.39
Catharista atrata	50.2	1.70
Aguila chrysatus	68.2	3.71
Hieraeus fasciatus	56.0	2.06
Helotarsus ecaudatus	51.2	2.10
Geranoatus melanoleucus	51.5	2.12
Circatus gallicus	53.3	1.66
Buteo bueto	40.4	1.03
Pernis apivorus	45.1	0.62
Pandion haliatus	49.6	1.11
Circus aeruginosos	41.3	0.68
Circus cyaneus (female)	37.4	0.472
Circus cyaneus (male)	33.9	0.331
Circus pygargus	35.9	0.237
Circus macrurus	35.7	0.386
Milvus milvus	50.7	0.927

The regression equation is: $\overline{\text{TotWeight}}$ = - 6.87 + 0.173 WingLen

Here is the scatterplot of the data. It appears that the relationship between total weight and wing length might be best described by a curve rather than a line.

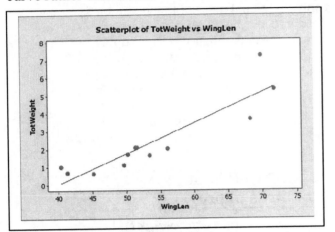

The residual plot brings out the nonlinear nature of the data:

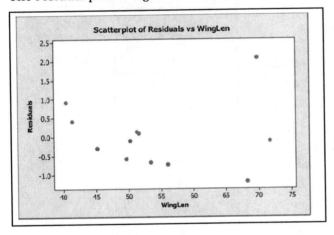

The residual plot clearly has a nonlinear pattern so a linear model is not the most appropriate way to summarize the data. One way to analyze the data is to transform the data to make the pattern linear. Two common transformations are exponential transformations and power transformations.

If we believe the relation is exponential we can "straighten the relationship" by substituting $y' = \log y$. Here is the scatterplot of the data and residual plot after an exponential transformation.

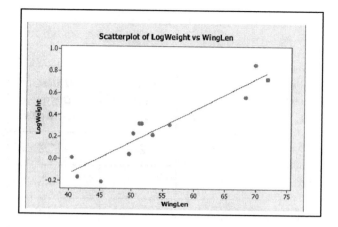

The regression equation for the transformed data is

$\overline{\text{LogWeight}} = -1.30 + 0.0290$ WingLen

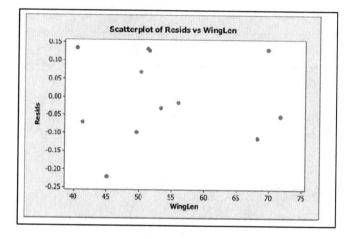

The transformation results in data that are linearly related. The random pattern in the residual plot for the transformed data (shown above) also confirms that an exponential model for the original data may be a more appropriate way to summarize these data.

EXAMPLE The home range of an animal (the area over which the animal typically travels) is a function of diet and energy consumption. The energy consumption is, in turn, typically a function of the animal's size. Data were gathered on 27 similar animal species and are presented in the table below.

Body weight vs. Home range for representative insectivorous mammalian species

Body Weight (g)	Home Range (Hectares)	Body Weight (g)	Home Range (Hectares)	Body Weight (g)	Home Range (Hectares)
1,000	180	142	0.1	210	1.2
3,900	110	100	20	58	0.3
11,600	87	66	0.1	45	0.1
2,400	4	21	2.9	700	10
780	31	20	0.4	100	0.3
1,260	240	5	0.2	3,800	25
270	30	10	0.4	4,500	78.5
3,400	64	3,300	3.5	20	0.4
8,300	100	200	5	1,200	1.8

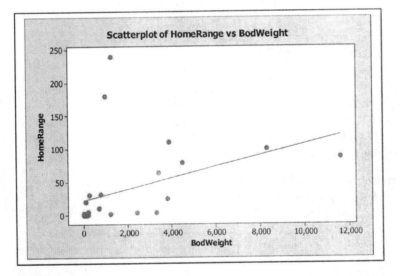

The scatterplot reveals that most of the animals have small values for body weight and home range, but a few have large values of body weight only or home range only. Thus, the linear model appears to be deficient. When distributions of both variables are skewed, a power transformation can be effective in "linearizing" the data. With a power transformation we make the following transformations:

$$x' = \log x$$

$$y' = \log y$$

Incredible as it might seem, these transformations result in a very nice linear fit:

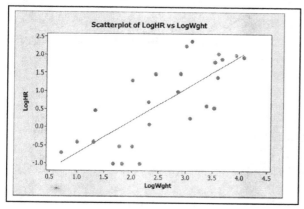

The regression equation is: LogHR = - 1.60 + 0.893 LogWght. The relation is not overly strong, as can be seen by considering the standard deviation of residuals and the coefficient of determination:

$$S = 0.734543 \quad R\text{-}Sq = 59.1\% \quad R\text{-}Sq(adj) = 57.4\%$$

However, the random scatter seen in the residual plot suggests modeling the relationship between log y and log x using a line is appropriate.

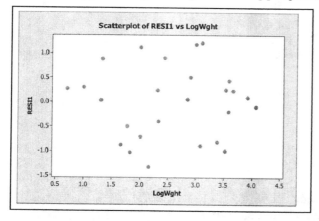

SUMMARIZING BIVARIATE DATA: STUDENT OBJECTIVES FOR THE AP EXAM

- You should be able to draw a scatterplot with labeled axes.
- Describe a scatter plot (form, direction, and strength in context).
- You should be able to interpret the sample correlation coefficient in context.
- You should be able to find the equation of the least squares regression line.
- You should be able to interpret the least squares regression line intercept (when appropriate) in context.
- You should be able to interpret the least squares regression line slope in context.

- You should be able to use the least squares regression line to make a prediction.
- You should be able to calculate and interpret a residual in context.
- You should be able to assess linearity of a data set using a residual plot.
- You should be able to interpret the coefficient of determination in context.
- You should be able to interpret standard deviation of the residuals in context.
- You should be able to read computer regression output.
- You should be able to use transformations to linearize data (specifically exponential and power transformations).

MULTIPLE-CHOICE QUESTIONS

1. Which of the following pairs of variables would be expected to have a negative relationship?
    (A) GPA and height
    (B) Number of chores and weekly allowance
    (C) Number of hours spent studying and test grade
    (D) Number of miles driven and amount of gas remaining in the gas tank
    (E) Age and IQ

2. Given $\bar{x} = 3.8, s_x = 1.2, \bar{y} = 2.7, s_y = 0.5,$ and $r = 0.6408.$ Find the equation of the least squares regression line.
    (A) $\hat{y} = -3.144 + 1.538x$

    (B) $\hat{y} = -.3526 + 1.538x$

    (C) $\hat{y} = 1.6854 + 0.267x$

    (D) $\hat{y} = 3.0791 + 0.267x$

    (E) $\hat{y} = 1.726 + 0.2563x$

3. Meg has a set of bivariate numerical data with a correlation coefficient of $r = 0.83.$ Elise has a data set with a correlation coefficient of $r = -0.83.$ What can you conclude about the two sets of data?
    (A) The scatterplots for these two data sets both display a strong linear pattern.
    (B) In Meg's data, 83% of the data points are closely related.
    (C) In Elise's data, 83% of the variability in y can be explained by the linear association with $x.$
    (D) Meg's data is more linear than Elise's data.
    (E) Nothing can be concluded about the two data sets without looking at a scatterplot.

4. Based on the scatterplot below, which of the following is true?

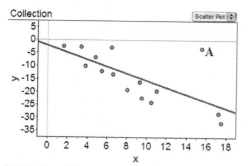

(A) Point A is an outlier and a potentially influential point.
(B) Point A is an outlier but not a potentially influential point.
(C) Point A is not an outlier but is a potentially influential point.
(D) Point A is not an outlier or a potentially influential point.
(E) None of these are true.

5. Given the following Minitab output, which of the following is false?

Regression Analysis: y versus x				
Predictor	Coef	SE Coef	T	P
Constant	-0.868	2.330	-0.37	0.716
x	-1.6914	0.2342	-7.22	0.000

S = 4.34426   R-Sq = 80.0%   R-Sq(adj) = 78.5%

(A) 80% of the variability in y is explained by the linear relationship with x.
(B) Since r = 0.898, the linear relationship between x and y is linear, strong, and positive.
(C) As x increases by one unit, y decreases, on average, 1.6914 units.
(D) The intercept of the least squares regression line is –0.868.
(E) The equation of the least squares regression line is
$\hat{y} = -0.868 - 1.6914x$.

6. Which of the following scatterplots could have created the following residual plot?

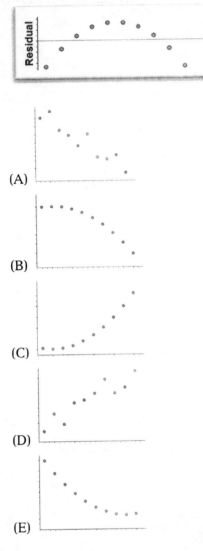

(A)

(B)

(C)

(D)

(E)

7. Two variables are exponentially related. The least squares regression line for transformed data is found to be ln $\widehat{y}$ = –0.041 + 0.0035x. What is the predicted value of y when x = 16.83?
(A) 2.49
(B) 1.018
(C) 12.98
(D) 16.74
(E) 54.15

Questions 8–12 refer to the following set of data:

The early human ancestors were similar in shape to most large primates. The data below are average male hind limb and forelimb lengths for different species of early hominids (humans and their ancestors). The relatively short hind limbs are thought to represent evolutionary specialization for vertical climbing.

Hindlimb and Forelimb lengths

Hindlimb Length (mm)	Forelimb Length (mm)
471	458
361	514
399	581
557	739
553	553
574	614
857	595
698	762

Computer output for a regression analysis relating the forelimb length to the hindlimb length is shown below:

```
The regression equation is
ForeLen = 436 + 0.297 HindLen


Predictor      Coef    SE Coef      T      P
Constant      435.8      135.9   3.21   0.018
HindLen      0.2974     0.2349   1.27   0.252


S = 99.9009    R-Sq = 21.1%    R-Sq(adj) = 7.9%
```

The scatterplot of the data is shown below:

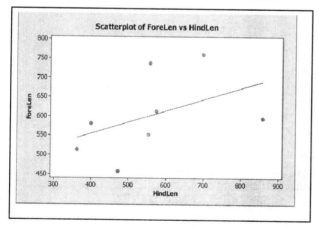

8. Which of the following is the correct interpretation of the slope of the least squares regression line?
   (A) An additional mm of forelimb length results in an increase in predicted hindlimb length of 435.8 mm.
   (B) An additional mm of hindlimb length results in an increase in predicted forelimb length of 435.8 mm.
   (C) An additional mm of hindlimb length results in an increase in predicted forelimb length of 0.297 mm.
   (D) An additional mm of forelimb length results in an increase in predicted hindlimb length of 0.297 mm.
   (E) None of the above are correct.

9. Which of the following is the correct interpretation of the intercept of the least squares regression line?
   (A) An additional mm of forelimb length results in an increase in predicted hindlimb length of 435.8 mm.
   (B) An additional mm of hindlimb length results in an increase in predicted forelimb length of 435.8 mm.
   (C) An additional mm of hindlimb length results in an increase in predicted forelimb length of 0.297 mm.
   (D) An additional mm of forelimb length results in an increase in predicted hindlimb length of 0.297 mm.
   (E) None of the above are correct.

10. Which of the following is a correct interpretation of the coefficient of determination?
    (A) 21.1% of the variability in forelimb length is explained by the linear relationship with hindlimb length.
    (B) 21.1% of the variability in hindlimb length is explained by the linear relationship with forelimb length.
    (C) 45.9% of the variability hindlimb length is explained by the linear relationship with forelimb length.
    (D) 45.9% of the variability in forelimb length year is explained by the linear relationship with hindlimb length.
    (E) The linear relationship between forelimb length and hindlimb length is strong and positive.

11. The predicted forelimb length of a hominid with a hindlimb length of 600 is
    (A) Approximately 400 mm.
    (B) Approximately 500 mm.
    (C) Approximately 600 mm.
    (D) Approximately 700 mm.
    (E) Approximately 800 mm.

12. Which of the following is false?
    (A) This is a good model; the residual plot would show no distinct pattern.
    (B) The slope of the line and the correlation coefficient are both positive.
    (C) The standard deviation indicates the scatterplot is curved.
    (D) The data point corresponding to (557, 739) is an outlier.
    (E) The correlation coefficient indicates that the linear relationship between forelimb length and hindlimb length is not very strong.

13. Suppose that the coefficient of determination between two variables x and y has a value equal to 1. Which of the following is NOT true?
    (A) There is a perfect linear relationship between x and y.
    (B) A scatterplot of x and y would show a linear relationship.
    (C) There is not a strong relationship between x and y.
    (D) 100% of the variability in y can be explained by the linear relationship with x.
    (E) 100% of the variability in x can be explained by the linear relationship with y.

14. A set of data has the following scatterplot. Which of the following is true?

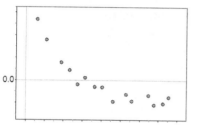

    I.   A line is not an appropriate model for the data.
    II. The least squares regression line has an intercept of 0.
    III. The least squares regression line has a slope of 0.
    (A) I only
    (B) II only
    (C) III only
    (D) I and III only
    (E) I, II, and III

15. The correlation between x and y is 0.84. What is the correlation between $3x + 2$ and $5 + y$?
    (A) 0.84
    (B) 0.48
    (C) –0.84
    (D) –0.48
    (E) Cannot be determined without the original data.

FREE-RESPONSE PROBLEMS

1. In a study of the growth of sparrows, data were gathered on the wing length (cm) and age (days) of 13 sparrows. A regression was performed with the following results:

Age (days)	Wing Length (cm)	Age (days)	Wing Length (cm)	Age (days)	Wing Length (cm)
3	1.4	9	3.2	15	4.5
4	1.5	10	3.2	16	5.2
5	2.2	11	3.9	17	5.0
6	2.4	12	4.1		
8	3.1	14	4.7		

The regression equation is
WingLength(cm) = 0.713 + 0.270 Age(days)

```
Predictor      Coef   SE Coef      T       P
Constant     0.7131   0.1479    4.82   0.001
Age(days)   0.27023   0.01349  20.03   0.000

S = 0.218405   R-Sq = 97.3%   R-Sq(adj) = 97.1%
```

(a) What is the equation of the least squares regression line that relates wing length and age for these birds?
(b) Interpret the slope of the least squares regression line. Does it make sense to interpret the intercept in this context?
(c) What is the value of the correlation coefficient? What does the correlation coefficient tell you about the strength of the relationship in this context?
(d) Do you feel the least squares regression line would be useful in predicting the wing length of sparrows that are 100 days old? Why or why not?
(e) There are no observations with a wing length with age = 7 days. Based on the regression analysis, what wing length would be expected for 7-day-old sparrows?

2. In an AP Statistics class, the teacher wanted to show the students that humans are not good at guessing or estimating quantities. She displayed a series of clear containers with different numbers of gumballs inside and asked each student to guess the number of gumballs in one of the containers. A large number of students participated in this study. The data was then used to produce the following computer regression output and residual plot.

Regression Analysis: Guessed Gumballs versus Actual Gumballs				
Predictor	Coef	SE Coef	T	P
Constant	28.52	46.80	0.61	0.543
Actual Gumballs	2.1690	0.1324	16.39	0.000
S = 140.344	R-Sq = 63.7%	R-Sq(adj) = 63.5%		

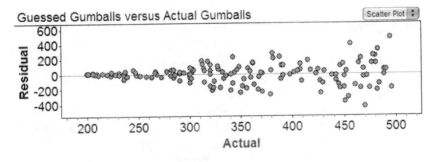

(a) What is the equation of the least squares regression line for predicting guessed number of gumballs from actual number of gumballs.
(b) Comment of the appropriateness of a linear model for these data.
(c) After making his guess, one student counted 325 actual gumballs in the container he was shown. For this data point

the residual was determined to be 76.555. What was the student's guess for the number of gumballs in the container?

Answers

MULTIPLE-CHOICE QUESTIONS

1. **D.** Choice (D) because as miles driven increase, the amount of gas left in the gas tank decreases (*Statistics: Learning From Data,* 1st ed. pages 182–197).

2. **C.** Solving for the slope we get: 0.6408*0.5/1.2=0.267. Solving for the y-intercept we get: 2.7-0.267*3.8=1.6854. So, the equation of the least squares regression line is =1.6854 +0.267x (*Statistics: Learning From Data,* 1st ed. pages 205–206).

3. **E.** You must look at scatterplots before you can assess the linearity of each data set (*Statistics: Learning From Data,* 1st ed. pages 182–197).

4. **A.** Point A is an outlier since it has a large residual. Point A is an influential point since it appears to be extreme in the x-direction (*Statistics: Learning From Data,* 1st ed. pages 96-98).

5. **B.** Since the slope of the least squares regression line is negative, the correlation coefficient must also be negative (*Statistics: Learning From Data,* 1st ed. pages 205–206).

6. **B.** The points farthest to the left and farthest to the right on the scatterplot will be below the least squares regression line (negative residuals), and the points in the middle of the scatterplot will be above the least squares regression line (positive residuals) (*Statistics: Learning From Data,* 1st ed. pages 214–217).

7. **B.** $\ln y = 0.017905$

   $y = 1.018$

   (*Statistics: Learning From Data,* 1st ed. pages 238–240)

8. **C.** Note the use of the word "predicted" in the answer (*Statistics: Learning From Data,* 1st ed. pages 205–206).

9. **E.** Note the use of the word "predicted" in the answer (*Statistics: Learning From Data,* 1st ed. pages 205–206).

10. **A.** Note the use of the word "linear" in the answer (*Statistics: Learning From Data,* 1st ed. pages 217–220).

11. **C.** Predicted forelimb length is: ForeLen = 435.8 + 0.2974(600) = 614.24 (*Statistics: Learning From Data,* 1st ed. pages 205–206).

12. **C.** The standard deviation of the residuals cannot provide any information about curvature in a scatterplot (*Statistics: Learning From Data,* 1st ed. pages 205–206).

13. **C.** When the coefficient of determination is equal to 1, there is a perfect linear relationship between *x* and *y* (*Statistics: Learning From Data,* 1st ed. pages 217–220).

14. **A.** The curved pattern in the scatterplot shows that a linear model is not the most appropriate model (*Statistics: Learning From Data,* 1st ed. pages 214–217).

15. **A.** The correlation coefficient does not change after this linear transformation on *x* and/or *y* (*Statistics: Learning From Data,* 1st ed. pages 185–187).

FREE-RESPONSE PROBLEMS

1. (a) predicted Wing Length = 0.713 + 0.270*Age

   (b) The slope of the least squares regression line is interpreted as meaning that on average, we expect the wing length to increase by 0.270 cm per day.

   We should be cautious about interpreting the intercept as wing length at 0 days (at hatching) because the smallest *x* value in the data set is 3 days.

   (c) r = 0.986; There is a strong, positive, linear relationship between wing length and age.

   (d) This equation should not be used to predict wing lengths far beyond the data. It is likely that in time growth would slow, and that a different model would be appropriate over longer time spans.

   (e) WingLength(cm) = 0.713 + 0.270 Age(days)

      WingLength(cm) = 0.713 + 0.270 * 7

      = 2.603 cm.

   (*Statistics: Learning From Data,* 1st ed. pages 185–187 and 205–206).

2. (a) predicted guess of number of gumballs = 28.52 + 2.169 × actual gumballs.

   (b) The residual plot does not show a discernable pattern, which provides strong evidence that a linear model is an appropriate model. The residual plot also shows that as the actual number of gumballs in a container increases, the least squares regression line is a less accurate predictor due to the increasing size of the residuals.

   (c) Since residual = actual – predicted, then actual = residual + predicted.

   actual = residual + predicted

   actual = 76.555 + (28.52 + 2.1690 × 325)

   actual = 76.555 + 733.445

   actual = 810

   The student guessed 810 gumballs in the container

   (*Statistics: Learning From Data,* 1st ed. pages 211–217).

4

COLLECTING DATA SENSIBLY

Data are used to make informed decisions in a variety of settings. But to be useful, data need to be collected in a sensible way. In this section, we look at data collections and we discuss the kind of conclusions that can be drawn from statistical studies.

OBJECTIVES

- Distinguish between an experiment and an observational study.
- Identify the most common sampling methods.
- Choose an appropriate sampling method for an observational study.
- Identify different types of bias that arise in sampling situations.
- Know the characteristics of a well-designed experiment.
- Design an experiment to investigate the effect of an explanatory variable on a response.

COLLECTING DATA SENSIBLY

(*Statistics: Learning From Data*, 1st ed. pages 13–23)

In order to make good statistical decisions, it is important to collect data that represents the group we want to make decisions about in our study. Representative data allow us to generalize the findings of our research to a larger population of interest. A well-designed experiment may allow us to identify causal relationships. A well-designed experiment with a representative sample allows us to talk about the likely causes and effects between the variables in the experiment and generalize to the population represented by the sample.

The first question to ask in the data collection process is, "what are we interested in knowing from the data?" There are two different approaches based on what type of information we want to glean and the type of data we need to gather. For example, if we want to know the average weight of 2-year-old calico cats, we would simply observe a collection of these cats and record the weights. This would be an *observational study* since we are only concerned with what already exists in the calico world. However, if we wanted to determine whether a new type of food would impact the weight of 2-year-old calico cats in a specific way, we would need to do more than observe. We would conduct an experiment using the new food type. We could discover how the new type of food impacts the weight of the cat by carefully studying the use of the new food type with our target population under very controlled conditions that allow us to rule out other influences on the weight of the cats besides the food itself. We would set up an experiment by manipulating the diet of the cats while attempting to keep values of other variables as similar as possible. We would need at least one comparison group of cats that are similar to the cats that are getting our food. Ideally they would be similar in every relevant respect except that they would not get the new food. Of course, it would not work to have cats in the control group eat nothing. We would then be studying the difference between cats that eat something and cats that eat nothing. That would not tell us how our new food compares to other cat foods in causing weight gain in cats. We would need a different food for our comparison group. These two different diets are known as the treatments that we would study, since we would be giving some cats one food and other cats a different food. This is an *experiment* because we are imposing a treatment on the experimental subjects.

SAMPLING

(Statistics: Learning From Data, 1st ed. pages 13–24)

Choosing a good representative sample is very important in the data collection process. It would be great if we could study the entire population. However, since this is typically not possible (affordable), and often not desirable, the researcher must rely on *sampling* in order to make decisions about the population. It is critical that we collect a sample in a way that makes it likely that the sample will be representative of the population. One way to achieve this is by selecting a random sample. Random sampling procedures help us minimize some types of bias.

BIAS

There are several types of bias whose presence could result in a sample that is not representative of the actual population of interest and thereby prevent us from knowing if the findings of the study would apply to the population at large. When a subgroup of a population is either systematically excluded or favored, this can introduce a form of bias called selection bias. Selection bias is just one type of bias that we should be concerned about. In selecting a sample, the researcher should be mindful of the following types of bias and attempt to minimize them.

SELECTION BIAS Selection bias occurs if some elements in a population are systematically favored or excluded from participation. For example, if we want to know about the senior class but only survey students in 12th grade English class on a particular day, then seniors who do not have 12th grade English or who were absent on that day will not be included in the sample. Our study would not tell us about the senior class—at best it tells us only about seniors in senior English class on that day.

MEASUREMENT/ RESPONSE BIAS Measurement bias can occur in a couple of ways. First, if the wording of a question favors a particular response or outcome, this introduces bias. In addition, if measuring equipment is working improperly, then it is possible that the measurements are inaccurate.

NONRESPONSE BIAS This type of bias can occur if a group that would have altered the overall results had they participated does not respond to a survey. This happens often in surveys when people choose not to submit their responses. Diligence should be used to ensure there is no systematic bias arising from the failure of individuals to respond to the survey.

VOLUNTARY RESPONSE BIAS Some individuals may tend to have strong opinions (for or against) on specific issue and they may tend to respond to surveys in a higher proportion. To help avoid this type of bias, it is important that individuals aren't simply allowed to volunteer to be participants.

AP Tip

When considering whether bias may have occurred, it is helpful to ask three questions:

1. Could the question wording tend to lead to a certain response?
2. Was the sample chosen in a way that would tend to exclude some parts of the population?
3. Could the results have been impacted by missing responses from those who did not respond?

If any of these problems occur, the results may not reflect characteristics of the population.

SAMPLE PROBLEM 1 Each of the following scenarios contains a potential source of bias. Identify and discuss the potential bias.

(a) Students in Algebra II were asked to read the course descriptions and choose their mathematics course for the following year based on the course descriptions below. The descriptions for all three courses are given.

   1. Analysis is for hard-working students who will do their homework.

   2. AP Statistics involves exciting data analysis and fun activities.

3. Finite Mathematics is for students who aren't planning to take other mathematics courses or go to college.

    For mathematics, I want to take _____.

(b) A phone survey was conducted to determine the need for after-school help sessions at a local student center. Phone numbers were generated at random and calls were made between the hours of 9 a.m.–11 a.m.

SOLUTION TO PROBLEM 1

(a) The question wording may lead to response bias. It makes the AP Statistics class sound more interesting than the others.

(b) The survey methodology may suffer from selection and nonresponse bias. Since the survey will be done by phone, it systematically excludes anyone who doesn't have a phone. In addition, since the calls will occur during daytime hours, families in which all adults work may be missed and these individuals may be more likely to want such a service.

RANDOM SAMPLING

A *simple random sample* is chosen in a way that ensures that every different possible sample of a given size n from the population has an equal chance of being selected. A random sample can be selected with replacement or without replacement. If a chosen individual is not eligible to be chosen again, this is called sampling *without replacement*. If, however, the individual is chosen, then placed back in the group and can be selected again, this is called sampling *with replacement*.

EXAMPLE Suppose that five prizes are available. A sample of five students will be selected to receive these prizes. The five winners will be chosen by pulling names out of a hat.

Without Replacement. If a person can only win one prize, once selected the name is removed and may not be drawn for additional prizes. This is sampling without replacement.

With Replacement. If we are going to pull a name out of the hat and then once the winner is given a prize we put the name back in the hat before drawing a name for the next prize, this is sampling with replacement.

Selecting a random sample can be done in a variety of ways. Some of the most common methods involve the use of a calculator, computer software, or a random digits table. No matter which method is used, it is important that each individual in the population be identified and then that each of them has an equal chance of being chosen.

Here are some examples of methods for selecting a random sample of size 12 from a group of 100 individuals who have each been assigned a unique number between 1 and 100.

Graphing Calculator. We would use the calculator to generate random integers between 1 and 100 and the first 12 distinct integers would be used to identify the individuals in the sample. On the TI-83 or TI-84 calculator, press **randint(1,100)[enter].** Continue hitting enter until we have 12 numbers. If we are sampling without replacement, we would ignore any duplicates that occur (until we have 12 numbers) since we don't want to choose the same person more than once.

Software. In this case, we will still generate 12 random integers but the manner in which this is done will be determined by the software package that is being used.

Random Digits Table. Since we need 100 numbers, we would number the participants 00–99 (or 001–100). Next, using any random digits table, go to any line on the table and read the digits in pairs to form two-digit numbers. The first 12 two-digit numbers you find on the table will indentify the participants. Again, any duplicates would be ignored and another two-digit number would be chosen if we are sampling without replacement.

AP Tip

The AP exam may require you to demonstrate that you know how to select a random sample using a random digit table. Be sure you know how to do this for the exam. Keep it simple and consider using the following strategies in choosing your participants.

10 or fewer participants—number 0–9. (For, say 6 participants, number 1–6 and ignore the digits 7–9 and 0 if they come up as well as any duplicates that occur.)

101–999 participants—number 000–999. (Again, ignore random numbers not needed or duplicates.)

1001–9999 participants—number 0001–9999.

Stopping Rule: How many participants will be picked?

(If given an option how to randomize on the AP Exam, the easiest way to randomize is to put "all the names in a hat, mix the names well, and then select n names)

SAMPLE PROBLEM 2 There are 863 seniors at a large high school.

(a) Explain how you would use a random number table to select a random sample of 20 seniors.

(b) Using the random digits below, select your first 5 participants using the scheme in (a).

56964	70827	35916	37811	97850
60854	11466	59330	18764	11778
68295	49267	12487	64350	09014

SOLUTION TO PROBLEM 2

(a) Number the seniors 001–863. Then, go to a randomly selected row of the random number table and select three-digit numbers from left to right. The seniors corresponding to the first 20 different three-digit numbers will be included in the sample. Any duplicates as well as numbers from 864–999 and 000 are to be excluded.

(b) The first 5 participants would be: 569, 647, 082, 735, and 378. See the circled numbers below to follow the scheme used. Notice the number 916 was skipped as it does not correspond to one of the seniors.

```
5 6 9 6 4    7 0 8 2 7    3 5 9 1 6    3 7 8 1 1    9 7 8 5 0
6 0 8 5 4    1 1 4 6 6    5 9 3 3 0    1 8 7 6 4    1 1 7 7 8
6 8 2 9 5    4 9 2 6 7    1 2 4 8 7    6 4 3 5 0    0 9 0 1 4
```

OTHER SAMPLING METHODS

There are other sampling strategies we might want to consider as alternatives to simple random sampling. For example, suppose a company has 58 male supervisors and 21 female supervisors. If five are going to be selected at random to be interviewed about job satisfaction, it is possible to get all males or all females, which may not provide a fair measure of job satisfaction. In such a situation, we may want to consider a different sampling method.

STRATIFIED RANDOM SAMPLING. When a population of interest can be divided into homogenous subgroups that don't overlap (such as male/female or freshman, sophomore, junior, and senior) stratifying may help to ensure that each subgroup, called *strata*, is represented in the sample. In this case, a set number of individuals will be selected at random from each subgroup (stratum) so that each subgroup is represented in the sample. Stratified sampling guarantees that all subgroups will be included in the sample.

CLUSTER SAMPLING. In cluster sampling, the population is divided into subgroups that are each as much like the population as possible. Instead of selecting a sample from each subgroup as in stratified sampling, a few of the subgroups are selected at random and the entire subgroups are included in the sample. For example, suppose that you want to know something about pillows that are packaged in crates of three dozen pillows per crate. If the manufacturer wanted to check the quality of nine dozen pillows, it would be easier to randomly select three of the crates at random and then study all of the pillows in these three crates.

SYSTEMATIC SAMPLING. This method works well if the population is easily arranged in a sequential order. We then randomly choose one of the first k individuals in the sequence and then we select every kth individual in the sequence from that starting point. For example, we might select one of the first 10 people to enter a store and then every 10th person that enters the store after that.

SAMPLE SIZE

It is a common misconception that a relatively small sample cannot reflect the overall population well if the population is very large. However, this is not true. If the researcher chooses a sample carefully and uses random selection, the samples will most likely reflect population characteristics. The key is to use a random sample.

OBSERVATIONAL STUDIES

(*Statistics: Learning From Data,* 1st ed. pages 9–10)

When a researcher collects information (data) about a population without imposing any type of treatment, the study is called an observational study. To use the earlier calico cat example, a researcher might select a sample of 2-year-old calico cats and record their weights. The researcher has not set out to manipulate any factor (or impose a treatment). Rather, the researcher is just observing the value of some variable in order to learn about the population of calico cats.

In observational studies, it is important to select a sample that is representative of the population. This might be done by selecting a simple random sample or by selecting a stratified random sample.

SAMPLE PROBLEM 3 A local school board wants to design a survey to see if more than 50% of the community would support increasing school funding by approving a local 1% tax on all restaurant purchases. The city has a large urban population as well as a smaller suburban population. How should the school board select a sample from this community?

SOLUTION TO PROBLEM 3 If the sample size is large, a simple random sampling plan would be fine and both groups (urban and suburban) should be represented in the sample. However, it would also make sense to stratify residents into urban and suburban strata and then select a random sample from each of these two strata.

EXPERIMENTAL DESIGN

(*Statistics: Learning From Data,* 1st ed. pages 24–44)

Unlike observational studies, in an *experiment* the researcher imposes some treatment in order to see if there is an effect on some response variable. Researchers performing experiments must not only consider what information they would like to gather, but also a multitude of other variables that may cloud interpretation of the results of the study.

There are several key components to a well-designed experiment that will allow the researcher to attribute differences in the response to the treatment and possibly generalize this finding to a broader population. A key difference between an experiment and an observational study is that in an experiment, the researcher assigns treatments to experimental units.

To use the calico cat example again, suppose the researcher wants to see if using a different type of food affects the weight of 2-year-old calico cats. The researcher will want to keep as many other outside variables as possible from changing to insure that only the controlled diet would be responsible for any observed differences in weight between the experimental groups. Once the sample of cats is chosen for participation in the experiment, they would need to be randomly assigned to two groups. One group would get the old diet (diet A) and the second group would get the new diet (diet B). The researcher would feed the cats and observe weight gain at the end of the study. This is a very basic design that includes comparing the results of only two treatment groups. Let's take a look at the key information that must be included for a complete description of an experiment of this type.

First, include an explanation of what the experimental units for the experiment are in your description. Next, the subjects should be divided into two groups: diet A and diet B. This assignment must be random and the method of assignment should be *clearly* explained. Once assigned to treatments, the cats will be weighed prior to being placed on the controlled diets and after they have been put on the diets. In demonstrating this for the AP exam, you can use a diagram to illustrate your design, but be sure to also include a written description of the important aspects of the design.

EXAMPLE Suppose there are 80 calico cats available for an experiment to see if diet has an effect on weight gain.

The diagram would look like this:

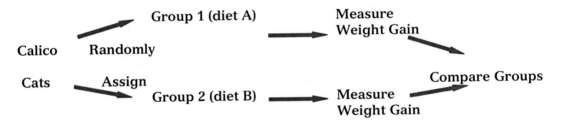

Here is an example of an accompanying statement:

> *Number the cats 01–80. Then, select two-digit numbers from a designated row in a random number table. The first 40 different numbers chosen will identify the cats who receive diet A and the rest will receive diet B. Any numbers that are repeated or 81–99 and 00 will be skipped. The cats will be weighed at the beginning of the experiment and again after a specified time on the diet. Mean weight gain for the two diets will be compared to determine if there is a difference for the two diets. (It is important for students to explain their design/picture on the AP Exam for full credit.)*

The plan for execution of the experiment is called the *design* of the experiment. The previous example illustrated a simple design, but sometimes more complicated designs are required. For example, an experiment may include *blocking* to create similar groups for comparison, such as males and females. A *placebo* may also be included with human subjects. First, we will offer a few definitions and then we will look at different ways to describe and to diagram these other experimental designs.

BLOCKING. Divide the subjects into blocks, or groups, that contain some key common factors that may be related to the response. For example, heart medication may have a different effect on men and women. With this noted, the subjects should first be divided into groups (called blocks) by gender and then subjects within each group should be randomly assigned to the treatments.

CONTROL GROUP. A control group is a group that exists to provide a comparison with treatment group(s). For example, if there is an existing heart medication, the control group would continue taking this and doing all things as usual. This helps determine if the new medication is actually causing a change to occur. Not all experiments need a control group, but all experiments need to control for outside variables entering the experiment.

PLACEBO. A placebo is given to disguise the identity of the treatment from subjects that might react to this knowledge. For example, one group might get a harmless sugar pill. This harmless pill is called a placebo. It looks just like the actual treatment. In the heart example, let's say there is not a current treatment for the particular condition of interest. In this case, a sugar pill of the same size and color would be given to the control group so the researchers could see if the new drug actually works or if subjects are reacting to the idea of receiving a treatment and this is what is affecting the response.

REPLICATION. Replication involves applying the treatments to a number of different subjects within the experiment in order to have any extraneous effects "even out" in the long run. This implies that we have at least two experimental units for each treatment (and usually you would want more than two).

AP Tip

Every description of an experiment should include the following components: random assignment (describe how this will be done); imposing a treatment, blocking (if needed); control of any extraneous variables that might also affect the response; and replication. Be sure the AP readers can see a clear indication of these four components in your written description of the experimental design. State what you will compare and use units if possible.

SAMPLE PROBLEM 4 A pharmaceutical company has just developed a new medication for use with adults who are experiencing mild memory loss. Researchers have identified 500 women and 600 men who are experiencing some memory loss and who are willing to participate in this study. Currently, there is an older product on the market that seems to result in minor improvement in most patients. Design an experiment to evaluate the effect of this new medication.

SOLUTION TO PROBLEM 4

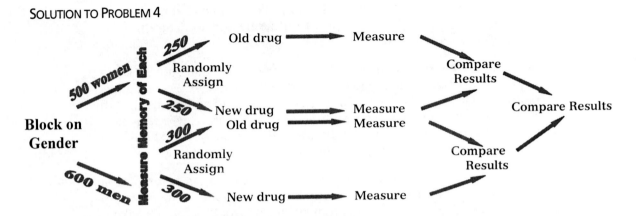

Since men and women may react differently to the drug, we will block on gender. Number the women 001–500. Then, go to the random number table and select three-digit numbers from a randomly selected row. The first 250 numbers chosen will determine which women are assigned to the new drug. The remaining 250 women will receive the old memory drug. Any numbers that are repeated or not in the range 001–500 on the random digits table will be skipped. Next, number the men 001–600. Again, start on a randomly selected line on the random number table and select three-digit numbers. The first 300 numbers will be the men assigned to the new drug while the remaining 300 will receive the old memory drug. Any repeats or numbers outside the range 001–600 on the random digits table will be discarded and another number will be chosen until we have the first 300 men assigned to the new treatment. The remaining 300 men will be assigned to the old drug treatment group. We will compare the mean memory loss between the two treatments.

CONFOUNDING VARIABLES

As the researcher attempts to ensure that only the treatment is causing any difference in response to occur, close monitoring for potential *confounding variables* must always be considered. A confounding variable is one that wasn't considered a factor in the experiment but may be affecting the response variable. The presence of a confounding variable clouds the interpretation of the results of an experiment. If there are confounding variables, any differences in response for the treatment groups cannot be attributed solely to the treatments. If there are potentially confounding variables that are known, we attempt to eliminate their potential effect by holding them constant. We count on the random assignment of subjects to treatments to even out the effects of other potentially confounding variables that we do not know about. This is why random assignment to treatment groups is so important in the design of an experiment. We randomize treatment so that the variation we see between treatments is due to chance and not some other factor.

MATCHED PAIRS

One way to avoid many potentially confounding variables is by having the same subject try both treatments. Sometimes it is possible to find pairs of subjects that are similar. For example, if two drugs are being compared, pairs of identical twins could be identified. One twin would be randomly assigned to drug X and the other to drug Y. Responses for each pair would be matched up and compared. Another type of matched pairs design is a before and after study, where the response is measured before and again after some treatment. Blocking/matching is a very powerful experimental tool in controlling for potentially confounding variables.

BLIND VERSUS DOUBLE BLIND

Finally, we consider the concept of blinding. *Blinding* occurs when the subjects or the person measuring the response do not know which treatment was received. As in the previous drug study, if the drugs look the same, then the participants receiving the old or new drug will not know which medication is the new drug or which drug they are receiving. *Double blinding* occurs when neither the subject nor the person measuring the response knows which treatment has been assigned. Any person involved in the evaluation process does not know which treatment is given. This can easily be accomplished by having the medications coded and given to the assistants that come in contact with the subject. They know which bottle to give the participant, but have no knowledge of which treatment is in the bottle.

SAMPLE PROBLEM 5 A local clinic has found 30 volunteers willing to participate in a study involving a new treatment for anxiety.

(a) Explain how the subjects in the experiment could be blinded.

(b) How could double blinding be accomplished?

(c) Could a placebo be added to this study? Explain.

SOLUTION TO PROBLEM 5

(a) Once the volunteers have been randomly assigned to either the old treatment or the new treatment, they would be given their medications each month. Both medications should look the same so the patients wouldn't know if they were getting the old or the new treatment.

(b) To make this double blind, it would be important that the individuals coming in contact with the patients and who evaluate the response to the treatment also not know which medication each patient received. This can easily be done by having the medications arrive with the patient's name on the bottle so the individual administering the treatment and measuring the response wouldn't know whether the patient was receiving treatment A or B. The researcher in charge of the experiment will know which patient received A or B and have all this information recorded.

(c) Yes. By simply adding a third group (that is, dividing the 30 subjects into 3 groups of 10) this could be done. The third group would also get a treatment but it would in fact be a harmless sugar pill that had no active ingredient.

AP Tip

Stratification is to		Blocking is to
Observational Studies	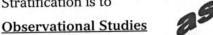	**Experimental Designs**
Strata are homogeneous groups, and because a random sample is selected from each stratum, all strata are represented in the sample.		Blocks are homogeneous groups and all treatments are tried within each block.

COLLECTING DATA SENSIBLY: STUDENT OBJECTIVES FOR THE AP EXAM

- You should be able to determine if a study is an observational study or an experiment.
- You should be able to identify potential sources of bias and provide an explanation of the bias.
- You should be able to describe an appropriate experimental design for a given study.
- You should be able to use blocking to improve an experiment when warranted.
- You should be able to describe an appropriate sampling plan for a given observational study.
- You should be able to describe how to select a simple random sample using a random digits table, your calculator, or another method such as drawing names/numbers from a hat.
- You should be able to incorporate blinding and double blinding into an experimental design when appropriate.

MULTIPLE-CHOICE QUESTIONS

1. A political polling group is interested in learning about the proportion of registered voters who favor a particular candidate for mayor. Which of the following sampling plans may have introduced bias?
   (A) individuals were selected randomly to participate from a list of registered voters.
   (B) every 50th person on a list of registered voters was contacted and asked to identify who they would vote for in the upcoming election.
   (C) a list of registered voters was first separated into two lists—one of males and one of females. Then individuals were selected at random from each of the two lists.
   (D) a survey was mailed to all registered voters and those who returned the survey make up the sample.
   (E) None of the above sampling plans creates the potential for bias.

2. Which of the following is not a <u>necessary</u> component of a well-designed experiment?
   (A) Imposing a treatment
   (B) Random assignment
   (C) Replication
   (D) Control of extraneous variables
   (E) Random selection of subjects

3. Which of the following is <u>always</u> problematic in a designed experiment?
   (A) No placebo is used
   (B) Systematic sampling is used instead of random sampling to select subjects
   (C) Subjects are not randomly assigned to treatments
   (D) No control group is used
   (E) Volunteers are randomly assigned to treatments

4. A garden club wants to use a new fertilizer on their prize petunia patch in hopes of increasing the number of petunias produced per patch. A club member suggests they set up an experiment to see if this fertilizer is better than the fertilizer they have used in previous years. The patch can be divided into eight sections in order to run this experiment. Four of the sections receive more sun while the other four sections receive more shade. Which would be the best way to assign fertilizers to the sections for this study?
   (A) Number the sections 1–8 and randomly pick four numbers from a hat to receive the new fertilizer and the remaining four would get the old fertilizer.
   (B) Number the four sections in the sun 1–4 and randomly assign the new fertilizer to two sections and the old fertilizer to the other two sections. Repeat this process with the shady sections.
   (C) Randomly assign one fertilizer to the sunny sections and the other fertilizer to the shady sections.
   (D) Put the new fertilizer down in all eight sections then grow petunias. Once these are harvested, repeat the process in all eight sections but now with the old fertilizer.
   (E) Choose either the shady or the sunny sections to experiment with and randomly assign one fertilizer to two sections and one to the other two sections.

5. What is the best reason to recommend the use of a placebo in an experiment?
   (A) When there are sufficient resources to warrant the use of a placebo
   (B) If the existing treatment is one that the researcher doesn't want to be included in the study
   (C) If the placebo can be used with animals for the appearance of an alternate treatment
   (D) When there isn't an existing treatment and you want to be sure that the effect seen is not due just to the fact that the subjects feel they are being treated
   (E) To make sure the placebo works before using it on other studies

6. Two variables are considered to be confounded if
   (A) they are both studied as part of the experiment.
   (B) their effects on the response variable cannot be separated.
   (C) more than one additional treatment group is included in the experiment.
   (D) the response variable is controlled by a placebo.
   (E) neither variable is related to the response variable.

7. The key difference between an observational study and an experiment is
   (A) the number of variables that are being studied.
   (B) the use of random selection of participants.
   (C) the ability to replicate the study.
   (D) the creation of groups of homogenous subjects to study.
   (E) the assignment of treatments to experimental units.

8. Which of the following doesn't result in some form of bias?
   (A) Choosing AP students to survey regarding their GPA in order to estimate the average GPA in a school
   (B) Polling a group of potential voters outside a Republican campaign office in order to estimate the proportion of city residents who favor a particular candidate
   (C) Surveying people who work in the marketing department of a large corporation in order to assess employee satisfaction with management policies
   (D) Choosing a stratified random sample of juniors and seniors at a high school in order to estimate the proportion of juniors and seniors who plan to go on to college
   (E) Asking shoppers leaving a grocery store if they think groceries should cost less in order to estimate the proportion of city residents who think groceries cost too much

9. A church board wants to survey church members to see which times they prefer to have services. The board wants to ensure that members who have families as well as members who are single adults are represented in the sample. Which method will yield the best chance of getting a representative sample of their members' opinions?
   (A) Send a survey to a random sample of 50 members and ask them to return it in a self-addressed envelope that was sent with the survey.
   (B) Separate church membership into two lists: families and singles. Select a random sample from each of the two lists and contact the selected members for their opinion via phone, email, or interview after a service.
   (C) After a service, have a designated person survey every kth person as they leave the service.
   (D) Call a meeting of members and survey those who attend the meeting.
   (E) Send an email survey to a sample of members who have published their email addresses in the church directory.

10. Which of the following are false statements?
    I.   Convenience samples are rarely useful.
    II. Every good experiment includes blocking.
    III. Every experiment should include a placebo.
    (A) I only
    (B) II only
    (C) I and II
    (D) II and III
    (E) I, II, and III

11. Why is double blinding used in an experiment?
    - (A) To keep knowledge of the treatment from influencing the response or the measurement of the response
    - (B) To insure that the treatments are randomly given to all subjects each time
    - (C) To keep all subjects unaware that they are participating in an experiment
    - (D) To avoid talking to the subjects at any time
    - (E) To easily replicate an experiment in other environments

12. A car manufacturer is concerned about the time it is taking for batteries to be installed on the assembly line. The use of a new tool has been suggested to reduce time spent on this part of the assembly process. Which of the following would be the best way to establish if this new tool reduces battery installation time compared to the old tool?
    - (A) Randomly assign one of two workers to use the old tool and the other worker to use the new tool.
    - (B) Have day shift workers use the new tool and night shift workers use the old tool.
    - (C) Have each worker use both tools for a day. Each worker would flip a coin to determine which tool would be used on the first day.
    - (D) Randomly assign four workers one task to be performed for a month and then change to the other one of the two tools to use for a month.
    - (E) Assign two different assembly lines of workers to use one of the tools and then compare the two lines.

13. Five thousand menopausal women are participating in a biofeedback regimen to see if this will help relieve the symptoms of menopause. Half of the women were randomly assigned to a group that received a biofeedback regimen while the other half were assigned to a control group and not given any treatment. This is an example of
    - (A) an observational study.
    - (B) a simple randomized comparison experiment.
    - (C) a single-blind randomized comparison experiment.
    - (D) a double-blind randomized comparison experiment.
    - (E) a matched-pairs, double-blind randomized comparison experiment.

14. The mathematics department wants to see if the use of a software program will improve scores in geometry. The department has decided to set up a small experiment with 10 sections of geometry. Two teachers have volunteered to have their students participate in the study. Ms. Smith is going to use the software alongside existing classroom materials while Mr. Jones is going to use the current classroom materials only. Each teacher will use 5 sections of geometry taught in periods 1 through 5. The students will be given a pretest as well as a posttest at the end of the course. What are the treatments in this experiment?
   (A) Students taking the course
   (B) The time of day of the course
   (C) The teacher
   (D) Software or no software
   (E) Posttest results

15. The mathematics department wants to see if the use of a software program will improve scores in geometry. The department has decided to set up a small experiment with 10 sections of geometry. Two teachers have volunteered to have their students participate in the study. Ms. Smith is going to use the software alongside existing classroom materials while Mr. Jones is going to use the current classroom materials only. Each teacher will use 5 sections of geometry taught in periods 1 through 5. The students will be given a pretest as well as a posttest at the end of the course. What variable is confounded with the treatments in this experiment?
   (A) Students taking the course
   (B) The time of day of the course
   (C) The teacher
   (D) Software or no software
   (E) Posttest results

FREE-RESPONSE PROBLEMS

1. An athletic company has seen declining sales of its weight lifting gloves. Company executives feel that a competing company has a glove that lasts longer and hence gyms and stores are selling more of the competitor's product. To regain their market share, they have developed a newer material that they believe will outlast their current product. You have been called in to advise the executives in setting up and carrying out an appropriate study to determine if gloves made of the new fabric last longer than gloves made with the fabric currently in use.
   (a) There are 84 men and 56 women who are willing to participate in this study. Design an experiment that the company could use to determine if the new gloves last longer.
   (b) Is there a way to make this study double blind? Justify your answer.

2. Ms. Partridge grows Japanese pears as well as Bosc pears for a living. She has learned of a new cross-pollination technique with another pear strain that may yield a larger, sweeter pear. Ms. Partridge has two of each type tree for use in this study and she

would like to know whether cross-pollination with the Japanese pear trees or cross-pollination with the Bosc pear trees will yield a better fruit.

(a) Describe how you would assign the four trees to four planting locations in a completely randomized design.

(b) The actual plot of land she will use for this study is shown below. You notice that a stream runs along the northern edge of the land. Explain how you would assign the four trees to the four planting locations shown in the diagram.

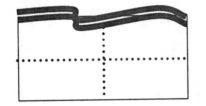

(c) What are some potential confounding variables that could affect the results of this study?

Answers

MULTIPLE-CHOICE QUESTIONS

1. **D.** This sampling method introduces the possibility of nonresponse bias (*Statistics: Learning From Data*, 1st ed. pages 19–20).

2. **E.** Although desirable, random selection of subjects is not a requirement for a well-designed experiment. Many experiments use volunteers. Random assignment of subjects to treatments is critical (*Statistics: Learning From Data*, 1st ed. pages 26–33).

3. **C.** If subjects are not randomly assigned to treatments, the possibility of confounding variables makes it impossible to be sure that any differences in response are actually due to the treatments (*Statistics: Learning From Data*, 1st ed. pages 26–33).

4. **B.** It would be best to block the experiment based on whether or not the petunias are in the sun. The club should try each type of fertilizer under both environmental conditions. This plan ensures that each fertilizer is evaluated in each environment (*Statistics: Learning From Data*, 1st ed. pages 25, 27, 28, and 33).

5. **D.** The purpose of a placebo is to offer a treatment that has no active ingredient but that makes the participants believe they may be receiving a treatment. This allows researchers to determine if the actual treatment being tested is working or if simply the idea of having a treatment is helping. Using a placebo is only important when the subjects in the experiment are people (*Statistics: Learning From Data*, 1st ed. page 34).

6. **B.** Confounding is what occurs when a study is performed and the effects of some variables on the response variable can't be distinguished from one another (*Statistics: Learning From Data,* 1st ed. page 25).

7. **E.** In an experiment, the researcher applies a treatment. However, in an observational study, we are just observing what is or is not happening. A well-designed experiment can lead to a cause and effect conclusion. An observational study cannot. (*Statistics: Learning From Data,* 1st ed. pages 9, 10, and 24).

8. **D.** This is the only choice without a clear bias in the initial design. Here, a sample of juniors and seniors will be chosen. The problem with **A** is the use of only an AP class as they may tend to have higher GPAs. **B** is biased as one would expect folks outside a campaign office to more closely follow the views of that political party. C only includes those who work in marketing and there are other groups of company employees that would be excluded from the study. Finally, **E** is biased as shoppers may have a sense of buyer's remorse after just paying for their groceries (*Statistics: Learning From Data,* 1st ed. pages 19–20).

9. **B.** This choice forms strata based on whether a person is single or has a family. It considers both types of members. Each of the other options potentially misses groups of individuals. Choices **A, D,** and **E** might have a problem with nonresponse bias. Choice **C** does not include members of the church who did not attend (*Statistics: Learning From Data,* 1st ed. pages 25–28 *and 33*).

10. **D.** Convenience samples are problematic and should be avoided. Not all experiments require blocking or a placebo group—it depends on the nature of the experiment (*Statistics: Learning From Data,* 1st ed. pages 18–20).

11. **A.** A double-blind experiment keeps both the subject and the person measuring the response from knowing which treatment a particular subject is receiving (*Statistics: Learning From Data,* 1st ed. page 35).

12. **C.** While both A and C incorporate random assignment of treatments, the matched pairs design in C also controls for the skill of an individual employee. Flipping a coin also provides for random assignment of treatments to the workers (*Statistics: Learning From Data,* 1st ed. pages 24 and 28).

13. **B.** This is an experiment where subjects are randomly assigned to one of two groups. Note that because one group receives a biofeedback regimen and the other does not, it is not possible to blind the subjects in this experiment (*Statistics: Learning From Data, 1st ed. page 35*).

14. **D.** There are two treatments in this experiment—software and no software—that are imposed on the subjects (*Statistics: Learning From Data,* 1st ed. page 24).

15. **C.** Because one teacher has all of the software sections and the other teacher has all of the no-software sections, it will not be possible to distinguish the effect of the software from a teacher effect (*Statistics: Learning From Data,* 1st ed. pages 24–25).

FREE-RESPONSE PROBLEMS

1. (a) Because men and women may lift weights differently, it makes sense to block on gender. It also makes sense in this setting to use a matched pairs design, with each participant using one glove of each type. For the block consisting of the 84 males, number the men from 01 to 84. Then use a table of random digits or a random number generator to select 42 of the men. These men will receive a pair of gloves that uses the new fabric in the right-hand glove and the old fabric in the left-hand glove. The other men will receive a pair of gloves with the old fabric used for the right-hand glove and the new fabric for the left-hand glove. This same process will be repeated for the women, except they will be numbered 01–56 and we will use random numbers to select 28 for one group (right new fabric and left old fabric) and the other 28 women will make up the other group (right old fabric and left new fabric). After 3 months, measure the difference in wear for each person.

   (b) Double blinding is possible. Blinding the subjects would be accomplished by making sure the subjects don't know which glove is made with the new fabric and which is made with the old fabric. This will work if the materials look and feel the same. Blinding the individual who is measuring the wear at the end of the three months is also possible. The gloves could be coded on the inside in a way that only the research team who will analyze the wear data would know what the code means (*Statistics: Learning From Data,* 1st ed. pages 24–41).

2. (a) One way to randomize the four trees would be to number these 1–4 with 1 and 2 representing the Japanese pear trees and 3 and 4 the Bosc pear trees. Go to a planting location and roll a die. Place the corresponding tree in that location (roll again if a 5 or a 6 appears on the die). Repeat for the next planting locations and remember that any numbers corresponding to trees that have already been assigned a location should be ignored along with any 5s and 6s.

   (b) Since there is a stream, there could be a difference in growing conditions close to the stream. In this case, you should block the two planting locations closest to the water (shaded sections) and randomly assign one of each type of tree to the locations in this block. Go to the top left shaded plot and flip a coin. If heads comes up, put in a Japanese tree. If tails comes up, put in Bosc. The other shaded plot will get the other type of tree. Repeat for the unshaded region.

Ex: 1st flip = tails. Bosc tree goes here, so the Japanese tree will go here.

(c) A variety of confounding variables could be given. A few examples are listed below.
(i) Soil condition
(ii) Unknown health of the trees that are planted
(iii) Any type of bug or animal that is a hindrance to one of the types of trees (that is, maybe local deer are drawn to the new Bosc cross-pollinated and they eat the fruit)

(*Statistics: Learning From Data,* 1st ed. pages 25–33).

5

PROBABILITY

What is the chance you will be admitted to your first choice university? How likely is it that Zach Johnson will win the Masters Golf tournament this year? What does it mean when the meteorologist on the evening news predicts rain with a 70% probability? Uncertainty surrounds us on a daily basis, but we still can make decisions based on available information. Understanding probability will help you to make well-informed decisions.

OBJECTIVES

- Be able to interpret probabilities as long-run relative frequencies.
- Understand the Law of Large Numbers in the context of probability.
- Be able to calculate the probability of an even using the complement, addition, and multiplication rules.
- Understand what it means to say that two events are independent.
- Be able to compute and interpret conditional probabilities.
- Be able to use simulation to estimate probabilities.

CHANCE EXPERIMENTS AND EVENTS

(*Statistics: Learning From Data*, 1st ed. page 273)

Consider a chance experiment where Sally is trying to decide what to have for dinner at the local diner. The limited menu includes three entrees: hamburger, pizza, and chef's salad; and two desserts: apple pie and carrot cake. Although we don't know what Sally's selections will be, if we know that Sally will choose one entrée and one dessert, we can describe the complete set of possible outcomes.

The collection of all possible outcomes of a chance experiment is the *sample space* for the experiment. The sample space for the menu selection example consists of six outcomes.

136

1. {Hamburger, Apple Pie}

2. {Hamburger, Carrot Cake}

3. {Pizza, Apple Pie}

4. {Pizza, Carrot Cake}

5. {Chef's Salad, Apple Pie}

6. {Chef's Salad, Carrot Cake}

The sample space can also be illustrated using a tree diagram (shown below).

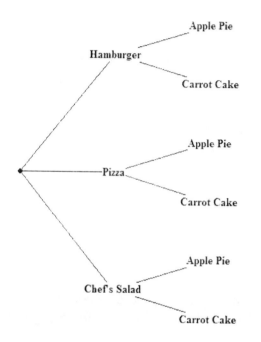

An *event* is any collection of outcomes from the sample space of a chance experiment. A *simple event* is an event consisting of exactly one outcome.

EXAMPLE Consider a chance experiment where every year eight basketball teams compete in a basketball tournament. Suppose we are interested in the tournament outcome—the team that wins the tournament. The teams competing are:

D = Duke Blue Devils   V = Vanderbilt Commodores

R = Rutgers Scarlet Knights   T = Texas Longhorns

M = Michigan Wolverines   S = Southern Mississippi Golden Eagles

G = Georgia Bulldogs   B = Bradley Braves

What is the sample space?

SOLUTION The sample space is {D, R, M, G, V, T, C, B}.

In the previous example, each outcome is a simple event. A compound event is an event that is made up of two or more simple events. Three common types of compound events are unions, intersections, and complements.

The event <u>not A</u> consists of all outcomes that are not in event A. Not A is sometimes called the *complement* of A and is usually denoted by A^C, A', or $\overline{A}$.

The event <u>A or B</u> consists of all outcomes that are in at least one of the two events, that is, A or B consists of all outcomes that are in A or in B or in both of these. A or B is called the union of the two events, and is often denoted by $A \cup B$.

The event <u>A and B</u> consists of all outcomes that are in both of the events A and B. A and B is called the intersection of the two events, and is often denoted by $A \cap B$.

EXAMPLE In the context of the basketball tournament, consider these events:

P = a private school wins the tournament

C = a school with a color in its name wins the tournament

A = a school with an animal in its name wins the tournament

List the outcomes in each of the following events:

(a) the complement of A

(b) A or C

(c) A and C

SOLUTION

(a) The complement of A (the same as not A) consists of all schools that do not have an animal in its name: {D,R,V,B}. This solution is also shown visually in the Venn Diagram below.

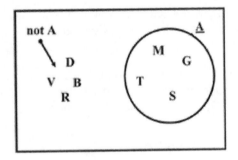

(b) The event A or C consists of all schools that have a color in its name, an animal in its name, or both a color and an animal in its name: {D, R, M, G, T, S}. This can be seen visually in the Venn Diagram below by looking at all schools in either of the two circles or in both circles.

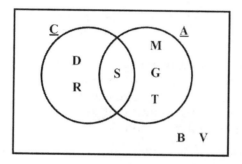

(c) The event A and C consists of all schools that have both a color and an animal in its name: {S}. This can be seen visually in the Venn Diagram above by looking at all schools that are in both circles.

Events that have no common outcomes (no overlap) are said to be *disjoint* or *mutually exclusive (mutually exclusive events are not independent of each other)*. The events P and A are disjoint events since the private schools are Duke and Vanderbilt and neither of these schools have an animal in its name.

DEFINITION OF PROBABILITY

(*Statistics: Learning From Data*, 1st ed. pages 267–289)

DEFINITION (CLASSICAL APPROACH)

If the outcomes in the sample space are equally likely, then for any event E, the probability of event E occurring, P(E), is

$$P(E) = \frac{\text{number of outcomes in event E}}{\text{number of outcomes in the sample space}}$$

P(E)= Events that work (successes)/Total Events

Note that it is only appropriate to use this approach to compute a probability if the outcomes in the sample space are equally likely.

EXAMPLE Paul practices the piano every morning. His repertoire consists of 8 classical pieces and 23 popular pieces. If he randomly selects one piece to play, what is the probability that it will be classical?

SOLUTION The sample space consists of the 31 pieces that Paul might play. Because he is selecting a piece at random, the outcomes in the sample space are equally likely. Let C = the event that a classical piece is chosen. Then since the number of outcomes in C is 8 and the total number of outcomes in the sample space is 8 + 23 = 31, $P(C) = \dfrac{8}{31}$.

LAW OF LARGE NUMBERS

As the number of repetitions of a chance experiment increases, the relative frequency (actual) of the occurrence of an event approaches the true, or theoretical, probability of the event.

As an illustration, when we repeatedly spin the spinner shown on the following page and calculate the relative frequency of landing in the shaded area, there is fluctuation over the short term but there is stability over the long term. Notice in the graph below that after a large number of spins, the relative frequency settles at about 0.75.

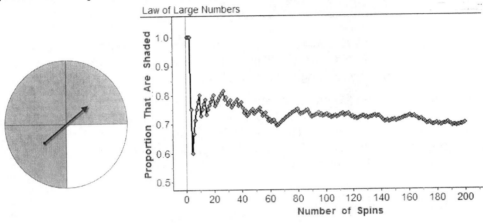

DEFINITION (RELATIVE FREQUENCY APPROACH)

For any event E, the estimated probability of event E is

$$P(E) \approx \frac{\text{number of times E occurs}}{\text{number of repetitions}}$$

The estimated probability should be close to the actual probability if the number of repetitions is large.

EXAMPLE Suppose we repeatedly flip a metal thumbtack to determine whether it will land with tip up or tip down. The results of the first 10 flips were:

Flip	1	2	3	4	5	6	7	8	9	10
Outcome	Down	Up	Down	Down	Up	Up	Up	Down	Down	Down

What is the estimated probability that the tack will land point up?

SOLUTION

Flip	1	2	3	4	5	6	7	8	9	10
Outcome	Down	Up	Down	Down	Up	Up	Up	Down	Down	Down
Cumulative Count of Ups	0	1	1	1	2	3	4	4	4	4
Relative Frequency of Ups	0	0.50	0.33	0.25	0.40	0.50	0.57	0.50	0.44	0.40

The estimated probability of event E is $P(E) = \dfrac{4}{10} = 0.40$. The relative frequencies fluctuate quite a bit over the 10 repetitions, so it does not appear that the relative frequency has settled down yet. The law of large numbers says that if we continue this process so that we had many/many more repetitions, the estimated (experimental) probability should be close to the actual (theoretical) probability.

Theoretical probability is calculated using mathematical formulas/models. Experimental probability is based upon running trials and observing our results. Probability can vary in the short term, but in the long run the experimental probability should approach the theoretical probability.

BASIC PROPERTIES OF PROBABILITY

(*Statistics: Learning From Data*, 1st ed. pages 271–273)

There are four basic rules of Probability. They are

1. For any event E, $0 \le P(E) \le 1$.

2. If S is the sample space for an experiment, P(S) = 1.

3. If two events are disjoint, then P(E or F) = P(E) + P(F).

4. For any event E, P(E) + P(not E) = 1.

   Two events are said to be disjoint if they cannot occur at the same time.

   [If two events are not disjoint, then P(E or F) = P(E) + P(F) – P(E and F).]

EXAMPLE A survey was taken in which students enrolled in AP Statistics at a particular high school were asked what type of music they preferred most. The resulting data is shown in the table below.

Genre	Rock	Rap	Country
Number of Students	32	18	25

If an AP Statistics student at this school is selected at random, what is the probability that the student's preferred type of music is Rap or Rock?

SOLUTION

$$P(\text{Rap or Rock}) = P(\text{Rap}) + P(\text{Rock}) = \frac{32}{75} + \frac{18}{75} = \frac{50}{75} = \frac{2}{3}.$$

EXAMPLE Brett counted the change he had placed in the console of his car and found 7 quarters, 10 dimes, 4 nickels, and 9 pennies. If he randomly selects one coin, what is the probability it was *not* a penny?

SOLUTION Using the complement rule (basic rule 4),

$$P(\text{not Penny}) = 1 - P(\text{Penny}) = 1 - \frac{9}{30} = \frac{21}{30} = \frac{7}{10}$$

AP Tip

When answering probability questions

1. Clearly define the event(s) of interest.

2. Use appropriate notation, i.e. $P(X > 3)$ or $P(A \cap B)$.

3. When appropriate, illustrate your method with a Venn diagram or a tree diagram.

4. Show all relevant calculations.

CONDITIONAL PROBABILITY

(*Statistics: Learning From Data*, 1st ed. pages 290–301)

In certain situations, knowing that one event has occurred will alter the probability that another event also occurred. For example, suppose your friend rolls one die. If your friend then tells you that the number showing is even, your assessment of the probability that the number rolled was a 2 would change from 1/6 to 1/3.

The *conditional probability* of event E given that event F has occurred, denoted by $P(E|F)$ is defined as

$$P(E \mid F) = \frac{P(E \text{ and } F)}{P(F)}.$$

EXAMPLE Suppose that you have torn a tendon and are facing surgery to repair it. The orthopedic surgeon explains the risks to you. Infection occurs in 3% of such operations, the repair fails in 14%, and both infection and failure occur together in 1%. Find the probability that a randomly selected patient from among those who have had this surgery has a failed repair given that an infection occurred.

SOLUTION Let I = person gets an infection and F = the repair fails. Then $P(I) = 0.03$, $P(F) = 0.14$, and $P(I \text{ and } F) = 0.01$.

$$P(F \mid I) = \frac{P(I \text{ and } F)}{P(I)} = \frac{0.01}{0.03} = 0.333$$

INDEPENDENCE

(*Statistics: Learning From Data*, 1st ed. pages 285–301)

It is important to understand that the concepts of mutually exclusive events and independence are two separate concepts. Mutually exclusive events are not independent.

If the Probability of A and B is zero then the events cannot be independent.

One way to show independence is $P(A \text{ and } B) = P(A) \times P(B)$

If events are mutually exclusive then the $P(A \text{ and } B) = 0$

Two events E and F are said to be *independent* if $P(E \mid F) = P(E)$. In other words, this means that the occurrence of event F has not changed the probability of event E. If E and F are not independent, they are said to be *dependent* events.

MULTIPLICATION RULE FOR TWO INDEPENDENT EVENTS

The events E and F are independent if and only if $P(E \cap F) = P(E) \times P(F)$.

GENERAL MULTIPLICATION RULE

Given two events, E and F, $P(E \cap F) = P(E) \times P(F \mid E)$.

EXAMPLE Determine if events A and B are independent.

(a) $P(A) = 0.5$, $P(B) = 0.7$, $P(A \text{ and } B) = 0.3$

(b) $P(B \mid A) = 0.25$, $P(B \mid A') = 0.3$, $P(A) = 0.4$

(c) $P(A) = 2/3$, $P(B) = 1/2$, $P(A \text{ or } B) = 5/6$

(d) $P(A) = 0.5$, $P(\text{not } B) = 0.4$, $P(A \text{ or } B) = 0.9$

SOLUTION

(a) $P(A) = 0.5$, $P(B) = 0.7$, $P(A \text{ and } B) = 0.3$. $P(A) \times P(B) = 0.35$. Since $P(A) \times P(B) \neq P(A \text{ and } B)$, events A and B are not independent.

(b) $P(B \mid A) = 0.25$, $P(B \mid A') = 0.3$, $P(A) = 0.4$. This means that the occurrence (or lack of occurrence) of A changes the probability of B occurring. So since $P(B \mid A) \neq P(B \mid A')$, events A and B are not independent.

(c) $P(A) = 2/3$, $P(B) = 1/2$, $P(A \text{ or } B) = 5/6$. $P(A) \times P(B) = 1/3$. $P(A \text{ and } B) = P(A) + P(B) - P(A \text{ or } B) = 2/3 + 1/2 - 5/6 = 1/3$. Since $P(A) \times P(B) = P(A \text{ and } B)$, events A and B are independent. Note: Start with the formula $P(A \text{ or } B) = P(A) + P(B) - P(A \text{ or } B)$ and solve for $P(A \text{ and } B)$ to find the formula used above.

(d) $P(A) = 0.5$, $P(\text{not } B) = 0.4$, $P(A \text{ or } B) = 0.9$. $P(B) = 1 - P(\text{not } B) = 0.6$. $P(A) \times P(B) = 0.3$.

$P(A \text{ and } B) = P(A) + P(B) - P(A \text{ or } B) = 0.5 + 0.6 - 0.9 = 0.2$. Since $P(A) \times P(B) \neq P(A \text{ and } B)$, events A and B are not independent.

SIMULATION

(*Statistics: Learning From Data*, 1st ed. pages 319–323)

Simulation is an artificial re-creation of a random process to study its long-run behavior. Random digits/calculators are often used to carry out a simulation.

Steps to perform a simulation using a random digit table:

1. State the assumptions you are making about how the real life situation works. Include any doubts you have about the validity of your assumptions.

2. Describe how you will use random digits to conduct one trial of a simulation of the situation. Show how you will assign a digit (or a group of digits) to represent each possible outcome. You can disregard some digits if necessary. Explain how you will use the digits to model the real-life situation. Describe what constitutes a single trial and what observation you will record.

3. Run the simulation a large number of times, recording the results in a frequency table.

4. Write a conclusion in context. Be sure to state that what you have computed is an <u>estimated</u> probability.

Think of
- Assumptions
- What is your scheme? (What will each of the numbers represent and why. Unlike choosing subjects you will be using repeat digits because you are simulating probability.)
- How will you define a trial?
- How many trials will you conduct?
- What is your conclusion from these trials?

EXAMPLE Suppose 70% of Beach High School students approve of a new uniform. You would like to estimate the probability that if five students are selected at random, all five approve of the new uniform. Use the list of random digits below to carry out 20 trials of simulation that would allow you to estimate this probability.

2039 2993 4362 6363 2914 4955 6364 5237 6456 5561

0176 2425 2968 3834 6077 4302 3499 9938 7231 2136

2161 1365 2764 7836 1584 2421 4247 2930 0783 9989

0407 1760 7048 1929 9034 0242 0753 4851 9465 0791

SOLUTION Each student in a sample of five will be represented by a single random digit. The digits 0 through 6 represent a student who approves of a new uniform and digits 7–9 represent a student who do not approve of the new uniform. A block of five random digits then represents a sample of five students. Count the number in the sample who approve (this is the number of digits in the block of five that are 0–6). Record whether or not all five students favor the new uniform. This is one trial. Repeat this process 19 more times to complete 20 trials. Notice that in this situation, repeated numbers are not ignored (in some problems, repeats symbolize a subject being chosen more than once and need to be ignored).

```
2039   2993   4362   6363   2914   4955   6364   5237   6456   5561
 NO     NO     YES    NO     NO     YES    NO     NO     YES

0176   2425   2968   3834   6077   4302   3499   9938   7231   2136
 NO     NO     NO     NO     NO     YES    NO     NO     YES

2161   1365   2764   7836   1584   2421   4247   2930   0783   9989
 YES    NO     NO     NO

0407   1760   7048   1929   9034   0242   0753   4851   9465   0791
```

The estimated probability that all five selected students approve of the new uniform is $\dfrac{\text{total number of yes trials}}{\text{total number of trials}} = \dfrac{6}{20} = 0.30$.

PROBABILITY: STUDENT OBJECTIVES FOR THE AP EXAM

- You should be able to interpret a probability in context.
- You should be able to interpret a long-run relative frequency as a probability.
- You should understand the concept of the law of large numbers.
- You should be able to calculate probabilities using the addition rule and the multiplication rule.
- You should be able to calculate and interpret a conditional probability.
- You should understand the concept of the independence.
- You should be able to determine mathematically whether or not two events are independent or not.
- You should be able to use simulation to estimate probabilities.

MULTIPLE-CHOICE QUESTIONS

1. In the fall of her senior year in high school, Carrie has begun to submit her college applications. Based on her research, she estimates the probability of her being admitted to the University of Georgia at 0.45, the probability of her being admitted to both the University of Georgia and Auburn University at 0.25, and the probability of her not being admitted to either at 0.10. If her assumptions are correct, what her probability of being admitted to Auburn University?
   (A) 0.90
   (B) 0.45
   (C) 0.70
   (D) 0.20

(E) 0.65

2. A recent survey of the employees at Aardvark Enterprises concerning their attitudes toward health care reform produced the following results.

		Type of Employee	
		Assembly Line Workers	Administrative Staff
Opinion on Health Care Reform	In Favor	127	13
	Opposed	42	31
	No Opinion	31	6

Considering only the employees who opposed health care reform, what is the probability that a randomly selected worker is on the administrative staff?

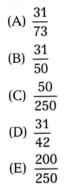

(A) $\dfrac{31}{73}$

(B) $\dfrac{31}{50}$

(C) $\dfrac{50}{250}$

(D) $\dfrac{31}{42}$

(E) $\dfrac{200}{250}$

3. The grade level breakdown at Bryans High School is as follows:

Freshman – 32%, Sophomores – 27%, Juniors – 23%, Seniors – 18%.

If a student at this school is selected at random, what is the probability the student is either a junior or a senior?
(A) 0.05
(B) 0.41
(C) 0.45
(D) 0.50
(E) 0.59

4. If $P(A) = 0.3$, $P(B) = 0.4$, $P(\text{not } A \text{ and not } B) = 0.42$, are the events A and B disjoint? Are the events A and B independent?
(A) disjoint; independent
(B) not disjoint; not independent
(C) disjoint; dependent
(D) not disjoint; independent
(E) cannot be determined

5. Each Saturday morning, Beau practices throwing pitches to his dad. His dad estimates the probability of any one pitch being a strike is 0.6. If pitches are independent, what is the probability that neither of Beau's next two pitches will be strikes?
   (A) 1.20
   (B) 0.48
   (C) 0.36
   (D) 0.16
   (E) 0

Questions 6–8 refer to the following information:

Suppose that 8% of all adults have a particular disease. Also suppose that a test for this disease returns a positive result for 96% of people who have the disease, but it also returns a positive result for 7% of people who do not have the disease.

6. What is the probability that a randomly selected adult has the disease and tests positive?
   (A) 0.0672
   (B) 0.0736
   (C) 0.0768
   (D) 0.08
   (E) 0.96

7. What is the probability that a randomly selected person tests positive for the disease?
   (A) 0.0768
   (B) 0.08
   (C) 0.1412
   (D) 0.144
   (E) 0.1536

8. If a person is selected at random and given the test and the test is positive, what is the probability that this person actually has the disease?
   (A) 0.0037
   (B) 0.0672
   (C) 0.0768
   (D) 0.1412
   (E) 0.5439

Questions 9–11 refer to the following information:

The 10,000 students at a polytechnic university were classified according to the type of program they were enrolled in (engineering, science or mathematics) and their degree objective (B.S., M.S. or Ph.D.), as shown in the table below:

	B.S.	M.S.	Ph.D.	Total
Engineering	2142	1890	2268	6300
Science	368	432	800	1600
Math	882	630	588	2100
Total	3392	2952	3656	10,000

9. What is the probability that a randomly selected student is in a M.S. program given that the student in not enrolled in a science program?
   (A) 0.0432
   (B) 0.0514
   (C) 0.2700
   (D) 0.3000
   (E) 1.0920

10. What is the probability that a randomly selected student is enrolled in a mathematics B.S. program?
    (A) 0.0882
    (B) 0.1606
    (C) 0.2600
    (D) 0.4200
    (E) 0.5492

11. What is the probability that a randomly selected student has a B.S. degree objective or is enrolled in an engineering program?
    (A) 0.2142
    (B) 0.3400
    (C) 0.6315
    (D) 0.7550
    (E) 0.9692

12. A weighted die is rolled 100 times and the results are given below. Find the estimated probability that, if this same die was rolled twice, the sum would be less than 4.

Number Shown	1	2	3	4	5	6
Frequency	15	10	6	9	30	30

(A) 0.0150
(B) 0.0300
(C) 0.0375
(D) 0.0525
(E) 0.2100

Questions 13–14 refer to the following information:
Many fire stations handle emergency calls for medical assistance as well as those requesting firefighting equipment. A particular station says that the probability that an incoming call is for medical assistance is 0.85.

13. What is the probability that a call is not for medical assistance?
    (A) 0
    (B) 0.1275
    (C) 0.1500
    (D) 0.8500
    (E) 1

14. Assuming independence, calculate the probability that exactly one of the next two calls will be for medical assistance.
    (A) 0.0450
    (B) 0.2550
    (C) 0.7225
    (D) 0.8500
    (E) 0.9775

15. When we say that a coin is "fair," what does this mean?
    (A) That every occurrence of a head must be balanced by a tail in one of the next two or three tosses.
    (B) If the coin is flipped many, many times, the proportion of heads will be approximately one half, and this proportion will tend to get closer and closer to one half as the number of tosses increases.
    (C) That regardless of the number of flips, half will be heads and half will be tails.
    (D) Generally, the flips will alternate between heads and tails.
    (E) If a head is observed on the first four tosses, the probability of a tail on the fifth toss is greater than ½.

FREE-RESPONSE PROBLEMS

1. At a certain high school, all students are required to take both physics and statistics. Suppose that the probability that a student earns an A in physics is one fifth and that the probability that a student earns an A in statistics is one fourth. What is the probability that a student will earn an A in at least one of physics or statistics if under each of the following conditions?
   (a) What is the probability that a student will earn an A in at least one of physics or statistics if the grades in each course are INDEPENDENT?
   (b) What is the probability that a student will earn an A in at least one of physics or statistics if the set of students who earn an A in physics is a subset of the set of students earning an A in statistics?

2. A company has seven mathematicians on its staff, two of whom are women. The president of the company wonders what the probability of two or fewer female mathematicians in a group of 7 randomly selected mathematicians would be. Suppose that 40% of mathematicians are female.
   (a) Using the random digits below, carry out five trials of a simulation that could be used to estimate the probability of two or fewer females in a group of seven randomly selected mathematicians.

   88565 42628 17797 49376 61762 16953 88604 12724

   62964 88145 83083 69453 46109 59505 69680 00900

   (b) Based on your five trials from part (a), what is the estimated probability of two or fewer females in a group of seven randomly selected mathematicians?
   (c) Do you think your estimate in part (b) is likely to be very accurate? Explain.

Answers

MULTIPLE-CHOICE QUESTIONS

1. **C.** Define the events: G = being admitted to Georgia, A = being admitted to Auburn. Since P(G) = 0.45 and P(G and A) = 0.25, then P(G and not A) = P(G)-P(G and A)= 0.45 -.25 = 0.20. P(G or A) = 1 - P(not G and not A)= 1 – 0.10 = 0.90. Therefore, P(A)=P(G or A)-P(G and not A) = 0.90 – 0.20 = 0.70 (*Statistics: Learning From Data* 1st ed. pages 273–290).

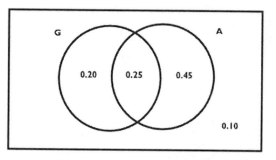

2. **A.** Define events: O = oppose health care, A = administrative staff. Since there are 73 employees who oppose health care (42 assembly line workers and 31 administrative staff) and 31 are on the administrative staff, P(A|O) = $\dfrac{31}{73}$ (*Statistics: Learning From Data,* 1st ed. pages 290–306).

3. **B.** Since the events junior and senior are disjoint, P(Junior or Senior) = P(Junior) + P(Senior) = 0.27 + 0.18 = 0.41 (*Statistics: Learning From Data,* 1st ed. pages 284 and 303).

4. **D.** Since P(not A and not B) = 0.42, P(A or B) = 1-0.42 = 0.58. Using the addition rule, P(A or B) = 0.58 = P(A) + P(B) – P(A and B) = 0.3 + 0.4 – P(A and B), which means that P(A and B) = 0.12. Since P(A and B) ≠ 0, A and B are not disjoint. Since P(A)· P(B) = P(A and B), A and B are independent (*Statistics: Learning From Data,* 1st ed. pages 284 and 303).

5. **D.** P(not strike) = .4. Since pitches are independent of one another, P(not strike and not strike) = P(not strike) × P(not strike) = .4 × .4 = .16 (*Statistics: Learning From Data,* 1st ed. pages 285–287 and 303).

In Problems 6–8, the following tree diagram may be helpful in keeping straight all of the possible probabilities.

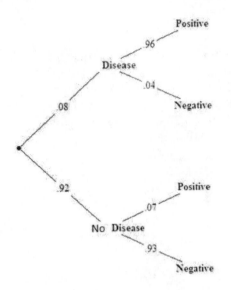

6. **C.** P(has disease and tests positive) = P(has disease) × P(tests positive | has disease) = 0.08 × 0.96 = 0.0768. Using the tree diagram, the top most branch represents P(has disease and tests positive).

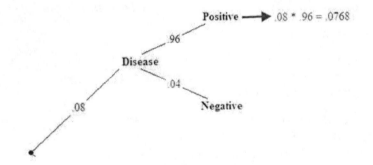

(*Statistics: Learning From Data*, 1st ed. pages 290–306)

7. **C.** P(tests positive) = P(tests positive and has disease OR tests positive and does not have disease) = P(tests positive and has disease) + P(tests positive and does not have disease) = (0.08 × 0.96) + (0.07 × 0.92) = 0.1412. Using the tree diagram, the two branches of interest end in a positive test result. This represents P(tests positive and has disease OR tests positive and does not have disease).

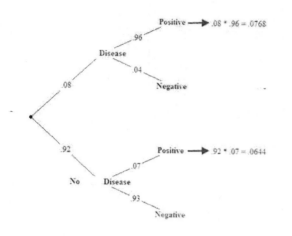

(*Statistics: Learning From Data,* 1st ed. pages 290–306)

8. **E.** P(has disease | tests positive) = P(has disease and tests positive) / P(tests positive) = 0.0768 / 0.1412 = 0.5439. The numerator and denominator here were found in the previous two problems (*Statistics: Learning From Data,* 1st ed. pages 290–306).

9. **D.** P(M.S. | not in science) = number of students in an M.S. program who are not enrolled in science / number of students at the university who are enrolled in engineering or math = (1890 + 630) / (6300 + 2100) = 2520 / 8400 = 0.3 (*Statistics: Learning From Data,* 1st ed. pages 290–306).

10. **A.** P(math and B.S.) = number of math B.S. students / total number of students at the university = 882/10000 = 0.0882 (*Statistics: Learning From Data,* 1st ed. pages 290–306).

11. **D.** P(B.S. or engineering) = number of B.S. students + number of engineering students – number of engineering students seeking a B.S / total number of students at the university = (3392 + 6300 - 2142) / 10000 = .7550 (*Statistics: Learning From Data,* 1st ed. pages 290–306).

12. **D.** P(sum less than 4) = P(1 and 1 OR 1 and 2 OR 2 and 1) = P(1 and 1) + P(1 and 2) + P(2 and 1) = 0.15 × 0.15 + 0.15 × 0.1 + 0.1 × 0.15 = 0.0525 (*Statistics: Learning From Data,* 1st ed. pages 284 and 290–306).

Using two dice

	1	2	3	4	5	6
1						
2						
3						
4						
5						
6						

Mark the places where the sum would be less than 4 and then calculate the probabilities for each of those choices.

13. **C.** P(call is not for medical assistance) = 1 – (call is for medical assistance) = 1 – 0.85 = 0.15 (*Statistics: Learning From Data,* 1st ed. page 271).

14. **B.** P(exactly one of two calls is for medical assistance) = P(first call is for medical assistance and the next call is not for medical assistance) + P(first call is not for medical assistance and the next call is for medical assistance) = (0.85 × 0.15) + (0.15 × 0.85) = 0.2550 (*Statistics: Learning From Data,* 1st ed. pages 284 and 290–306).

15. **B.** This is due to what is known as the Law of Large Numbers, which states that as the number of repetitions of a chance experiment (such as flipping a coin) increases, the likelihood that the relative frequency of occurrence for an event will differ from the true probability of the event by more than a very small number approaches zero. (*Statistics: Learning From Data,* 1st ed. pages 271 and 276).

FREE-RESPONSE PROBLEMS

1. Given P(Physics) = 0.2 and P(Statistics) = 0.25
    P(Physics or Math) = P(Physics) + P(Statistics) – P(Physics and Statistics)

    (a) Since the events are independent, P(Physics and Statistics) = P(Physics) × P(Statistics).

    P(Physics or Statistics) = P(Physics) + P(Statistics) – P(Physics and Statistics) = P(Physics) + P(Statistics) – (P(Physics) × P(Statistics)) = 0.2 + 0.25 – (0.2)(0.25) = 0.4

    (b) Since the students getting an A in Physics are a subset of those students getting an A in Statistics, all students that

make an A in Physics also make an A in Statistics. So
P(Physics and Statistics) = 0.2.

P(Physics or Statistics) = P(Physics) + P(Statistics) – P(Physics and Statistics) = 0.2 + 0.25 – 0.2 = 0.25 (*Statistics: Learning From Data,* 1st ed. pages 284, 303 and 304).

2. (a) Scheme: Since 40% of mathematicians are female, let the digits 0, 1, 2, and 3 represent selecting a female mathematician and digits 4, 5, 6, 7, 8, and 9 represent selecting a male mathematician. A group of seven randomly selected mathematicians will be represented by seven digits in the random digit table. Each trial is seven numbers to represent the seven openings. Count the number of female mathematicians and note whether two or fewer were women. The number of trials is five. Do this five times.

88565 42628 17797 49376 61762 16953 88604 12724
1 female ┆ 2 females ┆ 1 female ┆ 3 females ┆ 2 females ┆

(b) In the five trials, four resulted in two or fewer women, so the estimated probability is 4/5 = 0.80.

(c) Because the estimate in part (b) is based on only five trials, it is not likely to be very accurate. Many more trials would need to be included in order for the estimated probability to be considered a relatively accurate estimate of the theoretical probability.

(*Statistics: Learning From Data,* 1st ed. pages 319–323).

6

RANDOM VARIABLES AND PROBABILITY DISTRIBUTIONS

This chapter discusses the characteristics of random variables as well as probability distributions used to describe the behavior of random variables.

OBJECTIVES

- ■ Given a probability distribution, calculate probabilities for discrete and continuous random variables.
- ■ Find the mean and standard deviation of a discrete random variable.
- ■ Find the mean and standard deviation for linear combinations of random variables.
- ■ Calculate probabilities of interest for binomial, geometric, and normal random variables.

RANDOM VARIABLES

(Statistics: Learning From Data, 1st ed. pages 335–362)

A random variable is actually a function, one whose value depends on the outcome of a chance experiment. A *discrete* random variable is a function whose possible values consist of a set of individual points on the number line. For example, 2, 3, 4, 5 would be a set of discrete values. A *continuous* random variable is one whose possible values consist of an entire interval on the number line. For example, the interval from 2 to 3 includes 2, 2.001, 2.002, 2.003, and even values between 2.001 and 2.002. This interval includes every number between two boundaries. Sometimes the values of a continuous random variable are only measured to the nearest tenth or the nearest hundredth, but a continuous variable can theoretically take on any value in the interval. We will use lower case letters, like x and y, to represent random variables.

A *probability distribution* for a discrete random variable is typically represented as a table or a histogram showing the probabilities associated with each of the possible values of the variable. They describe a variable's long-run behavior and allow us to calculate probabilities based on the distribution. The probability distribution of a continuous random variable is represented by a curve and areas under the curve are interpreted as probabilities. We will use probability distributions to calculate probabilities for both discrete and continuous random variables.

PROBABILITY DISTRIBUTIONS FOR DISCRETE RANDOM VARIABLES

(*Statistics: Learning From Data*, 1st ed. pages 339–344)

Since the possible values of discrete random variables are individual points, the probability distribution specifies a probability associated with each possible value. For example, consider the number of decorative coasters in sets sold by a souvenir store. The store sells coasters in sets of 2, 4, 6, and 8. In this case, the random variable being considered is x = *number of coasters in the coaster set sold*. Suppose that over the course of any summer tourist season, the probability of selling each size of coaster sets is as shown in the table below.

x	2	4	6	8
$p(x)$	0.10	0.35	0.5	0.05

The table shows that the probability of selling a set with 6 coasters is 0.50, while the probability of selling a set of two coasters is 0.10. This means that in the long run, about 50% of coaster sets sold will contain 6 coasters and about 10% will contain 2 coasters. Another way to display this information is as a histogram where the x-axis shows the values for number of coasters in a set and the y-axis is the probability associated with each size coaster set.

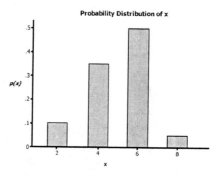

The histogram is an excellent visual representation of the probability distribution. Probability distributions for discrete random variables have the following properties:

1. for every value of x, $0 \le p(x) \le 1$

2. $\displaystyle\sum_{all\ x\ values} p(x) = 1$

   Note: the sum of the probabilities always adds up to one.

If the manager of the store wants to know the probability that a set of 2 or a set of 4 is sold, the manager would add the probabilities $p(2) + p(4) = 0.1 + 0.35 = 0.45$.

$P(x = 3)$ denotes the probability that the random variable x is equal to 3.

(Note: Sometimes capital letters are used to represent random variables.)

CONTINUOUS RANDOM VARIABLES

(*Statistics: Learning From Data*, 1st ed. pages 345–351)

Possible values of a continuous random variable consist of an entire interval along the number line. The height of an individual selected at random from some population is a continuous variable, even though we usually only measure it to the nearest inch or half of an inch. Since this is not just a set of discrete values, its behavior is described by a smooth curve which is defined by some function $f(x)$. This function is called a *density function*. A density function has the following characteristics:

1. $f(x) \geq 0$ for all x values

2. The total area under the density curve = 1

For a continuous random variable, the probability of observing a value in any particular interval is equal to the area under the density curve and above that interval. The inclusion or exclusion of the endpoints of the interval will not change the associated probability. Unlike the discrete case, where including a single value might change the probability, in the continuous case, this isn't the case. The diagram below shows a density function with a total area under the curve = 1. For this distribution, $P(1 < x < 2)$ is the total area under the curve and above the interval from 1 to 2. The probability of interest is the area of a rectangle with length 1 and height 0.4. It follows that $P(1 < x < 2) = 1(0.4) = 0.4$.

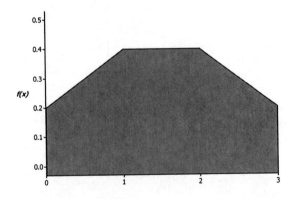

MEAN AND STANDARD DEVIATION OF RANDOM VARIABLES

(*Statistics: Learning From Data*, 1st ed. pages 352–362)

The mean and standard deviation of a random variable (or equivalently of a probability distribution) are computed differently for discrete random variables and continuous random variables. In the AP Statistics course, you are expected to be able to compute the mean and standard deviation given a discrete probability distribution. For continuous random variables, you do not need to be able to compute the mean and standard deviation (this involves the use of calculus), but you should be able to interpret the mean and standard deviation as measures of center and spread for values of the random variable.

DISCRETE RANDOM VARIABLES

For most discrete random variables, the mean and standard deviation are easy to calculate. These values are useful in describing the behavior of a discrete random variable.

For any discrete random variable x,

$$\mu_x = \sum xp(x)$$
$$\sigma_x = \sqrt{\sigma_x^2} = \sqrt{\sum (x - \mu)^2 p(x)}$$

The mean of a random variable is also called the expected value of the random variable and denoted as $E(x)$.

SAMPLE PROBLEM 1 Using the coaster set example, calculate the mean and standard deviation.

SOLUTION TO PROBLEM 1

$$\mu_x = \sum xp(x) = 2p(2) + 4p(4) + 6p(6) + 8p(8)$$
$$= 2(0.1) + 4(0.35) + 6(0.5) + 8(0.05)$$
$$= 5$$

$$\sigma_x = \sqrt{\sigma_x^2} = \sqrt{\sum (x - \mu)^2 p(x)}$$
$$= \sqrt{(2-5)^2 p(2) + (4-5)^2 p(4) + (6-5)^2 p(6) + (8-5)^2 p(8)}$$
$$= \sqrt{(-3)^2 (0.1) + (-1)^2 (3.5) + (1)^2 (0.5) + (3)^2 (0.05)}$$
$$= \sqrt{2.2}$$
$$= 1.48$$

Notice that the mean number of coasters in a set sold by the store would be 5; however, the store only sells them in sets of 4 or 6. As we have seen previously, the mean is an "average" value (and does not have to be one of the values in our distribution). The standard deviation of 1.48 describes the variability in x over a long sequence of observations.

Continuous Random Variables

Computing the mean and standard deviation of a continuous random variable requires calculus (and is not required in AP Stats). While you won't need to be able to do these calculations, you should be able to interpret the mean and standard deviation of a continuous random variable as measures of center and spread. As was the case with discrete random variables, the mean is the long-run average value and the standard deviation describes long-run variability in the values of the random variable.

Mean, Variance, and Standard Deviation for Linear Combinations

It is also possible to define new random variables that are functions of other random variables. One such random variable is a linear combination of other known random variables. If we wish to calculate the mean of a random variable constructed as a linear combination of two or more random variables, $y = a_1 x_1 + a_2 x_2 + \cdots + a_k x_k$, we would use the following formula.

$$\mu_y = a_1 \mu_1 + a_2 \mu_2 + \ldots + a_k \mu_k$$

where the x_i's are random variables, the a_i's are constants, and μ_i is the mean of the random variable x_i.

Example Suppose that the weight of a student book bag is a random variable with a mean of 26 pounds with a standard deviation of 4 pounds. What is the mean total weight of seven students' book bags? With x_i representing the weight of the ith book bag, the total weight y is $y = x_1 + x_2 + \cdots + x_7$. Then

$$\mu_y = \mu_{x_1} + \mu_{x_2} + \ldots + \mu_{x_7}$$
$$= 26 + 26 + \ldots + 26$$
$$= 182$$

Finding the standard deviation of a linear combination of random variables requires a little more work. If the random variables in a linear combination are independent, then the standard deviation of $y = a_1 x_1 + a_2 x_2 + \cdots + a_k x_k$ is

$$\sigma_y = \sqrt{a_1^2 \sigma_1^2 + a_2^2 \sigma_2^2 + \ldots + a_k^2 \sigma_k^2}$$

where the x_i's are independent random variables, the a_i's are numerical constants and σ_i^2 is the variance of x_i.

Note that the formula for the mean of a linear combination holds for any set of random variables, but the formula above for the standard deviation of a linear combination should only be used if the random variables are independent.

SAMPLE PROBLEM 2 A local high school requires seniors to take a comprehensive exam covering mathematics, science, and English prior to graduation. Each section is graded separately by a person who teaches the corresponding subject. The table below gives the mean and standard deviation for the time required to grade each section of the exam.

Subject	Mean Time per Test	Standard Deviation
Mathematics	5.2 min.	0.4 min.
Science	4.9 min.	0.3 min.
English	5.7 min.	0.5 min.

Since each exam contains all three sections, the total time required to score the exam is the sum of the times required to grade each section. Assuming that the time required to grade any one section is independent of the times required to grade the other two sections, what are the mean and standard deviation of y = total time to grade an exam?

SOLUTION TO PROBLEM 2

$y = x_M + x_S + x_E$, where x_M is the time to score the math section, x_S is the time required to score the science section, and x_E is the time required to score the English section.

$$\mu_{grading\ time} = 5.2 + 4.9 + 5.7 = 15.8 \text{ minutes}$$

$$\sigma_{grading\ time} = \sqrt{0.4^2 + 0.3^2 + 0.5^2} = \sqrt{0.16 + 0.09 + 0.25} = \sqrt{0.5} = 0.707$$

In this case, it will take an average of approximately 15.8 minutes to grade an exam. The standard deviation of total grading time is about 0.7 minutes, indicating that there is not much variability in exam grading times.

BINOMIAL AND GEOMETRIC DISTRIBUTIONS

(*Statistics: Learning From Data*, 1st ed. pages 363–372)

There are two discrete probability distributions of major interest—the binomial distribution and the geometric distribution. Both of these distributions arise in situations where the experiment consists of a sequence of trial and where each trial can result in one of two possible outcomes. The outcomes are usually called success (S) and failure (F). If a chance experiment meets the requirements for either of these distributions, then we are able to use the binomial or geometric distribution to compute probabilities.

BINOMIAL DISTRIBUTIONS

A binomial experiment has the following characteristics:

▪ The experiment consists of a fixed number of trials, n.
▪ Each trial results in one of two possible outcomes, success (S) or failure (F).
▪ The n trials are *all* independent.
▪ The probability of success, p, is the same for each trial.

The random variable defined as x = number of successes in a binomial experiment has a binomial distribution. Keep in mind that success doesn't necessarily mean "good;" it is the name given to the value of the outcome of the chance experiment in which we are interested.

For example, suppose we are interested in the number of students in a sample of 10 students who test positive for the H1N1 virus. Since a student will either test positive for the virus or not, we can think of the 10 students at 10 trials and each trial has two possible outcomes. To meet the independence criteria, we would need the events "student x tests positive" and "student y tests positive" to be independent. For this to be a binomial setting, we would also require a fixed probability that a randomly selected student has H1N1. Suppose that the county health department states that 18% of students will test positive for this virus. If we assume that 18% is a constant probability of a student testing positive, we have a binomial experiment.

Once we know that our random variable is binomial, a variety of probability questions can be considered.

The probability distribution of a binomial random variable is the *binomial probability distribution*. With

n = number of independent trials in a binomial experiment

p = constant probability that a trial results in a success

and the notation $\binom{n}{x} = \dfrac{n!}{x!(n-x)!}$

then

$p(x) = P(x$ successes among n trials$)$

$$= \binom{n}{x} p^x (1-p)^{n-x}$$

Suppose we want to know the probability that exactly 3 of the 10 students test positive for H1N1. Then $n = 10$, $p = 0.18$, and you can write B $(n,p) = B(10, 0.18)$

x = the number of students that test positive for H1N1 out of 10 students

$$p(3) = P(x = 3)$$

$$= \binom{10}{3} 0.18^3 (1 - 0.18)^{10-3}$$

$$= \frac{10!}{3!7!} 0.18^3 (0.82)^7$$

$$= 120(0.00583)(0.24929)$$

$$= 0.1744$$

The probability of exactly 3 of the 10 students testing positive is 0.1744.

What is the probability that 3 or fewer of the 10 students test positive for the virus? In this case, the event of interest would "happen" if 3, 2, 1, or 0 students test positive. Here, we would compute each of the individual probabilities of success and then add these probabilities. (We can add them because the separate events are disjoint.)

$$P(x \leq 3) = P(x = 0) + P(x = 1) + P(x = 2) + P(x = 3)$$

AP Tip

Remember that zero is a possible value (you can have zero successes) for a binomial random variable. To compute $P(x > 3)$, use $1 - (P \leq 3)$.

Binomial probabilities can also be calculated using a graphing calculator using one of the **binomialpdf** or **binomialcdf** functions. The arguments entered on your calculator for each are shown below:

$p(x = value)$ $p(x \leq value)$

use pdf command   use cdf command

binompdf(n, p, x)   **binomcdf**(n, p, x)

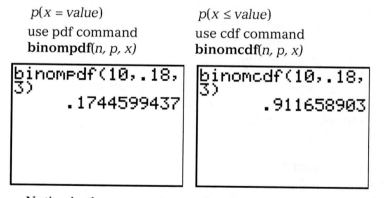

Notice both arguments require the n, p, x values. The difference is the binompdf calculates the probability for exactly three students with a positive test and binomcdf added the probabilities for 0, 1, 2, and 3 students testing positive.

AP Tip

The probability of "more than x" would be calculated on a graphing calculator using **1 – binomcdf**(n, p, x). Never use calculator syntax on the AP Exam. You should use the Binomial Probability formula, which is given on the formula sheet on the AP Exam.

Finally, if x has a binomial distribution, the mean and standard deviation of x are

$$\mu_x = np$$
$$\sigma_x = \sqrt{np(1-p)}$$

With the earlier example of students testing positive for H1N1 we can easily find the mean and standard deviation.

$$\mu_x = 10(0.18) = 1.8$$
$$\sigma_x = \sqrt{10(0.18)(0.82)} = 1.21$$

GEOMETRIC DISTRIBUTION

A geometric experiment has the same properties as a binomial experiment with one exception. Instead of a fixed number of trials, a geometric experiment continues until a success occurs.
- Each trial results in one of two possible outcomes, success (S) or failure (F).
- Trials are independent.
- Probability of success, p, is the same for each trial.

Then the random variable x = number of trials until the first success, including the success trial, has a geometric distribution. This is unlike the binomial distribution, which had a finite number of trials.

Using the example of the students testing positive for H1N1, in the geometric case we would continue to test students until we found one who tested positive. This is different from the binomial case because you could test one child and get a positive result or test many students before you had a positive result. For a geometric random variable, the geometric probability distribution is:

x = number of trials to first success

p = probability of success for each trial

$$p(x) = (1-p)^{x-1} p$$

Recall that for the H1N1 testing example, the probability that any one student tests positive is 0.18. Suppose that we plan to continue to test students until one student tests positive. What is the probability that we would have to test two or fewer students before we find one testing positive?

$$p(1) = (1 - 0.18)^{1-1}0.18 = 0.18$$
$$p(2) = (1 - 0.18)^{2-1}0.18 = 0.148$$
and the $P(x \le 2) = p(1) + p(2) = 0.18 + 0.148 = 0.328$

> ## AP Tip
>
> The geometric distribution formulas are not on the formula sheet that you get during the AP Exam, but these formulas tend to be easy to remember. One useful formula is :
>
> $$P(x > k) = (1 - p)^k \text{ for geometric settings}$$

NORMAL DISTRIBUTIONS

(*Statistics: Learning From Data*, 1st ed. pages 373–390)

Normal distributions are continuous probability distributions with a bell-shaped density curve. Normal distributions are described by a mean and a standard deviation. One special case is the standard normal distribution, which has a mean of 0 and a standard deviation of 1.

Review section 2 introduced a standardized score known as a *z* score. This standardized score allows us to see the relative standing of any value within a distribution. The *z* score separates the area under the curve that is below the *z* score from the area that is above the score. This area also represents the probability of observing a value less than or equal to *z* for a normal random variable.

For example, if *z* = –1.32, this would cut off an area below this *z* score. The shaded region in the picture is the area in question.

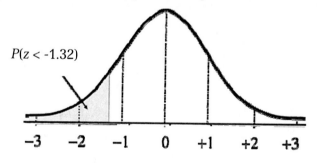

One way to calculate this area is to look up the *z* score of interest on a "standard normal" table. Use the following steps.

- Read down the first column until you find the whole number and first decimal value (that is, to the tenths place). In this case, find –1.3.
- Next, read across this row until you are in the column of the second decimal value (that is, the hundredths place). For this example, go over to the 0.02 column.
- This is the area to the left of the *z* score (0.0934), which is the probability of observing a value less than or equal to the value of *z* used.

(this is a portion of a *z* table)

z*	0.00	0.01	0.02	0.03	0.04
−1.5	0.0668	0.0655	0.0643	0.0630	0.0618
−1.4	0.0808	0.0793	0.0778	0.0764	0.0749
−1.3	0.0968	0.0951	0.0934	0.0918	0.0901
−1.2	0.1151	0.1131	0.1112	0.1093	0.1075
−1.1	0.1357	0.1335	0.1314	0.1292	0.1271

Another way to calculate this would be to use your graphing calculator. In this case, depending on your calculator, you would enter the arguments as follows into the calculator.

normalcdf(lower bound, upper bound, [μ , σ])

cdf for cumulative density function   lowest value cut off   highest value cut off   mean of normal distribution (if needed)   standard deviation of normal distribution (if needed)

For this example, since the default values are for a standard normal, we don't need to include the mean and standard deviation parameters. When entered, the calculator input returns the value shown below. The calculator command requires a lower and upper boundary. Since the data are in standardized form, we need a lower boundary that will include the entire tail area in standard units. We used −5 since that will include the left tail to −5 standard deviations. We could have used other values such as −10, −100, and so on, and this will not impact our final answer as there is very little "area" included in this lower tail beyond 5 SDs below the mean.

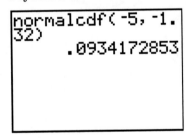

The value returned is the area under the standard normal curve and to the left of a *z* score of −1.32. See the shaded picture below.

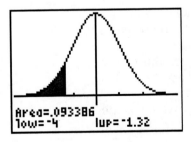

(This is created by **ShadeDraw**. See your calculator instructions as needed.)

You will notice the area under the curve corresponds to the area from the left tail of this distribution up to the z score value. We can also solve problems to find the area between two z scores or even above a z score value. The notation and calculations we would use are shown here.

$$P(a < x < b) = P(a^* < z < b^*)$$

where z is a variable from a standard normal distribution and

$$a^* = \frac{a - \mu}{\sigma} \text{ and } b^* = \frac{b - \mu}{\sigma}$$

AP Tip

You can also use the graphing calculator to compute area above an interval (a, b). By using the command shown here, your calculator will return the probability of the area between the two boundaries given. Never use calculator syntax on the AP Exam; use good statistical nomenclature.

normalcdf (lower bound, upper bound, [μ , σ])

You can either enter the lower bound and upper bound in terms of z scores (in which case you don't need to enter the values for μ and σ— the default values are 0 and 1) or you can use the original bounds (without transforming to z scores) and enter the appropriate values for μ and σ.

Here are a few things to keep in mind when solving probability problems:

1. Often questions will ask for the probability that involves finding an area under a curve. In these questions, always draw the distribution, mark off the interval of interest, shade the appropriate area, and then find the desired probability.

2. Binomial distribution questions may ask for the probability of observing one possible value or a probability that involves a collection of possible values. Watch for questions like $P(x < 3)$ vs $P(x \le 3)$. In the first case, you do not include $x = 3$ in the solution and in the second case you will include $x = 3$.

3. If the question involves some type of linear combination, you will need to first find the appropriate mean and standard deviation. The next sample problem involves this additional consideration.

SAMPLE PROBLEM 3 Recall the earlier example where time to grade an exam was the sum of the times required to grade each of three sections. Suppose that the time to grade each section has a normal distribution with the mean and standard deviation given earlier. Also suppose that the time to grade any one section of the exam is independent of the time to grade either of the other two sections. We saw earlier that

$$\mu_{grading\ time} = 5.2 + 4.9 + 5.7 = 15.8$$

$$\sigma_{grading\ time} = \sqrt{0.4^2 + 0.3^2 + 5^2} = \sqrt{0.14 + 0.09 + 0.25} = \sqrt{0.5} = 0.707$$

We also know that the sum of normal random variables also has a normal distribution. What is the probability that the time to grade a randomly selected exam is less than 14.9 minutes?

SOLUTION TO PROBLEM 3

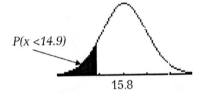

$P(x < 14.9)$

15.8

$$P(x \le 14.9) = P\left(z \le \frac{14.9 - 15.8}{0.707}\right) = P(z \le -1.273) = 0.1015$$

CONVERTING A z SCORE BACK TO AN X VALUE

Occasionally we would like to convert back to an x value if we know the z score or a related probability. This calculation is simple using the z score formula.

Since

$z = \dfrac{x - \mu}{\sigma}$, by using basic algebraic manipulation,

$x = z\sigma + \mu$

If you were interested in what score you need to get to fall in the 88th percentile on a test with a mean of 91 and standard deviation of 3, you would find the z score that goes with an area of 0.88 to the left of z and then solve the equation for x.

For an area of 0.88 to the left, look up the corresponding z score in the normal table or use

`InvNorm(0.88)` $= 1.175$

$1.175 = \dfrac{x - 91}{3}$,

$x = 1.175(3) + 91 = 94.5$

Since it is probably safe to assume that scores on the test are whole numbers, you would need to score a 95 to be at the 88th percentile for the test.

NORMAL APPROXIMATIONS

Sometimes the normal distribution is used to approximate a discrete distribution, such as the distribution of IQs or the distribution of the number of trucks at a truck stop. When we say that discrete data are approximately normally distributed, we can use a normal distribution to approximate probabilities. For example, suppose that the number of bags of groceries purchased on a military payday has a distribution that is approximately normal with a mean of 13 bags and a standard deviation of 2 bags. What is the probability that a randomly selected customer purchases 15 bags of groceries on a military payday? One way to answer this question is to approximate the probability of 15 bags using the area under the corresponding normal curve and between 14.5 and 15.5.

$N(13,2)$

$$z_{14.5} = \frac{14.5 - 13}{2} = 0.75 \quad \text{and} \quad z_{14.5} = \frac{15.5 - 13}{2} = 1.25$$

then

$$P(0.75 < z < 1.25) = 0.8944 - 0.7734 = 0.121$$

Another option would be to use the graphing calculator. The lower and upper bounds would be the 14.5 and 15.5, respectively.

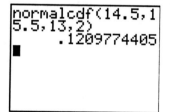

RANDOM VARIABLES AND PROBABILITY DISTRIBUTIONS: STUDENT OBJECTIVES FOR THE AP EXAM

- You will be able to calculate the mean and standard deviation of a discrete random variable.
- You will be able to use a given probability distribution to calculate probabilities.
- You will be able to find mean and standard deviations of a linear combination of random variables.
- You will be able to solve problems involving the binomial and geometric distributions.
- You will be able to use the normal distribution to compute probabilities.

MULTIPLE-CHOICE QUESTIONS

1. Air America Airways (AAA) tracks the number of delays per day for its jet fleet due to maintenance issues at a major airport. Suppose that the probability distribution of x = number of delays on a randomly selected day is given in the table below.

x	0	1	2	3	4	5
$P(x)$	0.42	0.31	0.12	0.08	0.04	0.03

   What is the probability that the airline has fewer than three delays due to maintenance at this airport on a randomly selected day?
   (A) 0.08
   (B) 0.12
   (C) 0.15
   (D) 0.85
   (E) 0.93

2. Which of the following random variables is not continuous?
   (A) The amount of gasoline in a car.
   (B) The time it takes to commute to work.
   (C) Number of goals scored by a hockey team.
   (D) Distance travelled by a police car in a day.
   (E) Lifetime of a AAA battery.

3. The probability distribution of x is given in the table below.

x	5	6	7	8	9
$P(x)$	0.1	0.15	0.45	0.25	0.05

   What is the probability that x is less than or equal to 7?
   (A) 0.45
   (B) 0.25
   (C) 0.15
   (D) 0.70
   (E) 0.30

4. What are the mean and standard deviation of a binomial distribution with $n = 20$ and $p = 0.25$?
   (A) $\mu = 2, \sigma = 0.1875$
   (B) $\mu = 2, \sigma = 3.75$
   (C) $\mu = 2, \sigma = 1.936$
   (D) $\mu = 5, \sigma = 0.1875$
   (E) $\mu = 5, \sigma = 1.936$

5. For a binomial distribution with $n = 20$ and $p = 0.25$, what is $P(x = 2)$?
   (A) 0.25
   (B) 0.0913
   (C) 0.0625
   (D) 0.0669
   (E) 0.1339

6. Two random variables x and y are independent. The mean and standard deviation of x are 78 and 6, respectively. The mean and standard deviation of y are 83 and 9 respectively. What is the standard deviation of $x + y$?
   (A) 15
   (B) 10.8
   (C) 7.5
   (D) 3.9
   (E) 3

7. Which of the following is not a characteristic of a binomial experiment?
   (A) A fixed number of trials.
   (B) Trials continue until a success is observed.
   (C) Trials are independent.
   (D) The probability of success is the same in all trials.
   (E) There are only two possible outcomes for each trial, success or failure.

8. Thirty percent of a school district's students receive advanced scores on their state end-of-course assessment in mathematics. If you choose 25 students at random, what is the probability that more than 10 of these students received advanced scores on their assessment?
   (A) 0.9022
   (B) 0.3231
   (C) 0.1894
   (D) 0.0978
   (E) 0.0442

9. The time to make coffee at a local convenience store is normally distributed with a mean of 4.5 minutes and standard deviation of 0.6 minutes. What is the probability that it will take less than 3.5 minutes for a worker to make a pot of coffee?
   (A) 0.0478
   (B) 0.0984
   (C) 0.4522
   (D) 0.5873
   (E) 0.9522

10. Two independent observations of a random variable are made. If the random variable has a normal distribution with a mean of 75 and standard deviation of 9, what is the probability that the sum of these two observations is greater than 168?
   (A) 0.1587
   (B) 0.0787
   (C) 0.0228
   (D) 0.0013
   (E) 0.0001

11. Which of the following are true statements?
    I.   A geometric experiment has a fixed number of trials.
    II. A normal distribution is defined by specifying μ and σ.
    III. The standard deviation of the sum of two independent random variables is found by taking the square root of the sum of the individual variances.
    (A) I only
    (B) II only
    (C) I and II
    (D) II and III
    (E) I, II, and III

12. What is the approximate value of the 93rd percentile of a normal distribution with a mean of 86 and standard deviation of 8?
    (A) 102
    (B) 100
    (C) 98
    (D) 96
    (E) 94

13. If the height of a randomly selected woman is normally distributed with $\mu = 64.5$ and $\sigma = 2.5$, approximately what proportion of women have heights between 65 and 67.5 inches?
    (A) 0.14
    (B) 0.16
    (C) 0.31
    (D) 0.58
    (E) 0.88

14. Suppose that 25% of all AP students in a school district have completed their homework. If AP students from this district are chosen at random and checked to see whether or not they have completed their homework, what is the approximate probability that more than five of these students will be checked before the district finds a student who has complete the homework?
    (A) 0.079
    (B) 0.237
    (C) 0.316
    (D) 0.763
    (E) 0.921

15. Two random variables X and Y are independent. The means and standard deviations are shown in the table below. What are the values of the mean and standard deviation of $2X + Y$?

	μ	σ
X	6	0.5
Y	8	0.6

    (A) $\mu_{2X+Y} = 14$, $\sigma_{2X+Y} = 1.17$
    (B) $\mu_{2X+Y} = 14$, $\sigma_{2X+Y} = 0.78$
    (C) $\mu_{2X+Y} = 20$, $\sigma_{2X+Y} = 1.1$
    (D) $\mu_{2X+Y} = 20$, $\sigma_{2X+Y} = 0.78$

(E) $\mu_{2X+Y} = 20$, $\sigma_{2X+Y} = 1.17$

FREE-RESPONSE PROBLEMS

1. A store catering to female pre-teens sells jeans in four sizes (0, 1, 2, and 3) corresponding to the waist measurements shown in the table below. These are the only sizes that the store carries.

Store Size	Waist Measurement
0	$20 \le$ waist < 22
1	$22 \le$ waist < 24
2	$24 \le$ waist < 26
3	$26 \le$ waist < 28

Suppose that for female pre-teens, waist measurement is normally distributed with a mean of 24.4 inches and a standard deviation of 2.5 inches.

(a) What proportion of female pre-teens will be unable to find jeans in this store?

(b) What proportion of female pre-teens wear a size 2?

(c) If 15 randomly selected female pre-teens are chosen, what is the probability that exactly 5 of them wear a size 2?

2. The time a car must wait for the light to change at an intersection depends on the type of intersection. Suppose that the wait time at major and the wait time at secondary intersections are approximately normal with the means and standard deviations given in the table below.

	Mean Wait Time (minutes)	Standard Deviation
Lights at Major Intersections	3.2 minutes	0.5 minutes
Lights at Secondary Intersections	2.5 minutes	0.7 minutes

(a) Kylie wants to know how likely it is that she will have to wait more than four minutes at a major intersection by her house. What is the probability that she will have to wait more than four minutes?

(b) Ella must travel through two major intersection lights and three secondary intersection lights on her way to work. Assuming that the wait times at these intersections are independent, what are the mean and standard deviation of her total wait time on the way to work?

(c) What is the probability that Ella's total wait time will be less than 12 minutes?

Answers

MULTIPLE-CHOICE QUESTIONS

1. **D.** Since x is a discrete random variable, the desired probability is the sum of the individual probabilities.
$P(x < 3) = P(x = 0) + P(x = 1) + P(x = 2) = 0.42 + 0.31 + 0.12 = 0.85$.
Notice that $P(x = 3)$ also would have been included if the question had asked for three or fewer delays (*Statistics: Learning From Data*, 1st ed. pages 352–356).

2. **C.** Possible values for number of hockey goals are discrete counts. The other four choices are continuous variables (*Statistics: Learning From Data*, 1st ed. pages 352–356).

3. **D.** Since this asks for $P(x \le 7)$, this is P(5) + P(6) + P(7) = 0.10 + 0.15 + 0.45 = 0.70. (*Statistics: Learning From Data*, 1st ed. pages 352–356).

4. **E.** μ = np = 20(.25) = 5. $\sigma = \sqrt{20(0.25)(0.75)} = 1.936$. (*Statistics: Learning From Data*, 1st ed. pages 438–443).

5. **D.** This is a binomial probability with p = 0.25 and n = 20. This probability can be evaluated using a graphing calculator by using binompdf(20, 0.25, 2) or by using the formula for binomial probabilities $\binom{20}{2} 0.25^2 (0.75)^{18} = 0.0669$ (*Statistics: Learning From Data*, 1st ed. page 367).

6. **B.** The standard deviation of $x + y$ is found by taking the square root of the sum of the variances. The calculation for this is $\sigma_{x+y} = \sqrt{\sigma^2_x + \sigma^2_y}$, in this case, $\sigma_{x+y} = \sqrt{6^2 + 9^2}$ (*Statistics: Learning From Data*, 1st ed. pages 352–356).

7. **B.** Continuing trials until a success is observed is a characteristic of a geometric experiment (*Statistics: Learning From Data*, 1st ed. pages 369–371).

8. **D.** The desired binomial probability is $P(x > 10)$, which can be found by computing $1 - P(x \le 10)$. On a graphing calculator this calculation would be 1-binomcdf(25,3,10) (*Statistics: Learning From Data*, 1st ed. page 367).

9. **A.** This is the area under the normal curve with mean 4.5 and standard deviation 0.6 to the left of 3.5 minutes. $P(x < 3.5) = 0.0478$. On a graphing calculator this calculation would be normalcdf(-E99, 3.5,4.5,0.6) (*Statistics: Learning From Data*, 1st ed. pages 376–380).

10. **B.**

$$\mu = 75 + 75 = 150$$

$$\sigma = \sqrt{9^2 + 9^2} = 12.728$$

$$P(sum \geq 168) = P\left(z \geq \frac{168 - 150}{12.728}\right) = 0.0787$$

On a graphing calculator normalcdf(168,E99,150,12.728) would calculate this probability.

(*Statistics: Learning From Data,* 1st ed. pages 352–356).

11. **D.** A geometric experiment continues until the first success occurs. (*Statistics: Learning From Data,* 1st ed. pages 369–371).

12. **C.** First find the z score that has an area to the left of 0.93 using the normal table. Then solve $1.48 = \dfrac{x - 86}{8}$ for x, which results in $x = 97.8 \approx 98$. Or the graphing calculator could be used to find the value using invNorm(0.93,86,8) = 97.8. (*Statistics: Learning From Data,* 1st ed. pages 160–164 and 376–377).

13. **C.** The area of interest is between the two x values so find the $P(x < 65)$ and subtract this from $P(x < 67.5)$. On a graphing calculator this calculation would be normalcdf(65.67,64.5,5) = 0.306. (*Statistics: Learning From Data,* 1st ed. pages 376–377).

14. **B.** This is a geometric experiment with $p = 0.25$. On a graphing calculator this calculation would be 1-geometcdf(0.25,5) = 0.237. (*Statistics: Learning From Data,* 1st ed. pages 369–371).

15. **E.** The mean is 2(6) + 8 = 20. The standard deviation $\sigma_{2X+Y} = \sqrt{2^2(0.5)^2 + (0.6)^2} = 1.17$ (*Statistics: Learning From Data,* 1st ed. pages 352–256).

FREE-RESPONSE PROBLEMS

1. (a) Anyone with a waist measurement < 20 or > 28 will not find jeans.

x = waist measurement

$$P(x < 20) + P(x > 28)$$

$$= P\left(z < \frac{20 - 24.4}{2.5}\right) + P\left(z > \frac{28 - 24.4}{2.5}\right)$$

$$= P(z < -1.76) + P(z > 1.44)$$

$$= 0.0392 + 0.0749$$

$$= 0.1141$$

(b)

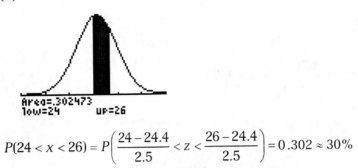

$$P(24 < x < 26) = P\left(\frac{24 - 24.4}{2.5} < z < \frac{26 - 24.4}{2.5}\right) = 0.302 \approx 30\%$$

(c) From part (b) we know that $P(\text{size } 2) = 0.30$. The desired probability is then a binomial probability with $n = 15$ and $p = 0.30$. Using the binompdf function on the graphing calculator binompdf(15, 0.3, 5) = 0.206. This could also be calculated as

$$P(x = 5) = \binom{15}{5}(0.3)^5(0.7)^{10} \quad (\textit{Statistics: Learning From Data}, \text{1st}$$

ed. pages 367 and 381–385).

2. (a) $P(x > 4) = P\left(z > \dfrac{4 - 3.2}{0.5}\right) = P(z > 1.6) = 0.0548$

(b) Ella's total wait time is the sum of the wait times for all five lights. Because the wait times are independent, the mean and standard deviation of Ella's total wait time are:

$$\mu_{time} = 2(3.2) + 3(2.5) = 13.9 \, \text{min}.$$
$$\sigma_{times} = \sqrt{0.5^2 + 0.5^2 + 0.5^2 + 0.7^2 + 0.7^2} = \sqrt{1.73} = 1.315 \, \text{min}.$$

(c) $P(x < 12) = P\left(z < \dfrac{12 - 13.9}{1.315}\right) = -1.45 = 0.0735$. There is about a 7.4% chance that Ella's total wait time will be less than 12 minutes (*Statistics: Learning From Data*, 1st ed. pages 352–356 and 367).

7

SAMPLING VARIABILITY AND SAMPLING DISTRIBUTIONS

In this review section, we will consider sampling distributions and see how they describe the behavior of a sample statistic.

OBJECTIVES

- Understand that a sampling distribution describes the behavior of a sample statistic.
- Know the general properties of the sampling distribution of a sample mean, $\bar{x}$.
- Know the general properties of the sampling distribution of a sample proportion, $\hat{p}$.

STATISTICS AND SAMPLING VARIABILITY

(Statistics: Learning From Data, 1st ed. pages 433–437)

What makes learning from sample data a challenge is that different samples from the same population give different results. For example, the average age at which a child learned to walk for one sample of 10 children would be different from the average age to walk for a different sample of 10 children. The value of the sample mean, $\bar{x}$, varies from sample to sample. To know what any one sample mean might tell us about the corresponding population, we need to understand this sampling variability. The sampling distribution of a sample statistic describes its behavior in repeated sampling.

THE SAMPLING DISTRIBUTION OF A SAMPLE MEAN

(Statistics: Learning From Data, 1st ed. pages 568–577)

To help understand the sample-to-sample variability in the sample mean, consider taking sample after sample from a particular population and looking at the resulting sample means. We took 100 different random samples for size 5 from a particular normal population distribution with mean $\mu = 0$ and computed the sample mean for each sample. These 100 sample means were used to construct the top histogram in the figure below. Notice there is variability in the sample means, but the sample means tend to cluster around 0, the value of the population mean.

We repeated this process with samples of size $n = 10$ to produce the second histogram in the figure, samples of size $n = 20$ to produce the third histogram, and samples of size $n = 40$ to produce the bottom histogram. These histograms are approximations of the sampling distribution of $\bar{x}$ for the given sample sizes.

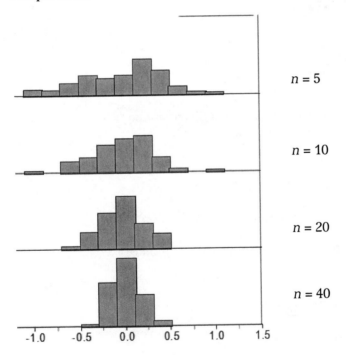

There are three characteristics of the graphs above that we should note. First, all four sampling distributions appear to have a roughly (approximately) normal shape. Second, the center for each histogram is approximately 0, the value of the population mean. Finally, the most interesting feature is that as the sample sizes increased, the overall spread decreased. In other words, as our sample size increased, $\bar{x}$ varies less from one sample to another and the $\bar{x}$ values tend to be closer to μ. We use x-bar to denote the sample mean (statistic) and mu to denote the population mean.

What if the population distribution is not normal? Would the sampling distribution of $\bar{x}$ still be described by these same three characteristics? Let's investigate by considering the population summarized in the histogram below. This population is skewed right and has a mean of $\mu = 1$.

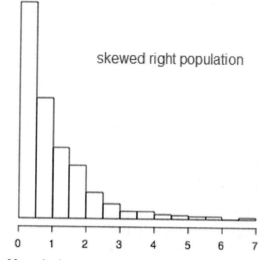

Now let's examine the approximate sampling distributions below. Each histogram was constructed using the sample means from 100 different random samples from the skewed population. Notice the distributions appear to be centered at $\mu = 1$. As n increases, the sampling distributions become less spread out, and for larger sample sizes, the sampling distribution is much less spread out than the population. Finally, although the shape of the sampling distribution is skewed for the small sample sizes, the sampling distributions become more symmetric (and less skewed) and for samples of size 30 the sampling distribution is approximately normal.

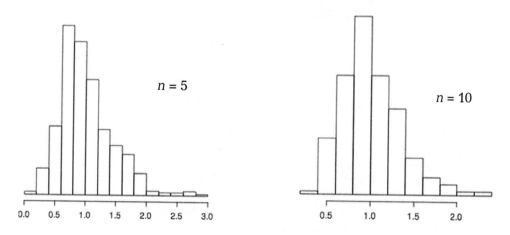

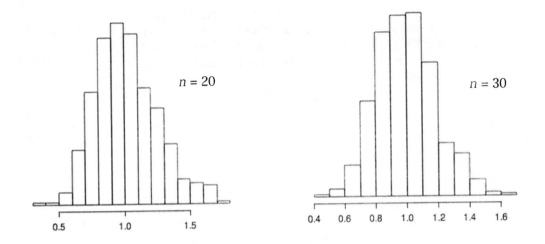

The fact that the sampling distribution of $\bar{x}$ is approximately normal in shape when the sample size is large, even if the population is not normal, is a consequence of a powerful result known as the *Central Limit Theorem*. This theorem states that if n is sufficiently large, the $\bar{x}$ distribution will be approximately normal even if the population is not normal.

There are two situations where we can count on the sampling distribution of $\bar{x}$ to be normal, or at least approximately normal:

1. If the population distribution is normal, the sampling distribution of $\bar{x}$ is (approximately) normal for any sample size.

2. If the population distribution is not terribly skewed, then the sampling distribution of $\bar{x}$ will be approximately normal if $n \geq 30$.

GENERAL PROPERTIES OF THE SAMPLING DISTRIBUTION OF $\bar{x}$

If $\bar{x}$ is the sample mean for a sample of size n from a population with mean μ and standard deviation, then

■ $\mu_{\bar{x}} = \mu$

■ $\sigma_{\bar{x}} = \dfrac{\sigma}{\sqrt{n}}$. This rule is exact if we have an infinite population; otherwise it's approximately correct with a finite population as long as no more than 10% of the population was included in the sample (population > 10n).

■ If the population distribution is normal, the sampling distribution of x is normal for any sample size n.

■ If the population distribution is not normal, the Central Limit Theorem states that for n sufficiently large, the sampling distribution is well approximated by normal curve. A sample size of 30 or more is generally considered large enough.

These properties imply that if n is large or the population distribution is normal,

$$z = \frac{\bar{x} - \mu_{\bar{x}}}{\sigma_{\bar{x}}} = \frac{\bar{x} - \mu}{\dfrac{\sigma}{\sqrt{n}}}$$

has a distribution that is approximately standard normal.

SAMPLE PROBLEM 1 Suppose that the age at which children begin to walk on their own is normally distributed with a mean of 12 months and a standard deviation of 1.5 months. A sample of four babies is observed and the age when each of these babies began to walk is recorded.

(a) Describe the sampling distribution of $\bar{x}$ for samples of size 4.

(b) What is the probability that the mean age to walk for a random sample of four babies is between 11 and 12.5 months?

SOLUTION TO PROBLEM 1

(a) $\bar{x}$ = the mean walking age for the children in the sample of four

μ = the mean walking age for all children

$\mu_{\bar{x}} = \mu = 12$

$\sigma_{\bar{x}} = \dfrac{\sigma}{\sqrt{n}} = \dfrac{1.5}{\sqrt{4}} = \dfrac{1.5}{2} = 0.75$

Because the population distribution of walking ages is normal, the sampling distribution of $\bar{x}$ is also normal.

(b) Because the sampling distribution of $\bar{x}$ is normal with a mean of 12 and a standard deviation of 0.75, we can use what we know about normal distributions to compute

$$P(11 \le \bar{x} \le 12.5) = P\left(\frac{11 - 12}{0.75} \le z \le \frac{12.5 - 12}{0.75}\right)$$

$$= P(-1.33 \le z \le 0.67)$$

$$= 0.7486 - 0.0918$$

$$= 0.6568$$

A graphing calculator can also be used to evaluate the desired probability:

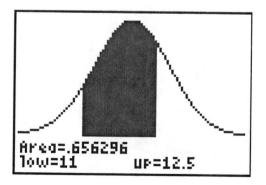

```
Area=.656296
low=11        up=12.5
```

Notice that the calculator gives a slightly different answer than the answer obtained using the normal table. This is due to rounding of the areas in the normal table, Normalcdf(11, 12.5, 12,.75).

THE SAMPLING DISTRIBUTION OF A SAMPLE PROPORTION

(Statistics: Learning From Data, 1st ed. pages 438–442)

When the variable of interest in a population is categorical with just two possible categories, such as gender or whether or not a manufactured part is defective, we usually want to learn about the value of a population proportion. In this case, the sample proportion, $\hat{p}$ is used to estimate the population proportion, p. The statistic $\hat{p}$ varies from sample to sample, just as was the case for $\bar{x}$ for numerical data. The sampling distribution of $\hat{p}$ describes the sample-to-sample variability in the value of $\hat{p}$. For example, consider the histograms below. The first histogram was constructed using the $\hat{p}$ values from 100 random samples of size 10 drawn from a population with a proportion of successes of 0.34. The other histograms were constructed using sample sizes of $n = 25$, $n = 50$, and $n = 100$. Notice that the $\hat{p}$ values tend to cluster around the population proportion of $p = 0.34$ and that the variability in the approximate sampling distributions decreases as the sample size increases. Also notice that for the larger sample sizes, the sampling distribution of $\hat{p}$ is more nearly symmetric and has a shape that more closely resembles a normal distribution.

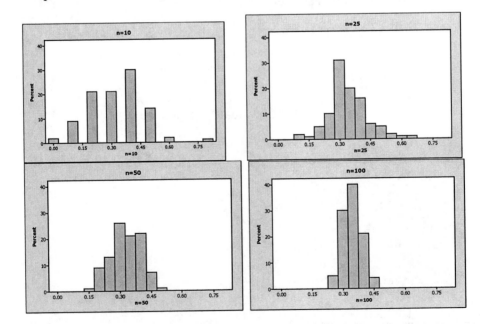

GENERAL PROPERTIES OF THE SAMPLING DISTRIBUTION OF $\hat{p}$

Let $\hat{p}$ be the proportion of successes in a random sample of size n from a population with a proportion of successes p. Then

- $\mu_{\hat{p}} = p.$

- $\sigma_{\hat{p}} = \sqrt{\dfrac{p(1-p)}{n}}$. This rule is exact if we have an infinite population; otherwise it's approximately correct with a finite population as long as no more than 10% of the population was included in the sample.
- When n is sufficiently large and p is not too near 0 or 1, the sampling distribution of $\hat{p}$ is approximately normal.

How large does n have to be in order for the sampling distribution of $\hat{p}$ to be approximately normal? It depends on the value of the population p. The further the population proportion is from 0.5, the larger the sample size needs to be in order for the sampling distribution of $\hat{p}$ to be well-approximated by a normal distribution. The sampling distribution of $\hat{p}$ is approximately normal as long as for a given value of p, n is large enough that

$$np \geq 10$$
and
$$n(1-p) \geq 10$$

SAMPLE PROBLEM 2 Suppose that approximately 4% of American children have Attention Deficit/Hyperactivity Disorder (ADHD). A random sample of 100 American children will be selected.

(a) Describe the mean and standard deviation of the sampling distribution of $\hat{p}$, where $\hat{p}$ is the proportion of children in the sample with ADHD.

(b) What is the smallest sample size that would be large enough for us to think that the sampling distribution of $\hat{p}$ would be approximately normal?

SOLUTION TO PROBLEM 2—For proportion, deal with categorical data. When defining a proportion, give a group and a trait.

p = the proportion of all american students that have ADHD

Population = all american students

$\hat{p}$ is a statistic (statistics have variation)

$\hat{p}$ = the proportion of american students from my sample that had ADHD (past tense)

(a) $\mu_{\hat{p}} = 0.04;$ $\quad \sigma_{\hat{p}} = \sqrt{\dfrac{0.04(1-0.04)}{50}} = \sqrt{\dfrac{0.04(0.96)}{50}} = \sqrt{\dfrac{0.0384}{50}}$

$$= \sqrt{0.000768} = 0.028$$

(b) For the sampling distribution of $\hat{p}$ to be approximately normal, we want

$np \geq 10$ and $n(1-p) \geq 10$. In this case, if we check both, we get

$n(0.04) \geq 10$, $n \geq 250$ and also $n(0.96) \geq 10$, $n \geq 10.42$ or 11. Since both conditions must be met, the smallest sample size is 250 children.

SAMPLE PROBLEM 3 Suppose that 30% of the seniors at a large urban school drink soda on a typical school day. If a random sample of 49 students is selected from the seniors at this school, what is the probability that more than 37% of the students in the sample drink soda on a typical school day?

SOLUTION TO PROBLEM 3 To solve this type of problem, we will

- Find the mean and standard deviation of the sampling distribution of $\hat{p}$.
- Verify that the sampling distribution of $\hat{p}$ is approximately normal.
- Use the normal distribution to calculate the probability of interest.
- p = the proportion of all students from this large urban school that drink soda on a typical school day.
- p = .3 $\hat{p}$ = .37 n = 49

$$\mu_{\hat{p}} = 0.30; \quad \sigma_{\hat{p}} = \sqrt{\dfrac{0.30(1-0.30)}{49}} = \sqrt{\dfrac{0.30(0.70)}{49}} = \sqrt{\dfrac{0.21}{49}} = \sqrt{0.0043} = 0.065.$$

Because $np = 49(0.3) = 14.7 \geq 10$ and $n(1-p) = 49(0.7) = 34.3 \geq 10$, the sampling distribution of $\hat{p}$ is approximately normal.

You can now evaluate the probability of interest, $P(\hat{p} \geq 0.37)$. Using a graphing calculator, we get normalcdf(0.37, 0.99, 0.3, 0.065) = 0.14.

The probability that more than 37% of the students in a random sample size of 49 drink soda on a typical school day is 0.14.

SAMPLING VARIABILITY AND SAMPLING DISTRIBUTIONS: STUDENT OBJECTIVES FOR THE AP EXAM

- You will be able to describe the sampling distribution of a sample mean.

⬛ You need to be able to explain a sampling distribution and what this means in a contextual setting.

⬛ You will be able to describe the sampling distribution of a sample proportion.

⬛ You will know when the sampling distribution of $\bar{x}$ is normal or approximately normal.

⬛ You will use properties of the sampling distribution of a sample mean to compute probabilities.

⬛ You will use properties of the sampling distribution of a sample proportion to compute probabilities.

MULTIPLE-CHOICE QUESTIONS

1. The mean of a population is 83 and the standard deviation is 5. What are the mean and the standard deviation of the sampling distribution of $\bar{x}$ for samples of size 40?
   (A) 83 and 5 respectively

   (B) 83 and $\dfrac{5}{40}$ respectively

   (C) 83 and $\dfrac{5}{\sqrt{40}}$ respectively

   (D) 83 and $\sqrt{\dfrac{5}{40}}$ respectively

   (E) 40 and $\sqrt{\dfrac{5}{40}}$ respectively

2. Fifty recent house sales were randomly selected from all sales in a large urban community. The average home sale price for these 50 sales was $372,000. $372,000 is the value of a
   (A) proportion.
   (B) population.
   (C) parameter.
   (D) statistic.
   (E) sample.

3. A population has $\mu = 78$ and a standard deviation $\sigma = 4.8$. What would be the mean and standard deviation of the sampling distribution $\bar{x}$ for random samples of size 100 from this population?
   (A) $\mu_{\bar{x}} = 78$, $\sigma_{\bar{x}} = 0.048$
   (B) $\mu_{\bar{x}} = 78$, $\sigma_{\bar{x}} = 0.48$
   (C) $\mu_{\bar{x}} = 78$, $\sigma_{\bar{x}} = 4.8$
   (D) $\mu_{\bar{x}} = 78$, $\sigma_{\bar{x}} = 48$
   (E) $\mu_{\bar{x}} = 78$, $\sigma_{\bar{x}} = 480$

4. Which of the following statements is true?
   I. The mean of any sample is always equal to the mean of the population.

    II. The sampling distribution of the sample mean is approximately normal when the sample size is sufficiently large.

    III. As the sample size increases, the standard deviation of the sampling distribution of the sample mean decreases.

(A) I only

(B) II only

(C) I and II

(D) I and III

(E) II and III

5. A population consisting of seven numbers has a mean of 8.5 and a standard deviation of 2.29. If random samples of size 3 are selected with replacement from this population, what can be said about the mean and standard deviation of the sampling distribution of $\bar{x}$?

(A) $\mu_{\bar{x}} = 8.5$ and the $\sigma_{\bar{x}} = 2.29$

(B) $\mu_{\bar{x}} = 8.5$ and the $\sigma_{\bar{x}} > 2.29$

(C) $\mu_{\bar{x}} = 8.5$ and the $\sigma_{\bar{x}} < 2.29$

(D) $\mu_{\bar{x}} > 8.5$ and the $\sigma_{\bar{x}} = 2.29$

(E) $\mu_{\bar{x}} < 8.5$ and the $\sigma_{\bar{x}} < 2.29$

6. Which of the following is a consequence of the Central Limit Theorem?

(A) When n is sufficiently large, the sampling distribution $\bar{x}$ is roughly approximated by the normal curve

(B) If the population distribution is normal, then the sampling distribution $\bar{x}$ will also be normal for all sample sizes n

(C) The shape of the sampling distribution will be the same as the shape of the population, but the sampling distribution will be less spread out than the population distribution.

(D) The shape of the sampling distribution will be the same as the shape of the population, but the sampling distribution will be more spread out than the population distribution.

(E) The sampling distribution will be centered at $\bar{x}$.

7. Each child in a random sample of 36 5-year-old children had his or her height measured. The sample mean height was 37.2 inches. Which of the following statements about the population of heights of 5-year-old children are true?

    I. The mean height of all 5-year-old children is 37.2 inches.

    II. The standard deviation of the heights of all 5-year-old children is six times smaller than the sample standard deviation.

    III. The standard deviation of the heights of all 5-year-old children is six times larger than the sample standard deviation.

(A) I only

(B) I and II only

(C) I and III only

(D) I, II, and III
(E) None of I, II, or III is true

8. In a simple random sample of 1,000 Americans who purchased a car in 2010, 61% were satisfied with the service provided by the dealer from which they bought their car. In a simple random sample of 1,000 Canadians who purchased a car in 2010, 58% were satisfied with the service provided by the dealer from which they bought their car. How do the standard errors associated with these statistics compare?
(A) The standard errors are about the same.
(B) The standard error is much smaller for the Canadian percentage because the population of Canada is smaller than that of the United States and so the sample is a larger proportion of the population.
(C) The standard error is much larger for the Canadian percentage because Canada has a lower population density than the United States. This means that Canadians tend to live further apart, which always increases sampling variability.
(D) The standard errors are equal.
(E) The standard errors cannot be compared because these are two different populations.

9. Suppose that the weights of packages of cheese are normally distributed with a mean of 8 ounces and a standard deviation of 0.2 ounces. A random sample of four packages of cheese is selected every 30 minutes and the sample mean weight is calculated. We would expect about 95% of the sample means to fall within what interval?
(A) 7.6 to 8.4 ounces
(B) 7.7 to 8.3 ounces
(C) 7.8 to 8.2 ounces
(D) 7.9 to 8.1 ounces
(E) 8.1 to 8.4 ounces

10. It is believed that currently 37% of American homes have at least one flat screen television. If the belief is true, which of the following is true about the sampling distribution of $\hat{p}$ for random samples of size 150 American homes?

(A) $\mu_{\hat{p}} = 0.37$ and $\sigma_{\hat{p}} = \sqrt{0.37(0.63)}$

(B) $\mu_{\hat{p}} = 0.37$ and $\sigma_{\hat{p}} = \dfrac{\sqrt{0.37(0.63)}}{150}$

(C) $\mu_{\hat{p}} < 0.37$ and $\sigma_{\hat{p}} = \sqrt{0.37(0.63)}$

(D) $\mu_{\hat{p}} < 0.37$ and $\sigma_{\hat{p}} = \sqrt{\dfrac{0.37(0.63)}{150}}$

(E) $\mu_{\hat{p}} = 0.37$ and $\sigma_{\hat{p}} = \sqrt{\dfrac{0.37(0.63)}{150}}$

11. Forty-three percent of the registered voters in a particular county are Democrats. A sample of 200 registered voters is

selected at random. The approximate probability that more than 100 of the people in the sample are Democrats is
(A) 0.463.
(B) 0.388.
(C) 0.038.
(D) 0.023.
(E) 0.012.

12. Suppose that 14% of the students at a university have a full-time job. If a random sample is to be selected from the students at this university, what is the smallest sample size for which it would be reasonable to think the sampling distribution of $\hat{p}$, the proportion in the sample who have full-time jobs, would be approximately normal?
(A) 10 students
(B) 11 students
(C) 12 students
(D) 71 students
(E) 72 students

13. A news magazine claims that 30% of all New York City police officers are overweight. To investigate, the police commissioner plans to select a random sample of 200 police officers and determine the proportion in the sample who are overweight. Assuming that the 30% figure is correct, which of the following statements must be *true*?
(A) The sample proportion will be 0.30.
(B) The sample proportion will be $\sqrt{\dfrac{(0.30)(0.20)}{200}}$.
(C) The sample size is not large enough to compute a sample proportion.
(D) The sample proportion will be $\dfrac{0.30}{\sqrt{200}}$.
(E) About 95% of all samples will have a sample proportion between 0.27 and 0.33.

14. A population has a mean of 27.3 and a standard deviation 20. What is the probability that the mean of a random sample of size 100 will be between 28.2 and 29.7?
(A) 0.2113
(B) 0.0500
(C) 0.0000
(D) 0.7887
(E) 0.9500

15. A population has a proportion of successes of 0.2. Let $\hat{p}$ be the proportion of successes in a random sample from this population. How large would n have to be in order for the standard deviation of $\hat{p}$ to be less than 0.01?
(A) 16
(B) 64
(C) 256
(D) 1600
(E) 4000

FREE-RESPONSE PROBLEMS

1. The amount of sugar in a one-gallon container of southern sweet tea is approximately normally distributed with a mean of 1.8 cups with a standard deviation 0.4 cups.
   (a) What is the probability that a randomly selected gallon container of this tea will have at least 2.3 cups of sugar in it?
   (b) What is the probability that the total amount of sugar in a sample of 10 gallon containers of this tea will be at least 23 cups?

2. A well-known food chain believes that 36% of their customers prefer to have the buns used to make their sandwich to be toasted. Suppose a random sample of 400 people is to be selected from the chain's customers.
   (a) What is the mean and standard deviation of the sampling distribution of $\hat{p}$, the proportion of customers in the sample who prefer toasted sandwich buns?
   (b) Is the sampling distribution of $\hat{p}$ approximately normal?
   (c) What is the probability that fewer than 32% of the customers in the sample will prefer to have their sandwiches on toasted?

Answers

MULTIPLE-CHOICE QUESTIONS

1. **C.** The mean of the sampling distribution is equal to the mean of the population and the standard deviation of the sampling distribution is equal to the population standard deviation divided by the square root of the sample size (*Statistics: Learning From Data,* 1st ed. pages 568–578).

2. **D.** The average of $372,000 is computed from a sample of 50 homes. A statistic is a quantity computed from values in a sample (*Statistics: Learning From Data,* 1st ed. page 13).

3. **B.** The mean of the sampling distribution is equal to the mean of the population and the standard deviation of the sampling distribution is equal to the population standard deviation divided by the square root of the sample size (*Statistics: Learning From Data,* 1st ed. pages 568–578).

4. **E.** Any particular sample mean is not necessarily equal to the population mean. However, the sampling distribution of the sample mean will be approximately normal when the sample size is sufficiently large, as the Central Limit Theorem states. Dividing by the square root of n makes the

standard deviation of the sampling distribution smaller as n gets larger. (*Statistics: Learning From Data,* 1st ed. pages 568–578).

5. **C.** The mean of the sampling distribution equals the mean of the population but the standard deviation of the sampling distribution will be smaller—it is the population standard deviation divided by the square root of the sample size (*Statistics: Learning From Data,* 1st ed. pages 568–578).

6. **A.** The Central Limit Theorem states that even in skewed populations, if the sample size, *n,* is sufficiently large (greater than 30), the sampling distribution will be approximately normal (*Statistics: Learning From Data,* 1st ed. pages 572–573).

7. **E.** The population mean and standard deviation are fixed and we don't expect the sample mean or the sample standard deviation to be exactly equal to μ and σ. In fact, in many repeated samplings, we expect to get several different values for the sample mean and standard deviation. None of the three statements are true (*Statistics: Learning From Data,* 1st ed. pages 568–578).

8. **A.** The two standard errors are close to each other (0.000236 and 0.000244), since the proportions of those who were satisfied are very close to one anoter and the two sample sizes are the same (*Statistics: Learning From Data,* 1st ed. page 456).

9. **C.** Using the empirical rule, about 95% of the sample means will be within two standard deviations. The standard deviation from the sampling distribution of size four = 0.1 (*Statistics: Learning From Data,* 1st ed. page 161).

10. **E.** Since $\mu_{\hat{p}} = \mu$ and $\sigma_{\hat{p}} = \sqrt{\dfrac{p(1-p)}{n}}$, $\mu_{\hat{p}} = 0.37$ and

$\sigma_{\hat{p}} = \sqrt{\dfrac{0.37(0.63)}{150}}$ (*Statistics: Learning From Data,* 1st ed. pages 460–467).

11. **D.** The related area under the curve for more than 100 of the sample to be Democrats is 100/200 = 0.50. Then this can also be found using Normalcdf(0.5, E99, 0.43,) (*Statistics: Learning From Data,* 1st ed. pages 375–385).

12. **E.** To ensure that $np \geq 10$, we need $n = \dfrac{10}{0.14} = 71.4$ students

and to ensure that $n(1-p) \geq 10$ we need $n = \dfrac{10}{0.86}$

= 11.6 students. In order to meet both conditions, at least

72 students should be included in the sample (*Statistics: Learning From Data,* 1st ed. pages 507–511).

13. **E.** The sampling distribution of the sample proportion will be approximately normal with mean 0.30 and standard deviation $\sqrt{\dfrac{(0.3)(0.7)}{200}} = 0.03$. About 95% of all possible random samples will have a value of $\hat{p}$ in the range 0.30 ± 0.03 (*Statistics: Learning From Data,* 1st ed. pages 467–480).

14. **A.** The standard deviation of the sampling distribution is $\dfrac{\sigma}{\sqrt{n}} = \dfrac{20}{\sqrt{100}} = 2$. Then $P(28.3 \le \bar{x} \le 29.7) = P(0.45 \le z \le 1.20) = 0.2113$ (*Statistics: Learning From Data,* 1st ed. pages 375–385).

15. **D.** This involves solving backward for *n*. Using the formula for the standard deviation of $\hat{p}$ gives the following

$\sigma_{\hat{p}} = \sqrt{\dfrac{p(1-p)}{n}}$, now substitute and solve for *n*.

$0.01 = \sqrt{\dfrac{0.2(0.8)}{n}}$

$0.01^2 = \dfrac{0.2(0.8)}{n}$

$n = \dfrac{0.16}{0.0001}$

(*Statistics: Learning From Data,* 1st ed. pages 352–363).

FREE-RESPONSE PROBLEMS

1. (a) Since this is just a single gallon container,

x =	the sugar content from one gallon of tea
xbar =	the mean sugar content from my sample of 10
u =	the mean sugar content from all the gallons of tea
population =	all the gallons of tea
parameter of interest =	the mean sugar content from all the gallons of tea

$P(x \ge 2.3) = P(z \ge 1.25) = 0.1056$.

(b) We can think of this as a question about $\bar{x}$, since the total sugar content will be 23 cups or more when $\bar{x} \ge 2.3$. Then

$$P(\bar{x} \ge 2.3) = P\left(z \ge \dfrac{2.3 - 1.8}{\dfrac{0.4}{\sqrt{10}}} \right) = P(z \ge 3.95) \approx 0.$$

(*Statistics: Learning From Data,* 1st ed. pages 375–385).

2. (a) $\mu_{\hat{p}} = 0.36$, $\sigma_{\hat{p}} = \sqrt{\dfrac{0.36(0.64)}{400}} = 0.024$.

(b) $400(0.36) \geq 10$, $400(0.64) \geq 10$. This means that the sample size is large enough for the sampling distribution of $\hat{p}$ to be approximately normal.

(c) $P(\hat{p} < 0.32) = P\left(z < \dfrac{0.32 - 0.36}{0.024}\right) = P(z < -1.67) = 0.0478$

(*Statistics: Learning From Data*, 1st ed. pages 375–385).

8

ESTIMATION USING A SINGLE SAMPLE

This review section will cover methods for estimating a single population parameter. Both point estimates (single numbers) and interval estimates (intervals of plausible values for a population parameter) will be considered.

OBJECTIVES

- Estimate a population mean or proportion using a point estimate.
- Construct and interpret a confidence interval estimate for a population proportion.
- Construct and interpret a confidence interval estimate for a population mean.
- Calculate the sample size needed in order to achieve a given bound on error.

POINT ESTIMATION

(Statistics: Learning From Data, 1st ed. pages 135 and 433)

When we want to estimate some population characteristic, we will often use sample data to compute a single value, such as $\bar{x}$, the sample mean. This value is called a point estimate of μ, the population mean. Given a different sample, the sample mean would most likely be different (but hopefully similar or close to the first value)—it is only an estimate of the value of μ. When estimating a population proportion, the usual point estimate is the sample proportion, $\hat{p}$. As an example, suppose a teacher wanted to estimate the proportion of student phone calls that are made to family members. Each student in a random sample of 100 students was asked whether the most recent call he or she made was to a family member. The teacher found that 29 of the students in the sample said the most recent call was to a family member. In this case, the sample proportion is $\hat{p} = 0.29$, and this is a point estimate of the proportion of calls made by students that are made to family members.

A statistic is said to be *unbiased* if the sampling distribution is centered at the true value of the population parameter. In other words, if the average value of the statistic for all possible random samples of size n is equal to the value of the population parameter.

LARGE-SAMPLE CONFIDENCE INTERVAL FOR A POPULATION PROPORTION

(Statistics: Learning From Data, 1st ed. pages 467–479)

Point estimates can be helpful in estimating a population characteristic; however, different samples would most likely yield different values. Instead of using just the point estimate, a better method is to present an entire interval of plausible values of the parameter. In this way, we are allowing for variation in the samples and yet offering a reasonable estimate of the population characteristic. In the previous phone call example, we might report that the proportion of calls made to family members by students to be some value in an interval extending from 0.20 to 0.38. In this case, all values within that range would be considered plausible. A narrow interval provides more precise information about the actual value of the population parameter, but at the same time, we want an interval for which we can be confident that the interval actually contains the correct value.

A *confidence interval* (CI) is an interval of reasonable values for the population characteristic of interest based on the sample data. This interval is constructed with a pre-chosen confidence level. The *confidence level* specifies the success rate for the method used to construct the interval. While in theory the confidence level can be any percent, the most commonly used confidence levels are 90%, 95%, and 99%. No matter what level of confidence is set by the researcher, the meaning of the confidence level is the same: if this method were used over and over again with different random samples, in the long run the confidence intervals generated would capture the true population characteristic the given percentage of the time. For example, if the confidence level was set at 90%, in repeated sampling about 90% of the intervals would capture the true value of the population characteristic.

Generating a confidence interval is relatively easy. We start with a point estimate. Let's assume we want a confidence level of 95%. We use the point estimate as the center of the confidence interval. We then determine how far the interval should extend on either side of the point estimate.

The general form of most confidence intervals includes three essential pieces.

$$\begin{pmatrix} \text{sample statistic} \\ \text{(point estimate)} \end{pmatrix} \pm (\text{critical value}) \begin{pmatrix} \text{standard deviation} \\ \text{of the statistic} \end{pmatrix}$$

In the case of proportions, $\hat{p}$, the sample proportion, is the sample statistic. If the sample size is large (check $n\hat{p} \geq 10$ and $n(1 - \hat{p}) \geq 10$), the sampling distribution of $\hat{p}$ is approximately normal. In this case, the critical value is found using the z distribution and depends on the desired confidence level. For a 95% confidence level, the critical value is found by finding the z value that corresponds to a central area of 0.95 under the z curve as shown below. For other confidence levels, adjust the central area. For a 95% confidence interval, the area for each of the tails is 0.025. Using the TI calculator invnorm(.025) will give you the critical z value for 95% confidence.

When you have an area and are looking for the z value you can use the invnorm (area) feature on your calculator.

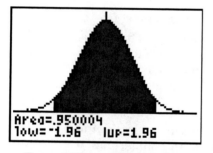

For a 95% confidence level, the z critical value is 1.96. It follows that a 95% confidence interval for a population proportion is:

$$(\hat{p}) \pm (1.96) \begin{pmatrix} \text{standard deviation} \\ \text{of } \hat{p} \end{pmatrix}$$

Finally, as seen in the last review section, the estimated standard deviation of the sampling distribution of $\hat{p}$ is $\sqrt{\dfrac{\hat{p}(1 - \hat{p})}{n}}$, resulting in $\hat{p} \pm (1.96)\sqrt{\dfrac{\hat{p}(1 - \hat{p})}{n}}$.

Returning to the example where 100 students were asked if the last phone call made was to a family member, a 95% confidence interval for the proportion of calls made to family members is

$$(0.29) \pm (1.96)\left(\sqrt{\frac{0.29(1 - 0.29)}{100}}\right) = (0.201, 0.379)$$

Note that the sample was a simple random sample and that the sample size is large enough $(n\hat{p} = 100(0.29) \geq 10, n(1 - \hat{p}) = 100(0.71) \geq 10)$.

These are the two conditions that must be checked in order for this confidence interval for a population proportion to be appropriate.

The 95% confidence interval is (0.201, 0.379). Interpreting the interval we would say that we are confident, in this case 95% confident, that the true proportion of students' phone calls that are to family members is between 0.201 and 0.379. Notice, instead of just having the point estimate of 0.29, we now have a range of reasonable values based on the sample data. This acknowledges the uncertainty in our original point estimate.

SAMPLE SIZE

The Margin of Error is defined to be the +/- part of the confidence interval. In other words, margin of error = (critical value)$\left(\sqrt{\dfrac{p(1-p)}{n}} \right)$.

Notice that the margin of error is related to sample size. Sometimes it is useful to determine the sample size needed to achieve a particular margin of error. For example, if we wanted to achieve a margin of error of 0.05 with 95% confidence, we would want to solve $0.05 = (1.96)\sqrt{\dfrac{p(1-p)}{n}}$ for n. But there is a catch—we don't know the value of p and we haven't yet selected a sample, so we can't just use $\hat{p}$ as an estimate. A conservative approach is to use $p = 0.5$ in this equation. This results in a sample size that is large enough to achieve the desired margin of error no matter what the actual value of p. (This is because p is always between 0 and 1 and $p(1-p)$ is largest when $p = 0.5$.)

EXAMPLE Suppose that a cell phone company wants to estimate the proportion of calls made by children that are made to contact family members. How large a sample would be needed to achieve a margin of error of 0.05 with 95% confidence?

$0.05 = (1.96)\left(\sqrt{\dfrac{0.5(0.5)}{n}} \right)$ Now, solve for n algebraically.

$0.05^2 = (1.96)^2 \left(\sqrt{\dfrac{0.5(0.5)}{n}} \right)^2$

$0.0025 = (3.8416)\left(\dfrac{0.5(0.5)}{n} \right)$

$0.0025 = (3.8416)\left(\dfrac{0.25}{n} \right)$

$n = (3.8416)\left(\dfrac{0.25}{0.0025} \right) = 384.16$

We would round this up to 385, and so to be 95% confident that the margin of error will be less than 0.05, we will need at least 385 in the sample. Always round up to the next integer when making a sample size calculation.

CONFIDENCE INTERVAL FOR A POPULATION MEAN

(*Statistics: Learning From Data*, 1st ed. pages 578–591)

A confidence interval for a population mean has the same form as the confidence interval for a population proportion:

$$(\text{sample statistic}) \pm (\text{critical value}) \left(\begin{array}{c} \text{standard deviaiton} \\ \text{of the statistic} \end{array} \right)$$

The sample statistic used is $\bar{x}$, the sample mean. When (1) the sample is a random sample and (2) either the sample size is large ($n \geq 30$) or the population distribution is approximately normal (these are the conditions that you need to check), the $\bar{x}$ sampling distribution is approximately normal with a standard deviation of $\dfrac{\sigma}{\sqrt{n}}$. There are two cases to consider depending on whether or not σ, the population standard deviation, is known.

If σ is known, a confidence interval for the population mean is

$$\bar{x} \pm (z \text{ critical value}) \frac{\sigma}{\sqrt{n}}$$

The z critical value depends on the desired confidence level and is found in the same way as it was for the confidence interval for a population proportion. Because it is rare that σ is known, the more commonly used interval is the one that follows.

If σ is not known, a confidence interval for the population mean is

$$\bar{x} \pm (t \text{ critical value}) \frac{s}{\sqrt{n}}$$

If the sample size is not large, we need to assume that the population distribution is approximately normal. This is reasonable when a dotplot or a boxplot constructed from the sample data is roughly symmetric and there are no outliers.

The t critical value in the interval above is found using a t distribution rather than the standard normal distribution. The t distributions have a mound shape, similar to the z distribution, but t distributions are more spread out than the z distribution. There is a family of t distributions that are distinguished from one another by a parameter called degrees of freedom (df). As df increases, the t distributions become more and more like the z distribution. For finding the critical value for a confidence interval for a population mean, df $= n - 1$.

The t critical value needed to compute the confidence interval can be found using the t distribution table (also provided during the AP Statistics exam) or generated on your calculator. To use the table provided by the College Board® on the day of the exam, find the degrees of freedom in the left-most column. Remember, df $= n - 1$. Then follow this row over to the column that contains the critical values for the desired confidence level. The partial *t table below* shows *t* critical values for df 8, 9, and 10. This is from the actual table you will be provided with on exam day. Keep in mind, the tail probability is ½ of the area left *outside* the interval you are seeking.

EXAMPLE If we want the *t* critical value for a 95% confidence interval and the sample size is 10, we would go down the first column to locate df = *n* − 1 = 9 and follow that row over to the column for a 95% confidence level.

Tail probability *p*							
Df	0.25	0.20	0.15	0.10	0.05	0.025	0.02
8	0.705	0.889	1.108	1.397	1.860	2.306	2.449
9	0.703	0.883	1.100	1.383	1.833	2.262	2.398
10	0.700	0.879	1.093	1.372	1.812	2.228	2.359
.	.	.	.	.	.		.
.	.	.	.	.	.		.
∞	0.674	0.841	1.036	1.282	1.645	1.960	2.054
	50%	60%	70%	80%	90%	95%	96%

Confidence Level C

(Notice, as the df increases, the critical values in any column of the table are getting smaller and closer to the critical value associated with a standard normal (*z*) distribution.)

An alternative to reading the table is to use the capabilities of your graphing calculator. The argument required by the calculator is the area under the curve in the left tail up to the critical value and the df. The calculator will give the correct critical value for either the left side or right side, based on the area you enter into the argument. The only difference below is that one yields the left critical value (the negative critical value) and the other yields the right critical value (the positive critical value) as shown in the two calculator displays. Many calculators still do not have this feature; if your calculator does not you can always use the table supplied on the AP Exam.

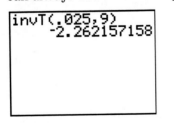
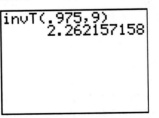

Once the critical value is determined, the confidence interval is computed using

$$\bar{x} \pm (t \text{ critical value})\left(\frac{s}{\sqrt{n}}\right)$$

where t critical value is based on $df = n - 1$

Before using this confidence interval, be sure to check the two necessary conditions: (1) random sample, and (2) large sample or normal population distribution.

EXAMPLE A cable TV company is interested in estimating the average time that customers must wait on hold when they call the company's customer support line. Twenty calls are selected at random from the calls made to the customer support line and the hold time is recorded for each call. The sample mean hold time was 10.5 minutes and the sample standard deviation was 1.1 minutes. A dotplot of the 20 hold times was approximately symmetric and there were no outliers. Estimate the mean hold time for customers calling the support line using a 90% confidence interval.

Because the population standard deviation is not known, we will consider using the t confidence interval. First we must verify that the conditions are met. The problem states that the sample is a random sample. The sample size is not large (20 < 30), so we must be willing to assume that the hold time population distribution is approximately normal. The actual sample data is not given here (if it were, we would construct a dotplot or a boxplot), but we are told that the dotplot was approximately symmetric and that there were no outliers.

With a sample size of $n = 20$, df = 19. For a 90% confidence level, the t critical value is 1.729. This gives

$\bar{x}$ = the mean wait time from the sample of 20.

$$\bar{x} \pm (t^* \text{ critical value})\left(\frac{s}{\sqrt{n}}\right)$$

$$= 10.5 \pm (1.729)\left(\frac{1.1}{\sqrt{20}}\right)$$

$$= 10.5 \pm (1.729)(0.246)$$

$$= 10.5 \pm 0.426$$

$$= (10.08, 10.93)$$

We can now say that, based on this sample, we are 90% confident that the true mean hold time is between 10.08 and 10.93 minutes. The 90% confidence level means that we used a method that will capture the true mean wait time about 90% of the time in repeated sampling.

As a final note, consider what happens if instead of using a 90% confidence level, we change to a 95% confidence level. Using the same information from the example above, the only change in the calculations would involve a change in the critical value. For a 95% confidence level, and df = 19, the *t* critical value is 2.093, resulting in

$$\bar{x} \pm (t^* \text{ critical value})\left(\frac{s}{\sqrt{n}}\right)$$

$$= 10.5 \pm (2.093)\left(\frac{1.1}{\sqrt{20}}\right)$$

$$= 10.5 \pm (2.093)(0.246)$$

$$= 10.5 \pm 0.515$$

$$= (9.98, 11.02)$$

Notice that this interval is wider than the 90% confidence interval. The higher confidence level comes at a price—a wider interval, which is a bit less informative in terms of pinning down the value of the population mean.

INTERPRETING AND COMMUNICATING THE RESULTS OF STATISTICAL ANALYSES

(*Statistics: Learning From Data*, 1st ed. pages 474–485)

Interpreting a confidence interval and the associated confidence level can be tricky. Here, wording is *everything*.

INTERPRETING THE CONFIDENCE INTERVAL

The confidence interval tells us that (based on the confidence level) we can be confident that the interval actually contains the actual value of the population characteristic. For example, for a 90% confidence interval for a population mean, you would say

> *I am 90% confident that the value of the population mean is contained within the interval (a, b).*

Of course, be sure to include appropriate context as well.

INTERPRETING THE CONFIDENCE LEVEL

A confidence level of 90% indicates that were we to perform *repeated sampling*, approximately 90% of the intervals constructed would contain the true value of the population characteristic being estimated. This does not mean that every 100 times we sample, we will get exactly 90 intervals of the 100 that contained the truth. Rather, this is saying that if we continued to select random samples and to generate confidence intervals, approximately 90% of the intervals would actually contain the true value. You would say something like

> *The method used to construct the interval will capture the true value about 90% of the time in repeated sampling.*

AP Tip

To interpret a confidence interval, say something like

I am ____% confident that the true value of the population _____ (proportion or mean) _____ (supply context here!) is contained within the interval (_____, _____).

To interpret the confidence level, say something like

The method that was used to construct the interval will capture the true _____ (proportion or mean) about _____% of the time in repeated sampling.

SAMPLE PROBLEM 1 A real estate company in Colorado has noticed the housing market seems to be doing better in some markets than others. In order to help new home sellers set a price, they selected a random sample of 20 recent sales in Colorado and found the average selling price was $142,000 and the standard deviation was $4,000.

(a) What conditions would need to be met in order to use a confidence interval to estimate the mean selling price of Colorado homes?

(b) Assuming the conditions are met, calculate a 95% confidence interval for the mean price in Colorado.

(c) Interpret the confidence interval.

(d) Interpret the confidence level of 95%.

SOLUTION TO PROBLEM 1

(a) Since the standard deviation is from the sample, a t-interval would be used. The conditions for the t confidence interval are that the sample is a simple random sample and that the sample size is large or the population distribution is approximately normal. We would need to assume these conditions are met to construct this confidence interval.

(b) $142,000 \pm 1.729 \left(\dfrac{4,000}{\sqrt{20}} \right)$, with df = 19.

(c) I am 95% confident that the true mean selling price of homes in Colorado is contained in the interval ($141,128, $143,872).

(d) The method I have used to generate the interval will capture the true mean selling price in Colorado about 95% of the time in repeated sampling.

ESTIMATION USING A SINGLE SAMPLE: STUDENT OBJECTIVES FOR THE AP EXAM

- You will be able to calculate a point estimate of a population mean or a population proportion.
- You will be able to calculate a confidence interval for a population mean.
- You will be able to calculate a confidence interval for a population proportion.
- You will be able to find the required sample size for a given margin of error.
- You will be able to correctly interpret a confidence interval in context.
- You will be able interpret the confidence level associated with a confidence interval.

MULTIPLE-CHOICE QUESTIONS

1. Which of the following is the z critical value for a 99% confidence level?
   (A) 2.326
   (B) 2.576
   (C) 2.807
   (D) 3.091
   (E) 3.291

2. Which of the following is a difference between a *t*-interval and a *z*-interval?
    (A) *z*-intervals are used if you know the population standard deviation and *t*-intervals are used when you do not know the population standard deviation.
    (B) *t*-intervals are used if you know the population standard deviation and *z*-intervals are used when you do not know the population standard deviation.
    (C) *t*-intervals are used when the sample size is large and *z*-intervals are used when the sample size is small.
    (D) *z*-intervals are used when the population is not normally distributed.
    (E) *t*-intervals can only be used when the sample size is large, whereas *z* intervals can be used for both large and small samples.

3. Assuming that the data came from a random sample, what other conditions must be met in order for the *z* interval to be an appropriate way to estimate a population proportion?
    (A) The sample size is greater than 30 or the population distribution is approximately normal.
    (B) The population size is greater than the sample size.
    (C) The population standard deviation is known.
    (D) The sample size must be large enough so that $n\hat{p}$ and $n(1-\hat{p})$ are both greater than or equal to 10.
    (E) All of the above conditions must be met.

4. A researcher wants to find a confidence interval estimate of the mean time it takes to complete an order over the phone at a call center for a large retail company. He has selected a random sample of 18 orders and recorded the time to complete the order. The sample mean time was 4.27 minutes and the sample standard deviation was 0.78 minutes. Assuming the conditions needed for inference are met, which of the following is the appropriate 98% confidence interval?

    (A) $4.27 \pm 2.326\left(\dfrac{0.78}{\sqrt{18}}\right)$

    (B) $4.27 \pm 2.054\left(\dfrac{0.78}{\sqrt{17}}\right)$

    (C) $4.27 \pm 2.552\left(\dfrac{0.78}{\sqrt{18}}\right)$

    (D) $4.27 \pm 2.567\left(\dfrac{0.78}{\sqrt{18}}\right)$

    (E) $4.27 \pm 2.567\left(\dfrac{0.78}{\sqrt{17}}\right)$

5. What size is needed to achieve a margin of error of 0.03 with 95% confidence when the population proportion of successes is 0.36?
   - (A) 984
   - (B) 983
   - (C) 256
   - (D) 255
   - (E) 250

6. Which of the following would not decrease the width of a confidence interval?
   - I. Increasing the sample size
   - II. Decreasing the degrees of freedom
   - III. Decreasing the confidence level
   - (A) I only
   - (B) II only
   - (C) III only
   - (D) I and II
   - (E) II and III

7. In a random sample of 64 children who received a new antibiotic for an infection, 52 of the children had positive results within 12 hours. Find a 95% confidence interval for the proportion of children with infections who will experience positive results within 12 hours when treated with this antibiotic.
   - (A) 0.81 ± 0.096
   - (B) 0.81 ± 0.080
   - (C) 0.81 ± 0.071
   - (D) 0.83 ± 0.079
   - (E) 0.83 ± 0.094

8. The treatment records of a small Army hospital were examined for 49 randomly selected days. The number of patients treated was determined for each of these days. For this sample, the mean was 11.2 patients and the standard deviation was 3.6 patients. Based on this sample, a confidence interval estimate of the mean number of patients treated per day was reported to be (10.3, 12.1). What confidence level is associated with this interval?
   - (A) 80%
   - (B) 85%
   - (C) 90%
   - (D) 95%
   - (E) 98%

9. What assumptions are necessary for a 95% t-interval with a sample size of 9 to be valid?
   - I. The sample was selected randomly from the population of interest.
   - II. The population standard deviation is known.
   - III. It is reasonable to assume that the population distribution is approximately normal.
   - (A) I only
   - (B) III only
   - (C) I and II only
   - (D) I and III only
   - (E) I, II, and III

10. A 98% confidence interval for a population mean is found to be 127 ± 18. Which of the following is a correct interpretation of this interval?
    (A) The probability that the true mean is contained in this interval is 0.98.
    (B) The probability that the sample mean is contained in this interval is 0.98.
    (C) The probability that another interval will give a true mean of 127 is 0.98.
    (D) If many different random samples of the same size are selected, about 98% of the confidence intervals constructed using this method will contain the value of the population mean.
    (E) If many different random samples of the same size are selected, about 98% of the confidence intervals constructed using this method will contain the sample mean of 127.

11. Which of the following statements about the width of a confidence interval is true if the sample size is doubled?
    (A) The confidence interval is wider.
    (B) The confidence interval is narrower.
    (C) It will have no effect on the width of the confidence interval.
    (D) The width of the confidence interval will double.
    (E) The confidence interval will be half as wide.

12. You have been asked to compute a 96% confidence interval for a population mean. If the population standard deviation is known to be 7 and the sample size is 40, what critical value would be used in computing the interval?
    (A) 1.751
    (B) 1.960
    (C) 2.054
    (D) 2.122
    (E) 2.125

13. A random sample of size 150 resulted in a sample proportion of 0.45. What is the approximate standard error of $\hat{p}$ for a sample of size 150?
    (A) 0.002
    (B) 0.02
    (C) 0.03
    (D) 0.04
    (E) 0.24

14. A large company wants to estimate the proportion of employees who would prefer a pay increase to an increase in retirement benefits using a 95% confidence interval. What sample size should be used in order to achieve a margin of error of 0.04?
    (A) 156 employees
    (B) 157 employees
    (C) 307 employees
    (D) 600 employees
    (E) 601 employees

15. Based on a random sample of 1,000 adult Americans, a consumer group states that 72% of adult Americans believe corporations are not concerned about public safety. They also reported a margin of error of 2 percentage points with 90% confidence. What does this mean?
    (A) If the poll were conducted again, the probability that 72% believe corporations are not concerned about safety is 0.90.
    (B) The probability that the proportion of adult Americans who believe corporations aren't concerned with public safety is between 70% and 74% is 0.98.
    (C) Between 88% and 92% of adult Americans believe corporations are not concerned with public safety.
    (D) About 90% of all random samples of 1000 adult Americans will result in a sample percentage that is within 2 percentage points of the actual population proportion.
    (E) Ninety out of every 100 samples of 1,000 adult Americans will have between 70% and 74% who believe that corporations are not concerned with public safety.

FREE-RESPONSE PROBLEMS

1. Many of the trees in a national forest suffer from a virus that attacks the bark of the tree. Trees with this virus should be removed in order to minimize the risk to nearby trees. To estimate the proportion of trees that have this virus, a random sample of 204 trees was selected. Each selected tree was inspected and it was found that 28% of the trees in the sample had the virus. A 95% confidence interval will be used to estimate the proportion of trees with the virus.
   (a) Verify that use of the large-sample z confidence interval is appropriate.
   (b) Calculate the 95% confidence interval.
   (c) Interpret the confidence interval and the associated confidence level.

2. Ninety minutes are allowed for students to complete the multiple-choice section of a national exam. A random sample of 28 students selected from the students at a large high school took a practice exam, and the time (in minutes) that it took each student to complete the multiple-choice section was recorded. The times are given below.

Time to complete multiple-choice section						
58	76	74	80	88	74	65
97	66	95	77	63	83	73
64	71	60	68	70	63	71
57	75	74	52	71	81	82

(a) Construct a 90% confidence interval for the mean time to complete the multiple-choice section for students at this school.

(b) Based on the confidence interval, do you think that 90 minutes is a reasonable amount of time to allow for the multiple-choice part of the test? Explain your reasoning.

Answers

MULTIPLE-CHOICE QUESTIONS

1. **B.** We need to locate the 99% confidence level at the bottom of the "t distribution critical values" table and select the value immediately above it (2.576) or use the z table or the calculator command invNorm(0.995) (*Statistics: Learning from Data*, 1st ed. pages 467–468).

2. **A.** A *t*-interval is used rather than a *z-interval* when we do not know the population standard deviation (*Statistics: Learning From Data*, 1st ed. page 473).

3. **D.** For proportions we need to have $np \geq 10$ and $n(1-p) \geq 10$

   $n(1-p) \geq 10$ (*Statistics: Learning From Data*, 1st ed. pages 507–511).

4. **D.** Since the standard deviation is from the sample, this is a *t*-interval. The *t* critical value be based on 17 df is 2.567. (*Statistics: Learning From Data*, 1st ed. page 581).

5. **A.** The calculation would be done in the following manner.

   $0.03 = 1.96\left(\sqrt{\dfrac{0.36(0.64)}{n}}\right)$. Therefore, $n = 984$. So n=984, since this

   is the smallest whole number that will result in a margin of error of 0.03 or smaller.

   (*Statistics: Learning From Data*, 1st ed. pages 460–467).

6. **B.** Fewer df correspond to a smaller sample size and a larger *t* critical value, making the interval wider (*Statistics: Learning From Data*, 1st ed. pages 467–468

7. **A.** $\dfrac{52}{64} \pm 1.96\sqrt{\dfrac{0.81(0.19)}{64}}$. or use the Zinterval command on the

   calculator and input 52 for x, 64 for n and 0.95 for C-Level.

   (*Statistics: Learning From Data*, 1st ed. pages 467–480).

8. **C.** The margin of error is (12.063-10.337)/2, which equals 0.863. Solving for t* gives t*=1.678. Since the degrees of freedom are 48, and t* for a 90% confidence interval with 50 degrees of freedom

equals 1.676, the answer is **C**. (*Statistics: Learning From Data,* 1st ed. pages 578–591).

9. **D**. If the population standard deviation is known, a z interval would be used. However, since the sample size is so small, both I and III will be needed for the interval to be appropriate (*Statistics: Learning From Data,* 1st ed. pages 467–468).

10. **D**. The method used produces an interval that contains the true mean value about 98% of the time in repeated sampling (*Statistics: Learning From Data,* 1st ed. pages 578–591).

11. **B**. Since the sample size is found in the denominator of the margin of error calculation, any increase in this value will make the width of the confidence interval smaller. (*Statistics: Learning From Data,* 1st ed. pages 578–591).

12. **C**. We need to locate the 96% confidence level at the bottom of the "t distribution critical values" table and select the value immediately above it (2.054) or use the z table or the calculator command invNorm(0.98) (*Statistics: Learning From Data,* 1st ed. page 581).

13. **D**. The estimated standard error is
$$\sqrt{\frac{\hat{p}(1-\hat{p})}{n}} = \sqrt{\frac{0.42(0.58)}{150}} = 0.0403\ ($$

(*Statistics: Learning From Data,* 1st ed. page 456).

14. **E**. The margin of error equals $0.04 = 1.96\sqrt{\frac{0.5(0.5)}{n}}$. Therefore, the sample size would be 601, since that is the smallest whole number that would result in a margin of error of 0.04 or smaller. Note that 0.5 is used as an estimate of the population proportion, because it gives a conservatively large value for the sample size. (*Statistics: Learning From Data,* 1st ed. pages 460–467).

15. **D**. The width of the confidence interval is the margin of error. (*Statistics: Learning From Data,* 1st ed. pages 467–480).

FREE-RESPONSE PROBLEMS

1. (a) There are two conditions that must be checked:
   (1) random sample: the problem states that the sample is a random sample of trees.
   (2) sample size is large enough. Because
   $n\hat{p} = 204(0.28) = 57.12 \geq 10$ and $n(1-\hat{p}) = 204(0.72) = 146.88 \geq 10$,
   the sample size is large enough.
   (b) **The 95% confidence interval is**
   $$0.28 \pm 1.96\sqrt{\frac{0.28(0.72)}{204}} = (0.218, 0.341).$$

(c) I am 95% confident that the true proportion of trees that have the virus is between 0.218 and 0.341. The method used to produce this estimate results in an interval that includes the true proportion about 95% of the time (*Statistics: Learning From Data,* 1st ed. pages 467–480).

2. (a) The population standard deviation is not known, so a *t* interval will be used. There are two conditions that must be checked:
   (1) Random sample: the problem states that the sample was a random sample.
   (2) large sample or normal population distribution: because the sample size is not greater than or equal to 30, we need to look at a plot of the sample data. Either a boxplot or normal probability plot would suffice. (Both are shown.)

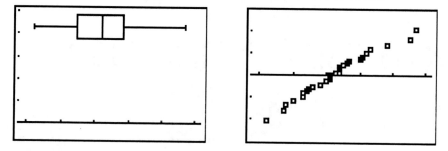

The boxplot is approximately symmetric with no outliers and the normal probability plot is relatively straight, so it is not unreasonable to think that the population distribution is approximately normal.

The confidence interval is

$$72.4 \pm 1.701 \frac{10.77}{\sqrt{28}} = 72.4 \pm 4.1 = (68.3, 76.6)$$

(b) We are 99% confident that the true mean time to complete the multiple-choice section is between 68.3 minutes and 76.6 minutes. Since 90 minutes is well above the upper end of the confidence interval, 90 minutes should be a reasonable amount of time to complete the exam (*Statistics: Learning From Data,* 1st ed. pages 578–591).

9

Hypothesis Testing Using a Single Sample

In this section, we will look at the basics of setting up and carrying out a hypothesis test using a univariate data set. Then we will use this information to draw conclusions about some unknown population parameter. Finally, anytime we make a decision based on sample data, there is a risk of error, so we will discuss what types of errors might be made when testing hypotheses.

Objectives

- Correctly set up and carry out a hypothesis test about a population mean.
- Correctly set up and carry out a hypothesis test about a population proportion.
- Describe Type I and Type II errors in context.
- Understand the factors that affect the power of a test.

Hypotheses and Test Procedures

(Statistics: Learning From Data, 1st ed. pages 591–604)

Making decisions based on sample data helps us evaluate claims about populations. Researchers and analysts use hypothesis testing methods in a variety of settings to choose between two competing claims about the characteristics of populations.

The first step in carrying out a hypothesis test is developing the null and alternative hypotheses. These are statements that will be used in the decision-making process. First, the researcher forms a hypothesis based on some initial claim. For example, suppose an auto manufacturer purchases off-road tires that are supposed to have a mean tread thickness of 0.3125 in. The auto manufacturer will assume

210

the new tires have been manufactured as specified. After all, the tire company wouldn't stay in business for long if it didn't provide what the initial claim about the mean tread thickness that the auto company believes to be fact. This initial assumption is called the null hypothesis. We write the null hypothesis as:

$H_0 : \mu = 0.3125$

where,

H_0 stands for "the null hypothesis"

μ is the population mean tread thickness for all tires of this type.

The auto manufacturer may suspect that there has been a change in the mean thickness of the tire tread, so they decide to check several of the tires. This leads the auto company to develop what is called an alternative hypothesis. The alternative hypothesis is a competing hypothesis and could be written in one of the following three ways:

$H_a : \mu \neq 0.3125$ in. *or*

$\mu < 0.3125$ in. *or*

$\mu > 0.3125$ in.

here,

H_a stands for "the alternative hypothesis"

Because the auto manufacturer suspects that the mean tread thickness has changed, but does not have a specific direction in mind, they would use $\mu \neq 0.3125$ in. as the alternative hypothesis (two-sided alternative).

No matter which alternative hypothesis the company uses, the hypothesis testing procedure only allows us to favor this alternative if there is strong evidence against the null hypothesis. This evidence would come from sample data. We would evaluate what we see in the sample to determine if the sample mean tire tread is just too far from what the null hypothesis specifies to be explained by just chance differences (normal sampling variability) from sample to sample. This same reasoning is used in all hypothesis tests considered in the AP Statistics course.

The null hypothesis is usually written as

H_0: *some population characteristic* = *the hypothesized value*

and the alternative hypothesis is written as one of the following:

H_a: *some population characteristic* $\neq$ *the hypothesized value*

H_a: *some population characteristic* $<$ *the hypothesized value*

H_a: *some population characteristic* $>$ *the hypothesized value*

EXAMPLE The marketing manager for an online computer game store targets the company advertising toward males because he believes that 75% of the company's purchases are made by men. The sales manager claims that the proportion of purchases made by females has increased. He believes that the proportion of purchases made by men is now less than what the marketing manager believes. What null and alternative hypotheses would be used to test the sales manager's claim?

The appropriate hypotheses are shown below. (Notice that we will now use p to represent the parameter, since the hypotheses are about a proportion.)

When defining a proportion, think about the group and trait and start with the first three words as the proportion of
p = the proportion of all customers that are males.

$$H_o : p_{males} = 0.75$$
$$H_a : p_{males} < 0.75$$

The null and alternative hypotheses are written in terms of population characteristics. In this example, the alternative is written as "less than" since the sales manager's claim is that the proportion of purchases made by males is less than what the marketing manager believes, 0.75.

ERRORS IN HYPOTHESIS TESTING

(*Statistics: Learning From Data*, 1st ed. pages 502–505)

Now that we have an understanding of how to generate the null and alternative hypotheses, a *test procedure* will be used to decide if we should reject the null hypothesis. Test procedures are considered in the next section. Once a decision is made after the test procedure is performed, there is a chance that the final decision is wrong. In other words, an error could have been made. There are two possible types of errors and they are called a *Type I* error and a *Type II* error.

Either types of error may occur when making a decision either to reject or to fail to reject the null hypothesis. For example, in the tire tread problem, if a decision is made to reject the null hypothesis, this could be a wrong decision that would cause the auto manufacturer to conclude that the tires did not meet specifications. However, if the decision was to *fail* to reject the null hypothesis, this could also be wrong and the auto manufacturer could end up using tires that do not meet specifications. In either case, a possible error exists that is potentially damaging in some way.

A *Type I* error is made if we reject the null hypothesis and the null hypothesis is actually true. Although the hypothesis test, based on probability, supports the decision, we are led to an incorrect inference about the population. This would amount to having strong enough evidence to conclude that the tires do not have a mean tread thickness of 0.3125 in. The company would decide to return the tires, causing the tire manufacturer to lose money. If the tires actually meet specifications, the tire manufacturer lost money due to the decision error.

A *Type II* error is made if we fail to reject the null and in reality the null hypothesis is not true and should have been rejected. This type of error would amount to not having enough evidence to say the tires did not meet specifications. In this instance, the company would unknowingly use these tires. This error could mean that customers receive cars with faulty tires and this could cause a lawsuit for the company and potentially even the risk of loss of life for the customer.

AP Tip

When giving a consequence for a certain type of error, you will need an "action" word. Use something that might happen as a result of the error. Many times students will just redefine the error and not give a clear consequence.

Avoiding these errors is, of course, desirable. However, it isn't always possible to avoid making an error because we make decisions based on looking only at a sample. What we can do is to try to keep the chance of these errors as small as possible.

We use the symbol α to denote the probability of a Type I error and the symbol β to denote the probability of a Type II error. We can control the value of α by the significance level we select for the test. Type II error is more problematic as it is something that we can't control easily. The values of α and β are related—the smaller we make α, the larger β becomes, all other things being equal. For this reason, we generally choose a significance level α that is the largest value that is considered an acceptable risk of Type I error. This will help control for the errors by keeping α small as well as controlling a bit for β.

LARGE-SAMPLE HYPOTHESIS TESTS FOR A PROPORTION

(*Statistics: Learning From Data,* 1st ed. pages 514–526)

Next, we take the hypotheses we developed and systematically test them to decide whether or not to reject the null hypothesis. This process is known as a test procedure and the same basic procedure is used in the many different hypothesis tests. However, depending on the type of data that we have and the question of interest, there are different hypothesis tests. The first test we consider is a large-sample hypothesis test for a population proportion.

In this case, we are looking at categorical data that come from a single sample, such as the data on the proportion of customers who are male. In this situation, the data consists of observations on a categorical variable with two possible values—male or female.

Just as was the case with a confidence interval, the hypothesis test is based on the properties of the sampling distribution of $\hat{p}$, listed on the next page. Recall that $\hat{p}$ is the sample proportion based on a random sample.

1. $\mu_{\hat{p}} = p$

2. $\sigma_{\hat{p}} = \sqrt{\dfrac{p(1-p)}{n}}$

3. When n is large enough, the sampling distribution of $\hat{p}$ is approximately normal.

It is important to check to make sure the sample size is large enough before carrying out a one-proportion hypothesis test. To verify that the sample size is large enough, check to make sure that

$$np \geq 10$$
$$n(1-p) \geq 10$$

Once we have verified that the sample size is large enough and that the sample is a random sample, we can proceed with the test. The sample size needs to be less than 10% of the population.

Using the properties of the sampling distribution of $\hat{p}$, we can form a z test statistic:

$$z = \frac{\hat{p} - p}{\sqrt{\dfrac{p(1-p)}{n}}},$$

where $\hat{p}$ = sample and p is the population proportion.

In the example where we wanted to test $H_0 : p = 0.75$ versus $H_a : p < 0.75$, we consider whether the sample proportion is enough smaller than the hypothesized proportion of 0.75 that the difference can't be explained just by sampling variability. To do this, we calculate the value of the z statistic using 0.75 (from the null hypothesis) as the value for p: $z = \dfrac{\hat{p} - 0.75}{\sqrt{\dfrac{0.75(1 - 0.75)}{n}}}$

If the null hypothesis is true, then this z statistic will have a standard normal distribution. If the value of the z statistic is something that would be "unexpected" for a standard normal variable, we regard this as evidence that the null hypothesis should be rejected.

EXAMPLE Suppose that the sales manager in the online computer game customer example selects a random sample of 423 previous customers and finds that 298 were males. The sample proportion is then $\dfrac{298}{423} = 0.70$. We can see that 0.70 is smaller than 0.75, but is it small enough to convince us that chance differences from sample to sample could not account for this difference? This is the question that is answered by a hypothesis test.

First, let's check the assumptions needed. The sample was a random sample of customers, so that condition is met. The second condition is that the sample size is large enough, so we check

$423(0.75) = 317.25 \geq 10$

$423(1 - 0.75) = 105.75 \geq 10$

Next, we calculate the value of the z test statistic.

$\hat{p} = \dfrac{298}{423} = 0.704$ so we can now substitute into our test statistic

$z = \dfrac{0.704 - 0.75}{\sqrt{\dfrac{0.75(1 - 0.75)}{423}}} = -2.1615$

Because the conditions were met, we know that if H_0 were true, z has an approximately normal distribution. Using what we know about the normal distributions, we know that getting a z score greater than +2 or less than -2 does not occur very often. In fact, we can compute the probability of observing a z value as small as -2.16 given that the distribution is standard normal:

$$P(z \leq -2.16 \text{ if } H_0 \text{ is true})$$

$$= \text{area to the left of } -2.16$$

$$= 0.0154$$

In other words, a sample proportion as small as or smaller than 0.704 would happen only about 1.5% of the time if the population proportion is really 0.75. Based on this sample, we have observed something that is unlikely if the null hypothesis is true, so we have enough evidence to reject the H_0 in favor of the H_a. The probability just computed is called a *P*-value.

Two key parts of a hypothesis test are the test statistic and the *P*-value.

Test statistic—a value computed from the sample data that is used to make a decision to either reject H_0 or fail to reject H_0

P-value—the probability of getting a test statistic value at least as extreme as what was observed for the sample, if the null hypothesis is actually true

If the *P*-value is small, we reject the null hypothesis. How small does the *P*-value have to be in order to reject H_0? This will depend on how large a risk of a Type I error we are willing to assume. This level of risk is preset by the researcher and is known as the significance level of the test. It is denoted by α. For example, if we set $\alpha = 0.05$, this means we know that a result as extreme as what we saw in the sample could happen as often as 5% of the time if the null is true, but we can live with a Type I error occurring for about 5 out of 100 of all possible random samples.

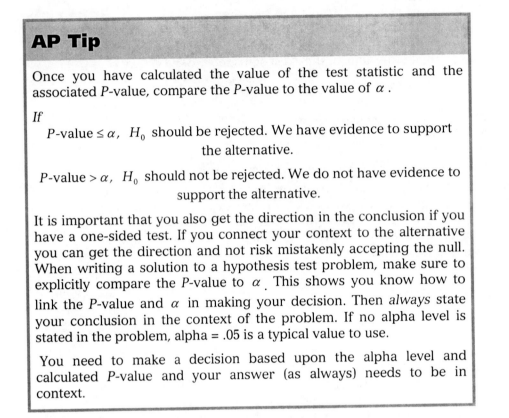

AP Tip

Once you have calculated the value of the test statistic and the associated P-value, compare the P-value to the value of α.

If

P-value $\leq \alpha$, H_0 should be rejected. We have evidence to support the alternative.

P-value $> \alpha$, H_0 should not be rejected. We do not have evidence to support the alternative.

It is important that you also get the direction in the conclusion if you have a one-sided test. If you connect your context to the alternative you can get the direction and not risk mistakenly accepting the null. When writing a solution to a hypothesis test problem, make sure to explicitly compare the P-value to α. This shows you know how to link the P-value and α in making your decision. Then *always* state your conclusion in the context of the problem. If no alpha level is stated in the problem, alpha = .05 is a typical value to use.

You need to make a decision based upon the alpha level and calculated P-value and your answer (as always) needs to be in context.

Finally, we note that there are three different computations of the P-value to consider. The computation chosen will depend on which inequality (<, >, or ≠) appears in the alternative hypothesis. Here are the three possible situations that can occur.

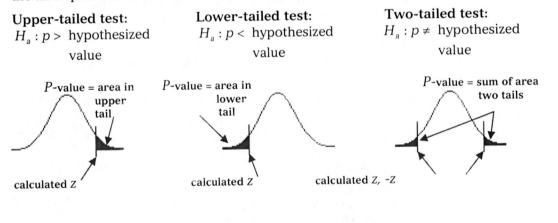

Upper-tailed test:
$H_a : p >$ hypothesized value

Lower-tailed test:
$H_a : p <$ hypothesized value

Two-tailed test:
$H_a : p \neq$ hypothesized value

P-value = area in upper tail

calculated z

P-value = area in lower tail

calculated z

P-value = sum of area two tails

calculated z, $-z$

> ## AP Tip
>
> Using incorrect notation can lower your score on free-response questions. Remember to keep the different "*p*" notations straight.
>
> *P*-**value** This is a probability computed from the value of the test statistic. *All* hypothesis tests involve the use of a *P*-value to make a decision.
>
> *p* This is the notation used to denote the population proportion when we have categorical data.
>
> $\hat{p}$ Read as p-hat, is used to denote the sample proportion.
>
> The H_0 and H_a are always written using the population parameters.

HYPOTHESIS TESTS FOR A MEAN

(*Statistics: Learning From Data,* 1st ed. pages 591–604)

Now that we have a handle on a hypothesis test for proportions, let's consider tests based on numerical data. With numerical data, we are usually interested in making inferences about a population mean. In Chapter 8 on confidence intervals, there were two types of intervals for estimating means, a *z interval* and a *t interval*. Which of these intervals is used in a particular situation is determined by whether we know the population standard deviation (σ) or whether we have only the sample standard deviation (s_x). In either case, we found that if n is large enough or if we know the population is roughly normally distributed, then either

$$z = \frac{\bar{X} - \mu}{\frac{\sigma}{\sqrt{n}}}, \text{ when we know } \sigma$$

has approximately a standard normal distribution or

$$t = \frac{\bar{X} - \mu}{\frac{s}{\sqrt{n}}}, \text{ when we DON'T know } \sigma$$

has a *t* distribution with $df = n - 1$

In either case, the forms of the hypotheses of interest look very much like those of the proportions test. The difference is that now the hypotheses are in terms of the population mean, so the null hypothesis is

$H_0 : \mu = $ hypothesized value

and the alternative hypothesis is one of the following:

$H_a: \mu \neq$ hypothesized value

$H_a: \mu <$ hypothesized value

$H_a: \mu >$ hypothesized value

The test statistic is based on either the z or the t statistic shown above, depending on whether or not we know σ.

If σ is known

$$z = \frac{\bar{x} - \text{hypothesized value}}{\dfrac{\sigma}{\sqrt{n}}}$$

P-value: computed as area under the z curve

If σ is NOT known

$$t = \frac{\bar{x} - \text{hypothesized value}}{\dfrac{s}{\sqrt{n}}}$$

P-value: computed as area under the t curve with $df = n - 1$

It is very rare that the population standard deviation is known, so we will focus mainly on the t statistic. Revisiting the earlier tire tread scenario, we can write the hypotheses, check assumptions, calculate the test statistic and *P*-value, and give a conclusion in context.

EXAMPLE Trustworthy Tires sells their leading high performance tire to a car manufacturer. The car manufacturer requires that the tires have a mean tread thickness of 0.3125. The car manufacturer thinks tires received from Trustworthy may not be meeting this requirement and that the mean tread may be greater than 0.3125, because they have been finding that in some cars the tires are hitting parts of the wheel well area on bumps. In a random sample of 32 tires, they found the tire treads to have a mean of 0.3625 in. and a standard deviation of 0.094 in. Since both these values came from the sample, we use the notation for sample statistics: $\bar{x} = 0.3360$ and $s_x = 0.094$. For this situation, let's look at the components that are common to all hypothesis tests.

HYPOTHESES

H_0: $\mu = 0.3125$

H_a: $\mu > 0.3125$ (since the company suspects the mean

tread is greater than the requirement of 0.3125)

set the significance level (α) at 0.05.

ASSUMPTIONS The assumptions that must be satisfied are the same as those with confidence intervals.

We must have a random sample and the sample size must be large or the population distribution of tread thickness must be approximately normal. The problem states that the sample was a random sample, so that condition is met. The sample size is large (n is greater than 30). Because both conditions are met, it is reasonable to proceed with the test.

TEST STATISTIC

$$t = \frac{\bar{x} - hypothesized\ value}{\dfrac{s}{\sqrt{n}}} = \frac{0.3360 - 0.3125}{\dfrac{0.094}{\sqrt{32}}} = 1.414$$

$$with\ df = n - 1$$
$$= 32 - 1$$
$$= 31$$
$$P - value = 0.08$$

Note that because the inequality in the null hypothesis was >, the *P*-value is the area to the right of 1.41 under the *t* curve with df = 31.

CONCLUSION In this example, with $\alpha = 0.05$ we see that the *P*-value is not less than α. We fail to reject the null hypothesis and conclude that there is not convincing evidence that the mean tread thickness of tires is greater than 0.3125 in.

AP Tip

When carrying out a test for means, whether or not you know the standard deviation of the population is what determines if you should use a *z* test or *t* test. Only use the *z* test when you are sure the standard deviation given is from the population, which almost never happens in the real world or on the AP Exam.

Note that the steps in a test about a population mean are the same as for the test for a population proportion. What distinguishes the two tests is the type of data (numerical for means and categorical for proportions), the specific assumptions that must be checked, and the test statistic used for the test.

SAMPLE PROBLEM 1 A well-known brand of pain relief tablets is advertised to begin relief within 24 minutes. To test this claim, a random sample of 18 subjects suffering from the same types of headache pain record when they first notice relief after taking the pain relief tablet. The data gathered from this study are shown.

Time to Pain Relief (in minutes)

25	24	27	23	25	22	25	25	24
27	24	23	26	28	25	24	26	23

Does the sample suggest the mean pain relief time is longer than the advertised time? Test the appropriate hypotheses using a 0.05 significance level.

SOLUTION TO PROBLEM 1 To solve this problem, we carry out a hypothesis test. All four parts of a hypothesis should be complete as shown.

HYPOTHESIS

With μ representing the population mean time to relief,

$H_o: \mu = 24$

$H_a: \mu > 24$

$\alpha = 0.05$ (*given*)

ASSUMPTIONS

Random sample: The problem states the 18 subjects are a random sample.

Large sample or normal population distribution: Since there were only 18 subjects, we need to be willing to assume that the distribution of relief times in the population is approximately normal. A dotplot of the sample relief times is shown below. Because the dotplot is approximately symmetric and there are no outliers, it is reasonable to think that the population distribution is approximately normal.

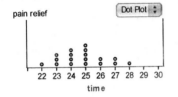

TEST STATISTIC

A *t* test will be used since we don't know the population standard deviation.

$$\bar{x} = 24.78, \quad s_x = 1.59, \quad df = 17$$

$$t = \frac{24.78 - 23}{\frac{1.59}{\sqrt{18}}} = 2.072$$

$$P\text{-value} = 0.027$$

CONCLUSION

Since $0.027 < 0.05$, we reject the null hypothesis in favor of the alternative hypothesis. There is convincing evidence that the mean time to pain relief is greater than 24 minutes.

As a final note, should you be in the unusual situation where the population standard deviation (σ) is given, you would use the *z* test statistic and the association *P*-value would be determined using the standard normal distribution. Otherwise, the process would be the same as the process illustrated in the example above.

AP Tip

For a hypothesis test, you will be required to provide complete answers to four key parts in every test. Get in the habit of thinking in terms of these four pieces. If you chunk the information under each of these categories, you will have a better chance of remembering all aspects of any hypothesis test. The four pieces are

State the Hypotheses and define any symbols used in the hypotheses.

H_0 and H_a (correctly written)

α level

Identify the test procedure by name or by formula. Check All Assumptions (or Conditions)

The assumptions vary by test, but you should always state AND CHECK the assumption appropriate for the test you are performing. Mathematically check the conditions that the sample size "is large enough."

Write the assumptions in the context of the problem and avoid giving a grocery list of general assumptions.

Test Statistic

Show your z or t calculation

Give the P-value associated with the value of the test statistic degrees of freedom.

Conclusion (be sure you state it in context too)

$P \le \alpha$, H_0 should be rejected.

$P > \alpha$, H_0 should not be rejected.

You need to make a decision based upon the alpha level and calculated p value and your answer (as always) needs to be in context.

POWER AND PROBABILITY OF TYPE II ERROR

(*Statistics: Learning From Data*, 1st ed. pages 526–531)

While the AP curriculum does not require you to calculate power, you are expected to know the factors that affect the power of a test. There are three factors that are generally considered when thinking about power. First, increasing α will raise the power. Although this seems like a fast fix, it is dangerous because the probability of a Type I error will also increase. Another way to raise power is simply to increase the sample size, although this isn't always practical. The other things that affect power aren't really things that we can control, but it is helpful to know that they do affect power. The variability in the population affects power, with power being greater when the variability in the population is small. Also, the difference between the actual value of the population characteristic and the hypothesized value affects power. The larger the difference, the greater the power of the test.

You want a powerful test. You hope to reject the null hypothesis if it is false.

SAMPLE PROBLEM 2 Consider the earlier scenario of the online computer game customers. The sales manager was interested in deciding if the proportion of males was less than the 0.75 claimed by the marketing manager. The appropriate hypotheses for this situation were:

$$H_0: p = 0.75$$
$$H_a: p < 0.75$$

(a) Identify the Type I error in this scenario and provide a possible consequence of this error.

(b) Identify the Type II error in this scenario and provide a possible consequence of this error.

(c) What can be done to increase the power of this test?

SOLUTION TO PROBLEM 2

(a) A Type I error would result if it were concluded that the proportion of customers who are male is less than 0.75, when in fact this proportion is 0.75. A possible consequence of this error would be that the company might change its strategy of targeting males in its advertising, which might result in a decrease in sales.

(b) A Type II error in this situation would occur by concluding there wasn't enough evidence to say that the proportion of customers who are male is less than 0.75 when this proportion really is less than 0.75. In this case, the company would probably continue to target males in its advertising, which might result in a loss of potential sales to female customers.

(c) One way to increase power would be to increase the significance level of the test. However, this will also increase the chance of a Type I error. Also, increasing sample size will increase power.

INTERPRETATION OF RESULTS IN HYPOTHESIS TESTING

(Statistics: Learning From Data, 1st ed. pages 496–501)

Once data has been gathered and an appropriate hypothesis test carried out, the findings are typically shared with others interested in the outcome. In communicating results in journals and newspapers, it is not common to provide the same level of detail that you would want to provide in a solution to a hypothesis testing question on the AP exam. Some of the important things to conclude when reporting results are:

- **Hypotheses:** In either symbols or words, you need to clearly state both the null hypothesis and the alternative hypothesis.
- **Test Procedures:** Clearly state what test you used (large sample z test for proportions, and so on) and mention any assumptions that are necessary in order for this test to be appropriate.
- **Test Statistic:** Be sure to report the value of the test statistic as well as the associated P-value. This will allow the readers to know if they would draw the same conclusion given the sample data.
- **Conclusion in Context:** Be certain you have provided a conclusion in terms of the originally posed research question. This needs to include a comparison of the P-value to α. Stating that you reject the H_0 is *not* sufficient.

In many cases, the reported results only include a statement such as P-value < 0.05. This is common and tells the reader that the results of the test yielded a P-value smaller than 0.05, hence statistically significant. Journals may also use a standard method of coding. * = significant, would mean their P-value was <0.05, ** = very significant, means P-value < 0.01.

As you review published reports, be sure to look for the four key components you would report and ask yourself some questions about these pieces. What were the hypotheses they tested? Did they use an appropriate test for these? What was the associated P-value and what significance level was used? Also, were the conclusions reached consistent with the results of the test?

AP Tip

In writing a conclusion for a hypothesis test, remember you are always either rejecting or failing to reject the null hypothesis. This means you either have convincing evidence in favor of the alternate hypothesis or that there is not enough evidence. You *never say that you accept the null hypothesis* because that implies strong evidence for the null hypothesis. ALWAYS WRITE YOUR ANSWER IN THE CONTEXT OF THE PROBLEM.

HYPOTHESIS TESTING USING A SINGLE SAMPLE: STUDENT OBJECTIVES FOR THE AP EXAM

- You will be able to write the null and alternative hypothesis for a test about a population mean or a population proportion.
- You will be able to describe Type I and Type II errors in context, how these errors are related, and how to minimize each type of error.
- You will be able to describe a possible consequence of each type error in context.

- You will be able to carry out a test of hypotheses about a population mean.
- You will be able to carry out a test of hypotheses about a population proportion.
- You will be able to interpret the result of a hypothesis test in context.

MULTIPLE-CHOICE QUESTIONS

1. A psychologist reports that the result of a hypothesis test was statistically significant at the 0.05 level. Which of the following is consistent with this statement?
   (A) The P-value calculated was smaller than the significance level of 0.05.
   (B) The P-value calculated was larger than the significance level of 0.05.
   (C) The significance level calculated was larger than 0.05.
   (D) The significance level calculated was smaller than 0.05.
   (E) There was not enough information to make a decision.

2. A concrete learner is a student who learns best when various types of hands-on or manipulative activities are used to illustrate abstract concepts. Researchers have long believed that 60% of all students remain concrete learners until they are between 16 and 21 years of age. Each student in a random sample of 32 students age 17 to 19 was evaluated, and it was found that 24 of the 32 were concrete learners. Would it be appropriate to use the z test for a population proportion to test to determine if the proportion of concrete learners in this age group is less than 0.60?
   (A) Yes. Since $32(0.6) = 19.2$ and $32(1 - 0.6) = 12.8$, and we can proceed with the test.
   (B) Yes. Since 32 is larger than 30, the sample is sufficiently large and we can proceed with the test.
   (C) Yes. Since we know from the sample was a random sample, we can proceed with the test.
   (D) No. Since $32(0.05) = 1.6$, we do not have a large enough sample to proceed with test.
   (E) No. While 32 is larger than 30, it is so close to 30 and we don't know if the population distribution is normal.

3. A Type I error occurs in which of the following situations?
   (A) H_a is rejected and the null hypotheses is true.
   (B) H_0 is not rejected and the null hypotheses is false.
   (C) H_0 is rejected and the null hypotheses is true.
   (D) The P-value is too small to reject the null hypothesis.
   (E) The α level is too small and so the null hypothesis is rejected.

4. A Type II error occurs in which of the following situations?
   (A) H_0 is rejected and the null hypotheses is true.
   (B) H_0 is not rejected and the null hypotheses is false.
   (C) H_a is not rejected and the null hypotheses is false.
   (D) The P-value is too small to reject the null hypothesis.
   (E) The α level is too large and so the null hypothesis is rejected.

5. A graduate student at a private university wanted to study the amount of money that students at his university carried with them. A recent study reported that the average amount of money carried by college students is \$31. He decides to collect data and carry out a test to see if there is evidence that the average is higher for students at his university. Which of the following describes a Type II error in this context?
   (A) This would lead to the incorrect idea that students at his university, on average, spend more money each month than students at other universities.
   (B) This would lead to the incorrect idea that students at his university carry, on average, more than \$31.
   (C) This would lead to the incorrect idea that students at his university carry, on average, less than \$31.
   (D) This would lead to the incorrect idea that there was no reason to believe that students at his university carry, on average, more than \$31.
   (E) This would lead to the correct idea that students at other campuses carry, on average, less than \$31.

6. An animal rights group has been very supportive of a new silicon product that caps the nails on cats as an alternative to surgically declawing the pets. The company that makes the caps claims they last for an average of 69 days before needing to be replaced. Before publically endorsing the product, the animal rights group plans to collect data to see if there is convincing evidence that the mean time before replacement is needed is actually less than what the company claims. Which of the following would be an appropriate pair of hypotheses for the animal rights group to test?
   (A) $H_o: \mu = 69$ days, $H_a: \mu > 69$ days
   (B) $H_o: \mu = 69$ days, $H_a: \mu < 69$ days
   (C) $H_o: \mu = 69$ days, $H_a: \mu \neq 69$ days
   (D) $H_o: \bar{x} = 69$ days, $H_a: \bar{x} > 69$ days
   (E) $H_o: \bar{x} = 69$ days, $H_a: \bar{x} < 69$ days

7. Neutering dogs is a common surgical practice. The mean time to recover from the general anesthetic used is 28 hours. A veterinarian believes that since changing to a new anesthetic, the mean recovery time is shorter than before. To investigate, she selects a random sample of 40 surgeries done with the new anesthetic and finds that the mean recovery time was 25 hours and the standard deviation was 2.5. She plans to use this sample data to test to see if there is evidence that the recovery time is shorter with the new anesthetic. Which of the following is the correct test statistic for this study?

(A) $z = \dfrac{25-28}{2.5}$, with $df = 39$

(B) $t = \dfrac{25-28}{\frac{2.5}{\sqrt{39}}}$, with $df = 39$

(C) $t = \dfrac{25-28}{\frac{2.5}{\sqrt{40}}}$, with $df = 39$

(D) $t = \dfrac{25-28}{\frac{2.5}{\sqrt{40}}}$, with $df = 40$

(E) $t = \dfrac{25-28}{\frac{2.5}{\sqrt{39}}}$, with $df = 40$

8. A recently published study reported that 63% of the nation's students have some type of structured homework study time. A school surveyed each student in a random sample of 83 students who attend the school and found that only 52% reported having a structured homework time. This data was used to carry out a hypothesis test to determine if there was evidence that the proportion of students at the school who had structured homework time was less that the proportion reported in the national study. Which of the following would be the test statistic for this test?

(A) $z = \dfrac{0.52-0.63}{\sqrt{\frac{0.52(0.48)}{83}}}$

(B) $z = \dfrac{0.52-0.63}{\sqrt{\frac{0.52(0.48)}{82}}}$

(C) $z = \dfrac{0.52-0.50}{\sqrt{\frac{0.52(0.48)}{83}}}$

(D) $z = \dfrac{0.52-0.63}{\sqrt{\frac{0.63(0.37)}{83}}}$

(E) $z = \dfrac{0.52-0.63}{\sqrt{\frac{0.63(0.37)}{82}}}$

9. Bicycles purchased from a discount store come unassembled. The assembly instructions that come with the bicycle claim that the average assembly time is 30 minutes. A consumer group has received complaints from people who say that the assembly time was greater than the time claims. They decide to purchase 40 of these bikes and have asked 40 different people to assemble them. The consumer group believed that it was reasonable to regard these 40 people as representative of the population of people who might purchase this bike. For this sample, they found that the assembly times had a mean of 34.2 minutes and a standard deviation of 8.6 minutes. Is there convincing evidence that the claimed average assembly time is too low at the 0.05 significance level?
   (A) No, $z = 0.49$ P-value $= 0.312$.
   (B) Yes, $t = 3.09$, df $= 39$, P-value $= 0.002$.
   (C) Yes, $t = 3.05$, df $= 39$, P-value $= 0.004$.
   (D) No, $t = 0.49$, df $= 39$, P-value $= 0.313$.
   (E) Yes, $t = 3.05$, df 40, P-value $= 0.002$.

10. The prom committee is thinking about changing the location of the prom. The new location is more expensive to rent, and for the increased cost to be reasonable, they would want to be fairly certain that more than 46% of the senior class would attend the prom. A survey of a random sample of 52 seniors found that 25 would attend if the site changed. Which of the following pairs of hypotheses should the prom committee test?
   (A) $H_o: \mu = 46\%$, $H_a: \mu > 46\%$
   (B) $H_o: \mu = 46\%$, $H_a: \mu \neq 48\%$
   (C) $H_o: p = 0.46$, $H_a: p > 0.46$
   (D) $H_o: p = 0.46$, $H_a: p \neq 0.46$
   (E) $H_o: p = 0.46$, $H_a: p < 0.46$

11. Which of the following is closest to the P-value associated with a two-tailed t test with 20 degrees of freedom if the value of the test statistic is 2.0?
   (A) 0.001
   (B) 0.01
   (C) 0.03
   (D) 0.06
   (E) 0.12

12. Which of the following statements are true?
   I.   The null hypothesis for test about a population proportion written as $H_0: \hat{p} =$ hypothesized value.
   II. For the z test to be an appropriate test for a population proportion, the following condition must be met: $np \geq 10$ and $n(1 - p) \geq 10$.
   III. The standard deviation of the statistic $\hat{p}$ is $\sigma_{\hat{p}} = \sqrt{np(1 - p)}$.
   (A) I only
   (B) II only
   (C) III only
   (D) II and III
   (E) I, II, and III

13. A local group claims that more than 60% of the teens driving after 10 p.m. are exceeding the speed limit. They plan to collect data in hopes that a hypothesis test will provide convincing evidence in support of their claim. Which of the following is true about the hypotheses the group should test?
    (A) The null hypothesis states that less than 60% of the teens are exceeding the speed limit.
    (B) The null hypothesis states that more than 60% of the teens are exceeding the speed limit.
    (C) The alternative hypothesis states that less than 60% of the teens are exceeding the speed limit.
    (D) The alternative hypothesis states that less than or equal to 60% of the teens are exceeding the speed limit.
    (E) The alternative hypothesis states that more than 60% of the teens are exceeding the speed limit.

14. A study by a geological research team found that a new piece of equipment designed to measure the forces of an earthquake is not effective. They based this conclusion on data from a sample of 40 pieces of equipment and they carried out a test with $\alpha = 0.05$. The manufacturer of the equipment claims this study was flawed and that their equipment is good. The research team is considering carrying out a second study with the intention of increasing the power of the test. Which of the following would ensure an increase in the power of the test?
    (A) Move the equipment to three randomly chosen new locations.
    (B) Change $\alpha = 0.05$ to $\alpha = 0.02$.
    (C) Carry out a two-sided test instead of a one-sided test.
    (D) Increase the sample size to 60 pieces of equipment being tested.
    (E) Decrease the sample size to 20 pieces of equipment being tested.

15. Suppose that the mean height of women in the United States is 64.5 in. with a standard deviation of 2.5 in. A clothing designer feels that women who use her products may actually be taller on average. She selects a random sample of 70 women from all women who have previously purchased her clothing. What is the population of interest, and what test would the designer use to test her claim?
    (A) The population is all women in the United States and the appropriate test is a *t* test with df = 70.
    (B) The population is all women in the United States and the appropriate test is a *z* test with σ = 2.5.
    (C) The population is all women who have previously purchased the designer's clothing and the appropriate test is a *t* test with df = 70.
    (D) The population is all women who have previously purchased the designer's clothing and the appropriate test is a *t* test with df = 69.
    (E) The population is all women who have previously purchased the designer's clothing and the appropriate test is a *z* test with σ = 2.5.

FREE-RESPONSE PROBLEMS

1. A bridal gown industry publication claims that nationwide the average amount spent for a wedding gown is $1,012. A local bridal shop in an urban community has noticed their more expensive gowns are not selling well. Instead, the brides seem to be selecting only lower priced gowns or clearance gowns. The shop wonders if the average amount spent for a wedding gown is less than $1,012 for their customers. To investigate, they selected a random sample of 50 wedding gown sales. They found a sample mean of $985 and a standard deviation of $235.

    Is there convincing evidence that the average amount spent on a wedding gown at this shop is less than the national figure? Test the relevant hypotheses using a 0.05 significance level.

2. A local school district believes that the proportion of seniors who are absent from school on the last day of school may be increasing. Over the past 5 years, 39% of the seniors have missed the last day. This year, the school district is considering a new reward program sponsored by local businesses where seniors who were at school on the last day would be entered in a drawing for an iPad. To see if this program might reduce the proportion of seniors who miss school on the last day, a random sample of 398 seniors from the school district was surveyed. Each student in the sample was asked if they planned to attend on the last day of class given the possibility of winning an iPad. Only 129 of the 398 seniors indicated that they would miss the last day of school. The school district would like to know if there is convincing evidence that the new program would reduce the number of seniors absent on the last day of school.
    (a) What hypotheses should the school district test?
    (b) Identify the appropriate test and verify that any conditions needed for the test are met.
    (c) Describe Type I and Type II errors in the context of this problem.

Answers

MULTIPLE-CHOICE QUESTIONS

1. **A.** When a researcher says the results were statistically significant, it means the P-value was less than the set significance level, making the outcome what is considered a rare event (*Statistics: Learning From Data,* 1st ed. page 504).

2. **A.** In a one-sample proportions test, one condition that is needed is $np \geq 10$ and $n(1 - p) \geq 10$. Notice these both use p and not $\hat{p}$ from the sample (*Statistics: Learning From Data,* 1st ed. pages 514–526).

3. **C.** By definition, a Type I error occurs when the null hypothesis is rejected when it should not be rejected. This might happen when

the *P*-value < significance level (*Statistics: Learning From Data,* 1st ed. pages 502–505).

4. **B.** A Type II error will occur anytime you fail to reject the null when in fact the null is false. This might occur if the *P*-value is greater than α (*Statistics: Learning From Data,* 1st ed. pages 502–505).

5. **D.** If a Type II error was made, this means he failed to reject the null hypothesis when in fact the null hyopotheses was false. In this case, that would amount to saying there was not convincing evidence that the mean for his university was greater than $31 when in fact this is incorrect and the mean really is greater than $31 (*Statistics: Learning From Data,* 1st ed. pages 502–505).

6. **B.** Since the group is concerned only if the caps last less than 69 days on average, the alternative hypothesis would be $\mu < 69$ days. Note that hypotheses always test parameters, not statistics. (*Statistics: Learning From Data,* 1st ed. pages 591–604).

7. **C.** Since we don't know σ, we must carry out a *t* test. Also, while *df* = 39 (n-1), *n* = 40 is the value that is used in the test statistic calculation (*Statistics: Learning From Data,* 1st ed. pages 591–604).

8. **D.** In calculating the test statistic, the denominator uses the hypothesized value of the parameter and sample size of n=83 is used in the calculation of the test statistic. (*Statistics: Learning From Data,* 1st ed. pages 514–526).

9. **B.** There are 39 degrees of freedom. We can use the table of "t distribution critical values". The t value of 3.09 lies between 3.030 and 3.385 on the line for 30 degrees of freedom (the largest number of df 39). This means that our P-value will be between 0.0025 and 0.001. We could also use the calculator *T*-test command, which gives a *t* value equal to 3.09 and a *P*-value of 0.0018. *t* test with 39 degrees of freedom would be used (*Statistics: Learning From Data,* 1st ed. pages 591–604).

10. **C.** This is a test about a population proportion. The question of interest is whether the population proportion *p* is greater than 0.46 (*Statistics: Learning From Data,* 1st ed. pages 507–511).

11. **D.** The table of "t distribution critical values" shows that for 20 df a t value of 2.0 lies between 1.725 and 2.086. This means that the P-value is between 2(0.05) and 2(0.025) or between 0.10 and 0.05. The calculator command 2*tcdf(2.0,E99,20) gives 0.0593 for the answer (*Statistics: Learning From Data,* 1st ed. pages 580–581).

12. **B.** Notice that choice I is written using $\hat{p}$ instead of *p*. All hypotheses are stated in terms of the population parameter, which would be *p*. The standard deviation of $\hat{p}$ is $\sqrt{\dfrac{p(1-p)}{n}}$ (*Statistics: Learning From Data,* 1st ed. pages 507–511).

13. **E.** Answer choices A, B, and D can be eliminated since the null hypothesis must state the equal to case. Because the group wants to show support for then claim that *more* than 60% are speeding, the alternative hypothesis would be $p > 0.60$ (*Statistics: Learning From Data,* 1st ed. pages 507–511).

14. **D.** The easiest ways to increase power are either to increase the sample size. Increasing the sample size will reduce the probability of a Type II error (β) and thus increase the power of the test (power = 1-β) (*Statistics: Learning From Data,* 1st ed. pages 527–530).

15. **D.** The designer is interested in the population of women who have previously purchased the designer's clothing and the standard deviation of this population is not known, so a t-test would be appropriate (*Statistics: Learning From Data,* 1st ed. pages 591–604).

FREE-RESPONSE PROBLEMS

1. **Hypothesis**

   $H_0: \mu = \$1012$

   $H_a: \mu < \$1012$

   $\alpha = 0.05$

   Assumptions
   The problem states this was a random sample of wedding gown sales.

   Since 50 > 30, the sample is large enough for the one sample *t* test to be appropriate.

   Test Statistic
   $\bar{x} = \$985$

   $s_x = \$235$

   $n = 50$

   $t = \dfrac{985 - 1012}{\dfrac{235}{\sqrt{50}}} = -0.81, \quad df = 49$

   $p = 0.21$

   Conclusion
   Since the *P*-value of 0.21 is greater than the α level of 0.05, we fail to reject the null hypotheses. There is not convincing evidence that the average amount spent on a wedding gown at this shop is less than the national average of $1,012
   (*Statistics: Learning From Data,* 1st ed. pages 591–604).

2. (a) $H_o : p = 0.39$ where $p =$ proportion of seniors skipping

   $H_a : p < 0.39$

(b) The appropriate test is a 1-*sample z* test where

$$z = \frac{\hat{p} - p}{\sqrt{\dfrac{p(1-p)}{n}}}$$

The conditions of this test are (1) random sample and (2) large sample. The problem states that the sample was a random sample of seniors. Checking the conditions for large sample: $(398)(0.39) = 155$, $(398)(0.61) = 243$. Since both $np \geq 10$ and $n(1-p) \geq 10$, the sample size is large enough.

(c) A Type I error would be that the iPad drawing would not actually reduce the proportion of seniors who miss the last day of school, but the school district thinks that that it will and implements the drawing.

A Type II error would be that the iPad drawing would in fact reduce the proportion of seniors who miss the last day of school, but the school district is not convinced of this and does not implement the drawing.

(*Statistics: Learning From Data,* 1st ed. pages 502–511).

10

Comparing Two Populations or Treatments

In this section, we will turn our attention to dealing with two populations or treatments. A typical question that arises involves how one treatment or population parameter compares to another: are they equal or is one greater than another? This chapter deals with the inference methods used to answer this question.

Objectives

- Correctly set up and carry out a hypothesis test for the difference in two population or treatment means using independent or paired samples.
- Correctly set up and carry out a hypothesis test for the difference in two population or treatment proportions.
- Construct and interpret a confidence interval for a difference in two means.
- Construct and interpret a confidence interval for a difference in two proportions.

233

INFERENCES CONCERNING THE DIFFERENCE BETWEEN TWO POPULATION OR TREATMENT MEANS USING INDEPENDENT SAMPLES

(Statistics: Learning From Data, 1st ed. pages 541–559 and 618–654)

To begin, let's take a look at inference methods based on two *independent* samples. Independent samples are samples where knowing the individuals selected for one sample does not tell you anything about which individuals are selected into the other sample. In contrast, in paired samples, an observation from one sample is paired in some meaningful way with an observation in the other sample.

As we saw in the previous review section, a hypothesis test can be consolidated into four key parts: hypotheses, test statistic and assumptions, computations, and a conclusion in context. Following this process, or *test procedure,* will be helpful in all other testing situations as well. The question of interest for a two independent sample procedure will be, "Are the two population means or proportions equal to each other or not?" However, since there are now two treatments or populations, it is important to define the symbols used to represent the populations or treatments. One way to manage this is by using subscripts to represent each population and sample separately. In the general notation shown, everything from the same population or treatment uses the same subscript.

Notation	Mean	Variance	Standard Deviation	
Population or Treatment 1	μ_1	σ_1^2	σ_1	
Population or Treatment 2	μ_2	σ_2^2	σ_2	
	Sample Size	Mean	Variance	Standard Deviation
Sample from Population or Treatment 1	n_1	$\overline{x}_1$	s_1^2	s_1
Sample from Population or Treatment 2	n_2	$\overline{x}_2$	s_2^2	s_2

When working with independent samples, hypotheses that compare the means would be written as one of the following:

$\mu_1 - \mu_2 = 0$, which is the same as saying $\mu_1 = \mu_2$

or if one mean were larger than the other, it would be written as

$\mu_1 - \mu_2 > 0$, which is the same as saying $\mu_1 > \mu_2$ or

$\mu_1 - \mu_2 < 0$, which is the same as saying $\mu_1 < \mu_2$

When the samples are independent, the sampling distribution of the difference in sample means, $\bar{x}_1 - \bar{x}_2$, has a mean that is equal to the difference of the two population means. The variance of the sampling distribution of $\bar{x}_1 - \bar{x}_2$ is the sum of the two population variances. Given this behavior, we get the following sampling distribution properties.

If the random samples are selected independently, then

1. $\mu_{\bar{x}_1 - \bar{x}_2} = \mu_{\bar{x}_1} - \mu_{\bar{x}_2} = \mu_1 - \mu_2$

   so this means the distribution will be centered around $\mu_1 - \mu_2$, making $\bar{x}_1 - \bar{x}_2$ an unbiased estimate for $\mu_1 - \mu_2$.

2. $\sigma^2_{\bar{x}_1 - \bar{x}_2} = \sigma^2_{\bar{x}_1} + \sigma^2_{\bar{x}_2} = \dfrac{\sigma^2_1}{n_1} + \dfrac{\sigma^2_2}{n_2}$ and $\sigma_{\bar{x}_1 - \bar{x}_2} = \sqrt{\dfrac{\sigma^2_1}{n_1} + \dfrac{\sigma^2_2}{n_2}}$

3. Finally, if both n_1 and n_2 are large or the population distributions are approximately normal, then both $\bar{x}_1$ and $\bar{x}_2$ are approximately normal and this implies that $\bar{x}_1 - \bar{x}_2$ is also approximately normal.

Simply interpreted, means that the general rules for two independent samples are

1. The mean value of a difference in means is the difference of the two individual mean values.

2. The variance of a difference in *independent* quantities is the *sum* of the two individual variances.

With this in mind, the statistic

$$z = \frac{(\bar{x}_1 - \bar{x}_2) - (\mu_1 - \mu_2)}{\sqrt{\dfrac{\sigma^2_1}{n_1} + \dfrac{\sigma^2_2}{n_2}}}$$

will have a standard normal distribution. However, because the values of σ^2_1 and σ^2_2 are rarely known, we will typically use a t statistic:

$$t = \frac{(\bar{x}_1 - \bar{x}_2) - (\mu_1 - \mu_2)}{\sqrt{\dfrac{s^2_1}{n_1} + \dfrac{s^2_2}{n_2}}}$$

$$**df = \frac{(V_1 + V_2)^2}{\dfrac{V_1^2}{n_1 - 1} + \dfrac{V_2^2}{n_2 - 1}}, \text{ where } V_1 = \frac{s_1^2}{n_1} \text{ and } V_2 = \frac{s_2^2}{n_2}$$

If you are using t table, the result for df should be rounded down (truncated) to get an integer for df.

**You are not required to know this formula for degrees of freedom on the AP exam. Instead, you are allowed to use the df that the calculator gives you. It is important that you are consistent with how you calculate the dfs for a two-sample problem. Either use the t table or just get the value from your calculator, which is acceptable on the AP Exam. When you give your test statistic, identify the df that matches that test statistic.

TEST PROCEDURES

Once we have established that the two samples are independently selected, the null hypothesis of interest would be of the general form

$$H_0 : \mu_1 - \mu_2 = \text{some hypothesized value (very often, this value = 0)}$$

We will follow the same basic hypothesis testing procedure as in previous sections. Two key differences when comparing two populations will occur in the procedures. First, in the hypotheses and test statistic sections, there will be two populations and two sets of sample statistics used in the calculations and claims. Second, in the assumptions section, it is necessary to check assumptions with each sample.

A summary of the two-sample t test for comparing two population means is outlined in the steps below.

1. State the hypothesis.

2. Be sure to identify your parameters of interest.

3. If you use different symbols then those stated in the problem and do not correctly identify your parameters, you will lose points on the AP test.

$H_0 : \mu_1 - \mu_2 = $ some hypothesized value

and the alternative hypothesis will be one of the following:

$H_a : \mu_1 - \mu_2 > $ some hypothesized value

$H_a : \mu_1 - \mu_2 < $ some hypothesized value

$H_a : \mu_1 - \mu_2 \neq $ some hypothesized value

state the α level

ASSUMPTIONS

Simple random sample: Check that *both* samples are random samples.

Large sample or normal population distributions: again, check *both* samples to insure that n_1 and n_2 are sufficiently large (typically greater than 30); if the samples sizes are not large, construct a plot to verify the plausibility that each population distribution is approximately normal. It is important that the two samples are independent of each other.

TEST STATISTIC

given $\bar{x}_1$, $\bar{x}_2$, s_{x_1}, s_{x_2}, n_1, n_2, then

$$t = \frac{(\bar{x}_1 - \bar{x}_2) - (\mu_1 - \mu_2)}{\sqrt{\dfrac{s^2_1}{n_1} + \dfrac{s^2_2}{n_2}}}$$

and

$$df = \frac{(V_1 + V_2)^2}{\dfrac{V_1^2}{n_1 - 1} + \dfrac{V_2^2}{n_2 - 1}}, \quad \text{where } V_1 = \frac{s_1^2}{n_1} \text{ and } V_2 = \frac{s_2^2}{n_2}$$

(Note: if a t table is used, remember to truncate the df to an integer. Or, a more conservative df can be used by using either $(n_1 - 1)$ or $(n_2 - 1)$, whichever is smaller.)

Be sure to state the df that you are using for your test statistic. If you use your calculator, give the df that your calculator identifies.

CONCLUSION Once the test statistic and P-value are calculated, the P-value can be compared to the significance level, α. If P-value $< \alpha$ there is sufficient evidence to reject the null hypothesis in favor of the alternative hypothesis. In other words, we would have sufficient evidence to say that the difference in the two means is not equal to the hypothesized value and favor whichever direction the alternative hypothesis claim takes.

Reject the null when you have evidence to support the alternative (when your P-value is small).

DO NOT reject the null when you DO NOT have evidence to support the alternative.

Statements are about MEANS. The Ho and Ha are defined using the population parameter mu for means.

SAMPLE PROBLEM 1 Students want to see if regular and low-fat chocolate chip cookies have the same number of chips in a cookie, on average. They suspect that the way low-fat cookies are made is by simply reducing the number of chips in each cookie. To test this theory, they selected random samples of cookies of each type and counted the number of chips in each cookie. Summary statistics are shown in the table below. Boxplots of the two samples were approximately symmetric and there were no outliers.

Cookie	Sample Size	Mean Chips	Standard Deviation
Regular	25	16.3	2.29
Low-Fat	25	15.2	3.25

Based on the data provided, is there evidence that the mean number of chips for regular cookies is greater than the mean number for low-fat cookies? Justify your answer.

SOLUTION TO PROBLEM 1   To answer this question, we will need to carry out an appropriate hypothesis test. The steps are shown here.

HYPOTHESES

$H_0 : \mu_r - \mu_l = 0$ where μ_r = mean number of chips for regular chocolate chip cookies

μ_l = mean number of chips for low-fat chocolate chip cookies

$H_A : \mu_r - \mu_l > 0$ to test to see if regular cookies have more chips, on average, than low-fat

$\alpha = 0.05$ (since no significance level is specified, we can choose a value.)

Be sure to identify your parameters of interest. If you use a symbol that is unusual and do not define your parameter you will lose credit.

TEST: two sample t test

ASSUMPTIONS

1. The problem states that the samples were random samples.

2. The samples sizes are not greater than 30, but the problem states that the two boxplots were approximately symmetric and there were no outliers. So, it is reasonable to assume that the two population distributions are approximately normal.

3. The two samples are independent of each other.

TEST STATISTIC

$$t = \frac{(\bar{x}_1 - \bar{x}_2) - (\mu_1 - \mu_2)}{\sqrt{\dfrac{s^2_1}{n_1} + \dfrac{s^2_2}{n_2}}} = \frac{(16.3 - 15.2) - (0)}{\sqrt{\dfrac{2.69^2}{25} + \dfrac{3.25^2}{25}}} = \frac{(16.3 - 15.2) - (0)}{\sqrt{0.289 + 0.423}} = 1.33$$

$$df = \frac{(V_1 + V_2)^2}{\dfrac{V^2_1}{n_1 - 1} + \dfrac{V^2_2}{n_2 - 1}} = \frac{(0.289 + 0.423)^2}{\dfrac{0.289^2}{24} + \dfrac{0.423^2}{24}} = \frac{0.50694}{0.01096} = 46.3$$

$$P - value = 0.095$$

CONCLUSION   Since $0.095 > 0.05$, we fail to reject the null hypothesis. In other words, there is not convincing evidence that the mean number of chocolate chips in regular cookies is greater than the mean number in low-fat cookies, based on the samples.

POOLED T TEST

The pooled t test is used when it is known that the two population variances are equal $(\sigma^2_1 = \sigma^2_2)$. However, this rarely is known. So, unless there is reason to believe the variances are the same, use an *unpooled* test. In the case of the AP Exam, we advise you to always use the unpooled test when testing hypotheses about the difference in means using independent samples.

CONFIDENCE INTERVALS

A confidence interval can be used to estimate a difference in means. Similar to the one-sample interval, the two-sample interval would be calculated in the following manner.

$$\bar{x}_1 - \bar{x}_2 \pm (t \text{ critical value})\sqrt{\frac{s_1^2}{n_1} + \frac{s_2^2}{n_2}}$$

with

$$df = \frac{(V_1 + V_2)^2}{\dfrac{V_1^2}{n_1 - 1} + \dfrac{V_2^2}{n_2 - 1}}, \quad \text{where } V_1 = \frac{s_1^2}{n_1} \text{ and } V_2 = \frac{s_2^2}{n_2}$$

The same assumptions used for the two-sample t test still apply and will need to be verified before computing the confidence interval. Remember, for a two-sample confidence interval for independent samples, both samples should be random samples and both sample sizes must be sufficiently large or the populations approximately normal for the interval to be appropriate.

INFERENCES CONCERNING THE DIFFERENCE BETWEEN TWO POPULATION OR TREATMENT MEANS USING PAIRED SAMPLES

(*Statistics: Learning From Data*, 1st ed. pages 633–644)

The two-sample t test just described is used with two independent samples. However, sometimes there are other variables that should be considered. For example, tracking weight loss of individuals who are exercising versus those who are not may be confounded by the individual's caloric intake as well. So just comparing a mean weight loss for those who exercise and those who don't may not be the best comparison. However, if the two groups were *paired* by caloric intake, then the weight loss difference in the pairs could be calculated. This would allow us to see if the exercise had an effect on weight loss for individuals who ate roughly the same number of calories per day.

If the two sample sizes are the same, think about whether or not the samples are paired. If the sample sizes are different, then the samples are not paired samples. To perform the paired t test, you will use the same steps as for the one-sample t test where

1. The hypothesis is based on the mean difference in the pairs (μ_d = hypothesized value).

2. The df is $n - 1$, where the n is the number of pairs we have.

The paired t test is summarized in these simple steps. If the two samples sizes are not the same, then it cannot be a paired test. However if the sample sizes are the same it may or may not be a paired test.

HYPOTHESES

$H_0 : \mu_d =$ hypothesized value, where $\mu_d = \mu_1 - \mu_2$

alternative hypothesis will be either

$H_a : \mu_d >$ some hypothesized value

$H_a : \mu_d <$ some hypothesized value

$H_a : \mu_d \neq$ some hypothesized value

state the α level

TEST: paired t test

ASSUMPTIONS

1. The samples are *paired*, and therefore, *not independent*.

2. The sample differences came from a random sample of population differences.

3. The sample size (number of differences) is large or it is reasonable to assume that the distribution of differences in the population is approximately normal.

TEST STATISTIC MEAN DIFFERENCE = PAIRED TEST

Given $\bar{x}_d =$ mean of the sample difference and $s_d =$ standard deviation of sample differences

$$t = \frac{\bar{x}_d - \text{hypothesized value}}{\frac{s_d}{\sqrt{n}}}$$

and

$df = n - 1$, where n is the number of sample differences.

CONCLUSION As before, if P-value $< \alpha$ there is sufficient evidence to reject the null hypothesis in favor of the alternative hypothesis. In other words, we would have sufficient evidence to reject the hypothesis that the mean difference equaled the hypothesized value. We use the same four steps as in other tests.

To better understand this procedure, let's reconsider the chocolate chip cookie scenario in the first part of the chapter. If you recall, we were given the mean number of chips in samples of 25 regular and low-fat cookies. Suppose that each student in the class counted the number of chips in one cookie of each type. Since everyone may count what constitutes a chip a little differently, a better way to analyze this data is to incorporate the pairing. The following sample problem uses the same set of data used for the two-sample test earlier, but instead of only having the means for each type of cookie, this data also contains information on the pairing. Notice how this test differs from the two-sample problem presented earlier.

SAMPLE PROBLEM 2 Twenty-five students each counted the number of chocolate chips in a randomly selected low-fat and a randomly selected regular chocolate chip cookie and recorded the counts as seen in the table below. Prior to counting, each student flipped a coin to decide if he or she would count the regular or the low-fat cookie first. Is there evidence that, on average, regular chocolate chip cookies contain more chips than low-fat cookies? Defend your answer.

R = regular chocolate chip cookies
L = low-fat chocolate chip cookies
Diff. = R − L

Person	R	L	Diff	Person	R	L	Diff
1	20	17	3	14	21	14	7
2	17	13	4	15	17	11	6
3	12	10	2	16	13	16	−3
4	15	13	2	17	16	16	0
5	16	12	4	18	17	16	1
6	20	23	−3	19	15	15	0
7	15	16	−1	20	18	20	−2
8	19	18	1	21	14	12	2
9	18	16	2	22	13	15	−2
10	17	15	2	23	21	19	2
11	15	19	−4	24	12	10	2
12	18	18	0	25	15	13	2
13	13	12	1				

$\bar{x}_d = 1.12$, $s_d = 2.68$

SOLUTION TO PROBLEM 2 To answer this question, we need to carry out a hypothesis test. Since the samples are paired, we will use a paired t test.

HYPOTHESES

$H_0 : \mu_d = 0$ where μ_d = difference in mean number of chips (regular − low-fat)

$H_a : \mu_d > 0$ since they want to know if regular have more chips

$\alpha = 0.05$

TEST: paired t test

ASSUMPTIONS
1. The samples are paired.
2. The differences can be regarded as a random sample of differences because the cookies were randomly selected.
3. The sample size is only 25, which is not large. A normal probability plot (below) is reasonably straight and so it is reasonable to assume that the distribution of differences is approximately normal.

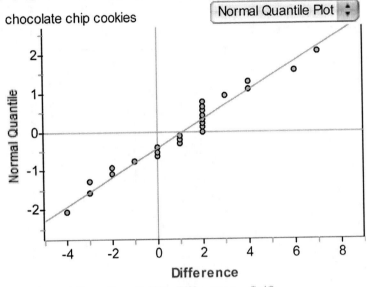

Normal Quantile = 0.373 Difference - 0.42

TEST STATISTIC

$$t = \frac{\bar{x}_d - 0}{\frac{s_d}{\sqrt{n}}} = \frac{1.12 - 0}{\frac{2.68}{\sqrt{25}}} = \frac{1.12}{\frac{2.68}{5}} = \frac{1.12}{0.536} = 2.09$$

and

$$df = n - 1 = 24$$

$$P-value = 0.024$$

CONCLUSION Since 0.024 < 0.05, there is sufficient evidence to reject the null hypothesis in favor of the alternative hypothesis. In other words, there is reason to believe that the mean difference in chips is greater than 0 so there does appear to be more chocolate chips, on average, in regular cookies than in low-fat cookies.

What is important to notice in the sample problem given is that when using the differences instead of two independent samples, our conclusion changed significantly. Both tests used the same samples; however, the paired *t* test took the pairing into account. This illustrates why it is important to use an appropriate test when the samples are paired.

LARGE-SAMPLE INFERENCES CONCERNING A DIFFERENCE BETWEEN TWO POPULATION OR TREATMENT PROPORTIONS

(*Statistics: Learning From Data*, 1st ed. pages 541–559 and 618–654)

As we saw in the previous chapters, categorical data involves working with proportions instead of means. We also learned that a hypothesis test and confidence interval could be used with this type of data. In this section, we will see how this concept can also be extended to compare two population proportions. For example, is the proportion of males who would be willing to pay an extra $300 to upgrade to a first class airline seat equal to the proportion of females who would be willing to pay extra to upgrade? This question and others that involve comparing two population proportions can be answered using a two proportions z test.

This test is based on the following properties of the sampling distribution of $\hat{p}_1 - \hat{p}_2$:

1. $\mu_{\hat{p}_1 - \hat{p}_2} = p_1 - p_2$, in other words, since the distribution of $\hat{p}_1 - \hat{p}_2$ is centered at $p_1 - p_2$, it is an unbiased estimator of $p_1 - p_2$

2. $\sigma^2_{\hat{p}_1 - \hat{p}_2} = \sigma^2_{\hat{p}_1} + \sigma^2_{\hat{p}_2} = \dfrac{p_1(1-p_1)}{n_1} + \dfrac{p_2(1-p_2)}{n_2}$

    making

    $\sigma_{\hat{p}_1 - \hat{p}_2} = \sqrt{\dfrac{p_1(1-p_1)}{n_1} + \dfrac{p_2(1-p_2)}{n_2}}$

3. When the sample sizes are both large, the sampling distribution of $\hat{p}_1 - \hat{p}_2$ is approximately normal. This MUST be checked by verifying all of the following:

    $n_1 p_1 \geq 10$ $n_2 p_2 \geq 10$

    $n_1(1-p_1) \geq 10$ $n_2(1-p_2) \geq 10$

AP Tip

It is not enough to just state the large sample conditions—you must show that you have actually checked these conditions and show your work.

Once we have verified that the sampling distribution is roughly normal and the samples were random samples, we can then form a test statistic:

$$z = \frac{(\hat{p}_1 - \hat{p}_2) - (p_1 - p_2)}{\sqrt{\dfrac{p_1(1-p_1)}{n_1} + \dfrac{p_2(1-p_2)}{n_2}}}, \text{ which simplifies to } z = \frac{\hat{p}_1 - \hat{p}_2}{\sqrt{\dfrac{p_1(1-p_1)}{n_1} + \dfrac{p_2(1-p_2)}{n_2}}}$$

when the null hypothesis specifies $p_1 - p_2 = 0$.

The actual test procedure uses a slightly different denominator. Because the values of p_1 and p_2 are not known, the denominator cannot be computed. Because the null hypothesis $H_0 : p_1 - p_2 = 0$ specifies that p_1 and p_2 are equal, we estimate this common value

p_c = combined estimate of the common population proportion,

using
$$p_c = \frac{n_1\hat{p}_1 + n_2\hat{p}_2}{n_1 + n_2} = \frac{\text{total S's in two samples}}{\text{total of the sample sizes}}$$

So given this information, we can now describe a two proportions z test.

HYPOTHESIS

$H_0 : p_1 - p_2 = 0,$

the alternative will again be one of

$H_a : p_1 - p_2 \neq 0$
$H_a : p_1 - p_2 > 0$
$H_a : p_1 - p_2 < 0$

ASSUMPTIONS

1. Independent random samples

2. Large sample sizes. Verify

$n_1\hat{p}_1 \geq 10$ $n_2\hat{p}_2 \geq 10$
$n_1(1 - \hat{p}_1) \geq 10$ $n_2(1 - \hat{p}_2) \geq 10$

TEST STATISTIC

TAKE NOTE OF THE STANDARD ERROR. P_c IS THE COMBINED PROPORTION.

$$z = \frac{\hat{p}_1 - \hat{p}_2}{\sqrt{\dfrac{p_c(1 - p_c)}{n_1} + \dfrac{p_c(1 - p_c)}{n_2}}}$$

This will allow us to find the related P-value using the standard normal distribution.

CONCLUSION As with other tests, the conclusion is based on comparing the P-value to α.

AP Tip

In any two-sample hypothesis test, you MUST be sure to check the standard assumptions for *both* samples. If you don't, you will not get credit for checking assumptions.

A great way to be sure you have checked everything is to use a table to insure that you have included a check for each of the samples on every assumption. Sketch something like the ones shown here and be sure you plug in the actual values or sketches as needed.

If you use the formula, make sure you have correctly entered the substitution.

MEANS		PROPORTIONS	
Sample 1	Sample 2	Sample 1	Sample 2
Random sample?	Random sample?	Random sample?	Random sample?
Is $n_1 > 30$?	Is $n_2 > 30$?	Check-	
If not, show a plot for normality	If not, show a plot for normality	$n_1 p_1 \geq 10$ $n_1(1 - p_1) \geq 10$	$n_2 p_2 \geq 10$ $n_2(1 - p_2) \geq 10$

EXAMPLE An airline wishes to know if the proportion of passengers who would pay $300 extra to upgrade to a first class airline seat is greater for international flights than for flights within the U.S. To investigate, they asked each person in a random sample of passengers on international flights and in a random sample of flights within the U.S. if they would pay extra. The resulting data are summarized in the following table:

Flight Type	n	# who would pay extra	Proportion who would pay extra
International	99	73	$\hat{p}_i = \dfrac{73}{99} = 0.74$
Within U.S.	91	56	$\hat{p}_w = \dfrac{56}{91} = 0.62$

let $\hat{p}_i$ = sample proportion of international passengers who would pay extra.

$\hat{p}_w$ = sample proportion of passengers on flights within U.S. who would pay extra.

Is there convincing evidence that the proportion of international passengers who would pay extra is greater than this proportion for passengers on flights within the U.S.? Use a significance level of 0.10.

Would the conclusion be different if a significance level of 0.05 had been used?

HYPOTHESIS

$$H_0 : p_i - p_w = 0$$
$$H_a : p_i - p_w > 0$$
$$\alpha = 0.10$$

ASSUMPTIONS

Sample 1

- problem states it's a random sample
- $100(0.74) \geq 10$
- $100(0.26) \geq 10$

So sample size is large enough

Sample 2

- problem states it's a random sample
- $92(0.62) \geq 10$
- $92(0.38) \geq 10$

So sample size is large enough

TEST STATISTIC

$$p_c = \frac{99(0.74) + 91(0.62)}{190} = 0.68$$

$$z = \frac{0.74 - 0.62}{\sqrt{\dfrac{0.68(0.32)}{99} + \dfrac{0.68(0.32)}{91}}} = 1.799$$

$$P\text{-value} = 0.072$$

CONCLUSION At $\alpha = 0.10$: since $0.072 < 0.10$, we would reject the null hypothesis and conclude that there is evidence that the proportion who would pay extra is greater for international passengers. However, at $\alpha = 0.05$: since 0.072 is not smaller than 0.05, we would fail to reject the null hypothesis.

AP Tip

If a question does not specify a significance level, you can select one. Just be sure to state what you have chosen and then use it to reach a conclusion. (Note: 0.05 is often selected.) You must show "linkage" by indicating that your p-value is > or < than your significance level.

A CONFIDENCE INTERVAL

Again, a confidence interval can be used to estimate the difference between two population proportions. The following confidence interval can be used when you have independent random samples and both sample sizes are large. The large sample conditions to be checked are the same as for the two proportions z test.

$$(\hat{p}_1 - \hat{p}_2) \pm (z \text{ critical value}) \sqrt{\frac{\hat{p}_1(1 - \hat{p}_1)}{n_1} + \frac{\hat{p}_2(1 - \hat{p}_2)}{n_2}}$$

Using our previous example of international passengers who would be willing to pay extra for a first class seat and passengers on flights within the U.S. who would pay extra, let's generate a 90% confidence interval estimate of the difference in population proportions. The assumptions were addressed in the hypothesis-testing example, so it is reasonable to proceed.

$$(\hat{p}_i - \hat{p}_w) \pm (z \text{ critical value}) \sqrt{\frac{\hat{p}_i(1 - \hat{p}_i)}{n_i} + \frac{\hat{p}_w(1 - \hat{p}_w)}{n_w}}$$

$$= (0.74 - 0.62) \pm 1.645 \sqrt{\frac{0.74(0.26)}{99} + \frac{0.62(0.38)}{91}}$$

$$= 0.12 \pm 1.645 \sqrt{0.0019 + 0.0026}$$

$$= 0.12 \pm 1.645(0.0671)$$

$$= 0.12 \pm 0.1104$$

$$= (0.01, 0.23)$$

Notice that the interval does not contain the value of 0. A difference of 0 corresponds to the case of no difference in the population proportions. We are 90% confident that the proportion who would pay extra for a first class seat is greater for international passengers than for passengers on flights within the U.S. by somewhere between 0.01 and 0.23. We used a method that will capture the true difference 90% of the time in repeated sampling.

INTERPRETING AND COMMUNICATING THE RESULTS OF STATISTICAL ANALYSES

(Statistics: Learning From Data, 1st ed. pages 541–559 and 618–654)

As we saw with the one-sample hypothesis tests, the conclusion is based on the *P*-value that is generated from our hypothesis test.

Two-sample tests are widely used as many studies are designed to compare two populations or two treatments. With this in mind, there are several things to remember when reading descriptions of statistical studies. Make sure you ask yourself the following questions.

- Are only two groups being compared? If there are more, we will need a different method.
- Are the samples independent or paired?
- What are the hypotheses being tested?
- Is the test appropriate? In other words, have all the needed assumptions been checked?
- If a confidence interval is calculated, has it been correctly interpreted?
- What is the *P*-value and does it lead us to reject the null hypothesis?
- Are the conclusions made consistent with the test results?

If you keep these questions in mind as you read descriptions of statistical studies or carry out two-sample hypothesis tests, you will avoid many common mistakes.

COMPARING TWO POPULATIONS OR TREATMENTS: STUDENT OBJECTIVES FOR THE AP EXAM

- You will be able to write the null and alternative hypothesis for two-sample tests.
- You will be able to carry out a two-sample hypothesis test for a difference in means.
- You will be able to carry out a two-sample hypothesis test for a difference in proportions.
- You will be able to distinguish between independent samples and paired samples.
- You will be able to construct and interpret a confidence interval for a difference in population means.
- You will be able to construct and interpret a confidence interval for a difference in population proportions.

MULTIPLE-CHOICE QUESTIONS

1. Insurance companies charge higher monthly premiums for male drivers than female drivers between the ages of 16–25. Their rationale is that male drivers have more traffic violations than females. A consumer group believes this rationale is no longer appropriate. To investigate, a random sample of male drivers and a random sample of female drivers in the target age groups were selected. Each person was asked how many traffic violations they had received. The resulting data were used to compute the given summary statistics. A 98% confidence interval for the difference in population mean number of traffic violations is (-1.3, 1.5). Based on this confidence interval, is there evidence that the population means are different?

Gender	n	Mean Violations	Standard Deviation
Male	73	4.2	1.8
Female	56	3.5	2.1

(A) Yes. Since the interval contains 0, there is strong evidence that the mean number of traffic violations is greater for males.

(B) No. Since the interval contains 0, it is plausible that the mean number of traffic violations is the same for males and females.

(C) Yes. Since the interval does not contain 0, there is strong evidence that the mean number of traffic violations is greater for males.

(D) No. Since the interval does not contain 0, there is insufficient evidence that the number of traffic violations is the same for males and females.

(E) Yes. Because the sample mean for males is greater than the sample mean for females.

2. In a random sample of 300 elderly men, 65% were married, while in a random sample of 400 elderly women, 48% were married. Which of the following is the 99% confidence interval estimate for the difference between the proportion of elderly men and the proportion of elderly women who are married?
    (A) 0.17 ± 0.073
    (B) 0.17 ± 0.096
    (C) 0.55 ± 0.067
    (D) 0.56 ± 0.067
    (E) 0.565 ± 0.096

3. Which of the following statements are *not* true?
    I.   When samples are paired, a two-sample t test is appropriate.
    II. The degrees of freedom for a paired t test is $n - 1$, where n is the sum of the two sample sizes.
    III. A two-sample z-test is used to test hypotheses about $p_1 - p_2$ with large independent samples.
    (A) I only
    (B) II only
    (C) III only
    (D) I and II
    (E) I and III

4. What would be the appropriate hypotheses for a research company that wants to see if there is a difference in the mean amount of vitamin D in a brand name multi-vitamin and generic brand multi-vitamin?
    (A) $H_0 : \mu_b - \mu_g = 0, \quad H_a : \mu_b - \mu_g > 0$
    (B) $H_0 : \mu_b - \mu_g = 0, \quad H_a : \mu_b - \mu_g < 0$
    (C) $H_0 : \mu_b - \mu_g = 0, \quad H_a : \mu_b - \mu_g \neq 0$
    (D) $H_0 : p_b - p_g = 0, \quad H_a : p_b - p_g > 0$
    (E) $H_0 : p_b - p_g = 0, \quad H_a : p_b - p_g \neq 0$

Question 5–6 refer to the following set of data:
When a virus is placed on a tobacco leaf, small lesions appear on the leaf. To compare the mean number of lesions produced by two different strains of virus, one strain is applied to half of each of eight tobacco leaves, and the other strain is applied to the other half of each leaf. The strain that goes on the right half of the tobacco leaf is decided by a coin flip. The lesions that appear on each half are then counted. The data are given below.

Leaf	1	2	3	4	5	6	7	8
Strain 1	31	20	18	17	9	8	10	7
Strain 2	18	17	14	11	10	7	5	6

5. What is the number of degrees of freedom associated with the appropriate t test for testing to see if there is a difference between the mean number of lesions per leaf produced by the two strains?
   (A) 7
   (B) 8
   (C) 11
   (D) 14
   (E) 16

6. Using a significance level of 0.01, is there statistical evidence that the mean number of lesions is not the same for the two strains?
   (A) Yes, $H_0 : \mu_1 - \mu_2 = 0, H_a : \mu_1 - \mu_2 \neq 0$, P-value $= 0.262$
   (B) No, $H_0 : \mu_1 - \mu_2 = 0, H_a : \mu_1 - \mu_2 \neq 0$, P-value $= 0.262$
   (C) No, $H_0 : \mu_1 = \mu_2, H_a : \mu_1 \neq \mu_2$, P-value $= 0.262$
   (D) Yes, $H_0 : \mu_d = 0, H_a : \mu_d \neq 0$, P-value $= 0.034$
   (E) No, $H_0 : \mu_d = 0, H_a : \mu_d \neq 0$, P-value $= 0.034$

7. A new classroom strategy uses interactive technology to review information for a test. A random sample of 40 classroom teachers has been asked to use the technology with one of their classes and not with another class. The average test score for each class is then recorded for each teacher. Which would be the appropriate test to run to see if the technology helped?
   (A) One-sample z test
   (B) One-sample proportions test
   (C) Two-sample t test
   (D) Paired t test
   (E) Two-sample z test

8. Two independent samples were studied, resulting in the following summary statistics:

$\bar{x}_1 = 10$, $s_1 = 2.1$, $n_1 = 108$ and $\bar{x}_2 = 15$, $s_2 = 2.9$, $n_2 = 78$.

Which of the following could be an appropriate test statistic for testing $H_0 : \mu - \mu = 0$?

(A) $z = \dfrac{10 - 15}{\sqrt{\dfrac{2.1}{108} + \dfrac{2.9}{78}}}$

(B) $z = \dfrac{10 - 15}{\sqrt{\dfrac{2.1^2}{108} + \dfrac{2.9^2}{78}}}$

(C) $t = \dfrac{10 - 15}{\sqrt{\dfrac{2.1}{108} + \dfrac{2.9}{78}}}$

(D) $t = \dfrac{10 - 15}{\sqrt{\dfrac{2.1^2}{107} + \dfrac{2.9^2}{77}}}$

(E) $t = \dfrac{10 - 15}{\sqrt{\dfrac{2.1^2}{108} + \dfrac{2.9^2}{78}}}$

9. A soup manufacturer is deciding which company to use for their mushroom purchases. Inspection of a random sample of 20 mushrooms for each company found 30% of the mushrooms from one company were damaged and 35% from the other company were damaged. What assumptions for the two proportions z test would be a concern?

   I.   We don't know that they are random samples from both companies.
   II. The samples may not be independent.
   III. The sample size is not large enough.

   (A) I only
   (B) II only
   (C) III only
   (D) I and II only
   (E) I, II, and III

10. In an experiment to test the effect of alcohol on fine motor skills, volunteers were randomly assigned to one of two groups, A and B. Everyone in group A drank 2 ounces of alcohol, and 20 minutes later everyone in both groups was timed on a manual dexterity test. The average completion times for the group A and B volunteers were 38 and 31 seconds, respectively. The 90% confidence interval estimate for the difference in means is (3,11). If μ_A and μ_B are the true mean completion times, respectively, for people who have and have not drunk 2 ounces of alcohol, how many of the following statements are reasonable conclusions?

I. $3 < \mu_A - \mu_B < 11$ with probability 0.90

II. There is a 0.90 probability that $3 < \mu_A - \mu_B < 11$.

III. The interval (3,11) was calculated using a method that includes the actual value of $\mu_A - \mu_B$ 90% of the time in repeated random sampling.

IV. We are 90% confident that $\mu_A - \mu_B$ lies between 3 and 11 seconds.

(A) None
(B) One
(C) Two
(D) Three
(E) Four

11. In a test of $H_0 : \mu_1 - \mu_2 = 0$ versus $H_a : \mu_1 - \mu_2 > 0$, the value of the test statistic was $t = 2.34$ and the P-value was 0.01. What conclusion would be appropriate if $\alpha = 0.05$?

(A) There was not a significant difference in the population means.
(B) There is convincing evidence that there is a difference in the population means.
(C) There is convincing evidence that the mean for population 1 is greater than the mean for population 2.
(D) The proportion of successes in population is greater than the proportion of successes in population 2.
(E) There is insufficient evidence to conclude that the proportions are different.

Questions 12–13 refer to the following set of data:
Ten men and ten women were given a supplement for weight loss, and the number of pounds lost by each person was measured at the end of one month. The data is shown below. The investigators would like to know if there is evidence that the mean weight loss for men is greater than the mean weight loss for women.

Subject	1	2	3	4	5	6	7	8	9	10
Males	5	8	12	7	9	11	10	16	8	14
Females	4	9	8	6	11	7	8	12	10	13

12. Which of the following is an appropriate set of hypotheses if $\mu_m = $ men, $\mu_w = $ women, $\mu_d = $ difference.
    (A) $H_0 : \mu_m - \mu_w = 0, \quad H_a : \mu_m - \mu_w \neq 0$
    (B) $H_0 : \mu_m - \mu_w = 0, \quad H_a : \mu_m - \mu_w > 0$
    (C) $H_0 : \mu_m - \mu_w = 0, \quad H_a : \mu_m - \mu_w < 0$
    (D) $H_0 : \mu_d = 0, \quad H_a : \mu_d \neq 0$
    (E) $H_0 : \mu_d = 0, \quad H_a : \mu_d > 0$

13. Assuming that the conditions for inference are met, what would be an appropriate conclusion for a significance level of 0.10?
    (A) With $t = 1.62$, $df = 9$, and $p = 0.140$, we would fail to reject the null hypothesis. There may be no difference in men's and women's mean weight loss.
    (B) With $t = 1.62$, $df = 9$, and $p = 0.140$, we would reject the null hypothesis. There is a significant difference in men's and women's mean weight loss.
    (C) With $t = 1.62$, $df = 18$, and $p = 0.140$, we would fail to reject the null hypothesis. There may be no difference in men's and women's mean weight loss.
    (D) With $t = 0.87$, $df = 17$, and $p = 0.197$, we would fail to reject the null hypothesis. There may be no difference in men's and women's mean weight loss.
    (E) With $t = 0.87$, $df = 17$, and $p = 0.197$, we would reject the null hypothesis. There is a significant difference in men's and women's mean weight loss.

14. A school district wishes to estimate the difference in the proportion of girls who purchase a school lunch and the proportion of boys who purchase a school lunch. Assuming that the conditions for inference are met, which of the following confidence intervals would be appropriate?
    (A) A one-sample t interval
    (B) A two-sample t interval
    (C) A paired t interval
    (D) A one proportion z interval
    (E) A two proportions z interval

15. Time spent studying in a typical week was recorded for each student in a random sample of juniors and a random sample of seniors in a large school district. A 95% confidence interval for the difference in mean time (junior–senior) was (7.4, 14.7). Which of the following is a correct interpretation of this interval?

	n	Mean Time (minutes)	Standard Deviation
Juniors	152	168	12.5
Seniors	108	157	15.8

(A) I am 95% confident that the sample mean difference is between 7.4 and 14.7 minutes.
(B) I am 95% confident that the true mean difference in time spent studying is between 7.4 and 14.7 minutes.
(C) I am 95% confident that the difference in the proportion of students who study is between 7.4 and 14.7.
(D) I know that 95 out of 100 times the mean difference in time is between 7.4 and 14.7 minutes.
(E) 95 of the sample differences were between 7.4 and 14.7 minutes.

FREE-RESPONSE PROBLEMS

1. Two types of fertilizer for roses are being considered by a housing community for their landscaping needs. The community decided to test the fertilizer on 170 bushes to see if one yielded more rose growth than the other. Each rose bush was assigned at random to one of the two fertilizers. The average growth, in centimeters, for each fertilizer was recorded. Fertilizer A is less expensive and will be used unless there is convincing evidence that mean growth is greater for Fertilizer B. Carry out an appropriate hypothesis test using $\alpha = 0.05$, and make a recommendation as to which fertilizer should be used.

Fertilizer	Sample Size	Mean Growth (cm)	Standard Deviation
Type A	87	12.7	1.5
Type B	83	13.3	2.2

2. A recent study reported that in a random sample of 248 women, 58 had changed their political affiliation since the last election. It also reported that 120 in a random sample of 387 men had changed political affiliation. The researchers would like to know if these data provide convincing evidence that the proportion changing political affiliation is greater for men than for women.
   (a) State the hypotheses of interest.
   (b) Identify the appropriate test and verify the conditions that must be met.
   (c) Is there convincing evidence that the proportion changing political affiliation is greater for men than for women? Use a significance level of 0.05.

Answers

MULTIPLE-CHOICE QUESTIONS

1. **B.** When this two-sample interval was computed, the value of 0 was within the interval, which means there is not sufficient evidence to conclude at the $\alpha = 0.02$ level of significance that the two population means are not the same (*Statistics: Learning From Data,* 1st ed. pages 649–651).

2. **B.** The difference between 0.65 and 0.48 is 0.17, making A or B the solution. Answer choice A, however, is for a 95% confidence interval (*Statistics: Learning From Data,* 1st ed. pages 649–651).

3. **D.** If the samples are paired, the appropriate test is a paired t test, and df = $n - 1$ where n is the number of pairs (*Statistics: Learning From Data,* 1st ed. pages 633–644).

4. **C.** The researchers want to know if there is a difference in the means of the brand name and the generic brand multi-vitamins. The alternative hypothesis would be two-sided (*Statistics: Learning From Data,* 1st ed. pages 633–644).

5. **A.** This is a paired t test so the df is calculated as $n - 1$, where n is the number of sample pairs (*Statistics: Learning From Data,* 1st ed. pages 633–644).

6. **E.** This would require a paired t test. Each virus is being tested on the same tobacco leaf and that makes the samples paired. When $P - value < \alpha$ there is strong evidence in favor of the alternative hypothesis. A, B, and C are incorrect. The t statistic for these paired samples is $t = \dfrac{4 - 0}{\dfrac{4.31}{\sqrt{8}}} = 2.62.$ The table of "t distribution critical values" shows that for 7 degrees of freedom, the P-value is between 2(0.02) and 2(0.01) or between 0.04 and 0.02 (the calculator gives a P-value of 0.034). Since this value is > the significance level of 0.01, we do not have sufficient evidence to conclude that the mean difference does not equal zero. We fail to

reject the null hypotheses (*Statistics: Learning From Data,* 1st ed. pages 633–644).

7. **D.** Each teacher is being assigned a class to use the technology and a class to not use the technology. The two sample means for each teacher are paired (*Statistics: Learning From Data,* 1st ed. pages 633–644).

8. **E.** This is a two-sample *t* test. The denominator must contain both the sample variances and the sample sizes (*Statistics: Learning From Data,* 1st ed. pages 633–644).

9. **C.** The sample size is not large enough to meet the conditions of the two proportions *z* test; $n_1 p_1 = (0.3)(20) = 6$, which is not greater than or equal 10 and $n_2 p_2 = (0.35)(20) = 7$, which is not greater than or equal to 10 (*Statistics: Learning From Data,* 1st ed. pages 514–526).

10. **C.** Both choice III and IV are reasonable statements based on the confidence interval. Correct interpretations should not refer to probability (*Statistics: Learning From Data,* 1st ed. pages 649–651).

11. **C.** Since the p-value is less than 0.05, the null hypothesis should be rejected in favor of the alternative $\mu_1 > \mu_2$ (*Statistics: Learning From Data,* 1st ed. pages 633–644).

12. **B.** The samples are independent since there is no indication that the male and female subjects are paired in any meaningful way. The alternative is $\mu_M - \mu_F > 0$ since we are asking if the mean for men is greater than the mean for women (*Statistics: Learning From Data,* 1st ed. pages 633–644).

13. **D.** The appropriate test is based on two independent samples with the alternative being that mean loss for men is greater than mean loss for women. The table of "t distribution critical values" shows that the P-value is slightly less than 0.20 for 9 degrees of freedom (the calculator give a p-value of 0.197). Since this P-value is greater than the significance level, we fail to reject the null hypotheses (*Statistics: Learning From Data,* 1st ed. pages 580–581).

14. **E.** A test of the difference in population proportions would use a *z* test (*Statistics: Learning From Data,* 1st ed. pages 514–526).

15. **B.** We are interested in the true mean difference (*Statistics: Learning From Data,* 1st ed. pages 633–644).

FREE-RESPONSE PROBLEMS

1. To answer this question, we will need to carry out an appropriate hypothesis test. In this case, it would be a two-sample *t* test. The steps are shown here.

    ### HYPOTHESES

    $H_0 : \mu_A - \mu_B = 0$

    (Note: this could have been written as $H_0 : \mu_A = \mu_B$)

    $H_A : \mu_A - \mu_B < 0 \quad \alpha = 0.05$

    ### ASSUMPTIONS

    1. The rose bushes were randomly assigned to treatments.

    2. Both sample sizes are greater than 30.

    TEST STATISTIC   Since the population standard deviations are not known, we will use a two-sample *t* test.

    $$t = \frac{(\bar{x}_1 - \bar{x}_2) - (\mu_1 - \mu_2)}{\sqrt{\dfrac{s^2_1}{n_1} + \dfrac{s^2_2}{n_2}}} = \frac{(12.7 - 13.3) - (0)}{\sqrt{\dfrac{1.5^2}{87} + \dfrac{2.2^2}{83}}} = \frac{(12.7 - 13.3) - (0)}{\sqrt{0.026 + 0.058}} - 2.07$$

    $$**df = \frac{(V_1 + V_2)^2}{\dfrac{V_1^2}{n_1 - 1} + \dfrac{V_2^2}{n_2 - 1}} = \frac{(0.026 + 0.058)^2}{\dfrac{0.026^2}{86} + \dfrac{0.058^2}{82}} = \frac{0.007056}{0.000049} = 143$$

    $p - value = 0.02$

    ** Again, we recommend you use the *df* from the calculator.

    CONCLUSION Since 0.02 < 0.05, we reject the null hypothesis in favor of the alternative hypothesis. There is convincing evidence that the mean growth for Fertilizer B is greater than for Fertilizer A (*Statistics: Learning From Data*, 1st ed. pages 633–644).

2. (a) The hypotheses would be:

    $H_0 : p_w - p_m = 0$

    where p_w = proportion of women changing affiliation

    p_m = proportion of men changing affiliation

    $H_a : p_w - p_m < 0$

    (b) The appropriate test is a two-sample proportions *z* test. The samples were independently selected and both were random samples. Check to see that the samples sizes are large enough.

    $\hat{p}_w = \dfrac{58}{248} = 0.23$ $\hat{p}_m = \dfrac{120}{387} = 0.31$

    248(0.23) >10   387(0.31) > 10

    248(0.67) > 10   387(0.69) > 10

    (c) $z = -2.09$, *P*-value = 0.018. Since 0.018 < 0.05, we reject the null hypothesis. There is convincing evidence that the proportion of men who changed political affiliation is greater than the proportion of women who did so (*Statistics: Learning From Data*, 1st ed. pages 514–526).

11

THE ANALYSIS OF CATEGORICAL DATA AND GOODNESS-OF-FIT TESTS

This section will explore inference for categorical data using chi-square test procedures. These procedures are used with univariate as well as bivariate data sets if the variables are categorical.

OBJECTIVES

■ Carry out a chi-square goodness-of-fit test.
■ Carry out chi-square tests for homogeneity of proportions and for independence.

CHI-SQUARE TESTS FOR UNIVARIATE DATA

(*Statistics: Learning From Data*, 1st ed. pages 694–708)

In this section, we extend techniques for analyzing univariate categorical data sets. Here we can consider questions about variables that involve two or more categories. A chi-square test (χ^2) for univariate data allows us to test hypotheses about the proportions falling into the different categories for a categorical variable. We do this by looking at the frequencies observed in each of the categories and comparing them to what would be expected if a null hypothesis were true. The hypothesized proportions can be equal or they can be different for each category.

The goodness-of-fit test is used when we are testing against some claim and there is only one sample. The data are displayed in one row.

EXAMPLE A new donut shop plans to sell plain, strawberry, blueberry, and cinnamon donuts. They wonder if there is a preference for one of these types of donuts or if each type is preferred by the same proportion of customers. If we let

p_p = proportion of customers preferring plain donuts

p_s = proportion of customers preferring strawberry donuts

p_b = proportion of customers preferring blueberry donuts

p_c = proportion of customers preferring cinnamon donuts

We are interested in knowing if $p_p = p_s = p_b = p_c = 0.25$ or if there is evidence that these proportions are not all the same. To answer this question, a random sample of 60 customers is surveyed, and each person in the sample is asked which of the four donut types they prefer.

Instead of running a separate two-sample proportions test for each of the possible pairs of proportions, a chi-square test will allow us to decide if the proportions we have observed in our sample are significantly different from the hypothesized proportions. The chi-square goodness-of-fit test will allow us to test the following hypotheses:

$H_0 : p_p = p_s = p_b = p_c$

$H_a :$ not all of the proportions are equal to 0.25

or at least one of the proportions is not equal to 0.25

(For chi-square tests, it is acceptable, and usually easier, to state hypotheses in words on the AP Exam.)

To make the decision to reject or fail to reject the null hypothesis, the chi-square test will compare the number of responses observed in each category to what we would expect to see in each category if the null hypothesis is true. If the difference is too large, we reject the null hypothesis. Otherwise, we fail to reject the null hypothesis.

Data on a single categorical variable is usually summarized using a one-way frequency table. Returning to the previous example, suppose that the sample of 60 customers resulted in the data summarized in the table below. The table entries are observed frequencies or counts.

Types of Donut Preferred

	Plain	Strawberry	Blueberry	Cinnamon
Observed Count	13	12	16	19

These counts represent the number of times a person in the sample selected each particular donut type. *Expected* counts are calculated using the hypothesized proportions from the null hypothesis. In this example, all four hypothesized proportions are equal to 0.25, so if the null hypothesis is true, we would expect to see the same number preferring each type. We calculate expected counts by multiplying the sample size by each hypothesized proportion:

$$np_p = 60(0.25) = 15$$
$$np_s = 60(0.25) = 15$$
$$np_b = 60(0.25) = 15$$
$$np_c = 60(0.25) = 15$$

The expected counts can now be entered into the table as shown here.

Types of Donuts Sold

	Plain	Strawberry	Blueberry	Cinnamon
Observed Count	13	12	16	19
Expected Count	60(0.25) = 15	60(0.25) = 15	60(0.25) = 15	60(0.25) = 15

Now we are ready to calculate the value of the test statistic. This test statistic is called the *goodness-of-fit* statistic because it considers how well the observed values fit to what we expected to see. Calculating the value of a chi-square $\left(\chi^2\right)$ test statistic is relatively easy.

$$\chi^2 = \sum \frac{(\text{observed cell count} - \text{expected cell count})^2}{\text{expected cell count}}$$

which is often abbreviated

$$\chi^2 = \sum \frac{(O - E)^2}{E}$$

For the example problem, the value of this test statistic is

$$\chi^2 = \frac{(13-15)^2}{15} + \frac{(12-15)^2}{15} + \frac{(16-15)^2}{15} + \frac{(19-15)^2}{15}$$
$$= \frac{(^-2)^2}{15} + \frac{(^-3)^2}{15} + \frac{(1)^2}{15} + \frac{(4)^2}{15}$$
$$= \frac{4}{15} + \frac{9}{15} + \frac{1}{15} + \frac{16}{15}$$
$$= \frac{30}{15}$$
$$= 2$$

To find the associated P-value, we use a chi-square distribution with df = k - 1, where k is the number of categories of a categorical variable. The Chi-Square distribution will become more normally distributed as the sample size/Dfs become larger but will always be at least somewhat skewed to the right. The P-value is the area to the right of the computed test statistic value (the chi-square goodness-of-fit test is an upper tail test). This area is found by referring to the chi-square table or by using a graphing calculator or other technology.

Some of the older TIs do not have a goodness-of-fit test. The student will need to use the distribution menu for chi square to find the p value (or the table feature of their calculator).

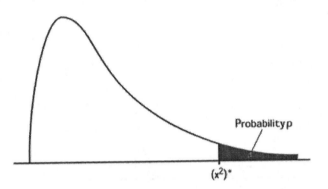

Probability p

$(x^2)^*$

The chi-square table is similar to the t table. You locate the df in the leftmost column then read across the row to find the value of the χ^2 test statistic. If the value is between two columns, then the P-value is between the two corresponding tail probabilities.

Chi-Square Tail Probability (χ^2)						
df	0.25	0.20	0.15	0.10	0.05	0.025
1	1.32	1.64	2.07	2.71	3.84	5.02
2	2.77	3.22	3.79	4.61	5.99	7.38
3	4.11	4.64	5.32	6.25	7.81	9.35
4	5.39	5.99	6.74	7.78	9.49	11.14
⋮	⋮	⋮	⋮	⋮	⋮	⋮

Since our $\chi^2 = 2.00$ and this value is smaller than 4.11, we know the tail probability is greater than 0.25. Using any reasonable α level will result in a failure to reject the null hypothesis.

The only things that we have not yet discussed are the overall assumptions required for the chi-square test. The first assumption is that we have a random sample of observations of a categorical variable. In addition, the sample size must be large enough that the following conditions are met:

1. No expected counts are < 1,

2. All of our expected counts should be ≥ 5 and if they are not,

3. No more than 20% of the expected counts are < 5.

Returning to the donut example:

HYPOTHESIS

$H_0 : p_p = 0.25, \ p_s = 0.25, \ p_b = 0.25, \ p_c = 0.25$

$H_a : H_0$ is not true; in other words, not all of the proportions are equal

We will use a significance level of 0.05 for this test.

TEST AND ASSUMPTIONS

Test: Chi-square goodness-of-fit test

Assumptions:

1. The sample was a random sample of customers

2. All expected counts are ≥ 5, so the sample size is large enough.

TEST STATISTIC As seen earlier, the calculated test statistic for this test would be

$$\chi^2 = \frac{(13-15)^2}{15} + \frac{(12-15)^2}{15} + \frac{(16-15)^2}{15} + \frac{(19-15)^2}{15}$$
$$= \frac{(-2)^2}{15} + \frac{(-3)^2}{15} + \frac{(1)^2}{15} + \frac{(4)^2}{15}$$
$$= \frac{4}{15} + \frac{9}{15} + \frac{1}{15} + \frac{16}{15}$$
$$= \frac{30}{15}$$
$$= 2$$

with $df = 3$

Based on $df = 3$ and $\chi^2 = 2$, the P-value is greater than 0.25.

CONCLUSION Since P-value $> \alpha$, there is insufficient evidence to reject the null hypothesis. In other words, there is not convincing evidence that the four types of donuts are not equally preferred.

AP Tip

For all chi-square tests, make sure to provide the expected counts and verify that they are large enough. On the AP Exam you should show calculations for at least two cells. You do not need to show the math for all four categories in this case.

If you specify the chi-square test, be sure to define the correct type.

Goodness-of-fit—one sample, one row, testing against a claim

Test for homogeneity—at least two samples, one category (at least two rows)

Test for independence—one sample, at least two categories (at least two rows)

SAMPLE PROBLEM 1 A music producer is interested in marketing a new artist via ads in movie theaters. The target age group is teenagers from 14–18 years of age. The company developing the advertisement has offered to run the ad in conjunction with the following types of movies: 20% of the time with comedies, 50% with dramas, and 30% with action films. However, the movie producer is not sure these percentages reflect the types of movies that teens attend. To investigate, each person in a random sample of teens was asked what type of movie they had seen most recently, resulting in the following data:

Movie Most Recently Watched By 100 Teens

	Comedy	Drama	Action
Observed Count	41	35	24

Do these data provide convincing evidence that the proportions of teens watching the different types of movies are different than the proportions proposed by the company developing the ad? Use a significance level of 0.05.

SOLUTION TO PROBLEM 1

HYPOTHESIS

p_c = proportion of teens who watched a comedy

p_d = proportion of teens who watched a drama

p_a = proportion of teens who watched an action films

$H_0 : p_c = 0.2; \quad p_d = 0.5; \quad p_a = 0.3$

H_a : the null hypothesis is not true (at least one of the proportions of teens watching the various film types is not equal to the hypothesized proportion).

$\alpha = 0.05$

ASSUMPTIONS The problem states the teens were a random sample.

All the expected counts are greater than or equal to 5 (see table below).

Movie Most Recently Watched By 100 Teens

	Comedy	Drama	Action
Expected Counts	100(0.2) = 20	100(0.5) = 50	100(0.3) = 30

TEST STATISTIC

χ^2 Goodness-of-fit test:

$$\chi^2 = \frac{(41-20)^2}{20} + \frac{(35-50)^2}{50} + \frac{(24-30)^2}{30}$$

$$= \frac{(21)^2}{20} + \frac{(-15)^2}{50} + \frac{(-6)^2}{30}$$

$$= 22.05 + 4.5 + 1.2$$

$$= 27.75 \quad \text{with } df = 2$$

CONCLUSION The *P*-value associated with $\chi^2 = 27.72$ and $df = 2$ is approximately 0.000. Since 0.000 < 0.05, there is strong evidence to reject the null hypothesis. In other words, based on this sample there is strong evidence that at least one of the proportions is not equal to the proportion of movies watched stated by the ad company.

Just as with other hypothesis tests, it is important to address all parts of the test. Clearly conveying your understanding of the procedure will include

- explaining the notation used,
- addressing all assumptions that are required for the test,
- demonstrating correct mechanics,
- and finally writing a conclusion in context.

As with other tests, the mechanics can be done using a graphing calculator, but it is strongly recommended that you still show the initial set-up. This lets the AP Reader know you understand how to compute the value of the test statistic.

TESTS FOR HOMOGENEITY AND INDEPENDENCE IN A TWO-WAY TABLE

(Statistics: Learning From Data, 1st ed. pages 708–723)

A chi-square test procedure can also be used with bivariate categorical data, which is usually displayed in a two-way frequency table. There are two different types of investigations that arise from this type of data. One type involves inferences about association between two different categorical variables being observed on a single population. The other type involves comparing two or more populations or treatments when a single categorical variable is observed. The calculation procedure is the same for both types of investigation, but the primary question of interest is different.

TEST FOR HOMOGENEITY

In a chi-square test for homogeneity, we are interested in whether the proportions falling into each of the possible categories of a categorical variable are the same for all of the treatments or populations studied. In this case, the null hypothesis is that the distribution of the categorical variable is the same for each population or treatment.

EXAMPLE A snack manufacturer produces three types of chips in two different locations (Location A and Location B). Sometimes bags of chips are damaged in the packaging process. Each bag in a random sample of 45 bags of chips packaged at Location A and in a random sample of 30 bags of chips packaged at Location B was classified into one of three categories: no damage, minimal damage, and severe damage. The resulting data is summarized in the table below. The manufacturer was interested in determining if there was sufficient evidence to conclude that the proportions falling into each of the three damage categories is not the same for the two locations.

	No Damage	Minimal Damage	Severe Damage	Row Totals
Location A	15	18	12	45
Location B	8	12	10	30
Column Totals	23	30	22	75

To answer the question of interest, the hypotheses would be

H_0: There is no difference in the proportions falling into each damage category for the two locations

H_a: The proportions falling into each damage category are not the same; there is a difference for the two locations

The assumptions that need to be are (1) the sample must be a random sample and (2) all expected counts must be at least 5.

The expected counts for a chi-square test of homogeneity are calculated using the following formula:

$$\frac{(\text{row total})(\text{column total})}{\text{table total}}$$

For example, the expected count for the cell of no damage and Location A is

$$\frac{(53)(23)}{76} = 16.04$$

A way to record these expected counts is either in a separate two-way table of just the expected counts or in parenthesis beside each observed count as shown in the table below. All of the expected counts are greater than or equal to 5.

	No Damage	Minimal Damage	Severe Damage	Row Totals
Location A	19 (16.04)	20 (18.13)	14 (18.83)	53
Location B	4 (6.96)	6 (7.87)	13 (8.17)	23
Column Totals	23	26	27	76

Degrees of freedom for the chi-square test of homogeneity is computed as follows:

$$df = (r-1)(c-1),$$

where r = number of rows and c = number of columns

Notice that the total column and row are not used in computing the degrees of freedom.

The chi-square test statistic that was used in the goodness-of-fit test is also used here.

χ^2 test for homogeneity

$$\chi^2 = \frac{(19-16.04)^2}{16.04} + \frac{(20-18.13)^2}{18.13} + \frac{(14-18.83)^2}{18.83} + \frac{(4-6.96)^2}{6.96} + \frac{(6-7.87)^2}{7.87} + \frac{(13-8.17)^2}{8.17}$$
$$= 6.5$$

with $df = (2-1)(3-1) = 2$

Using the graphing calculator technology, you first enter a matrix of the observed counts and then run a chi-square test from the test menu. The calculator will calculate the expected counts and store them in matrix[B].

```
MATRIX[A]  2 x3
[ 19    20    14    ]
[ 4      6    13    ]
```

```
X²-Test
 Observed: [A]
 Expected: [B]
 Calculate Draw
```

```
X²-Test
 X²=6.534131993
 P=.0381181018
 df=2
```

Finally, since $P < \alpha$ (0.038 < 0.05), we reject the null hypothesis. There is strong evidence that the proportions falling into each of the three damage categories is not the same for both locations.

TEST FOR INDEPENDENCE

The chi-square test for independence, also known as the chi-square test for association, is used to investigate if there is an association between two categorical variables within one population. The calculations will proceed in the same manner as the test for homogeneity; however, we are actually looking to see if knowing the value of one variable provides information about the value of the other variable.

EXAMPLE A car manufacturer has two production lines building three types of cars. An engineer is wondering if there is an association between the type of car and the production line that made the vehicle for cars that are found to have major defects. Each car in a random sample of 75 cars selected from all cars found to have major defects was classified according to the type of car and the production line that produced the car. The resulting data is given in the table below.

	Sedan	Wagon	Truck
Line A	13	9	12
Line B	18	12	11

The question of interest is whether there is an association between car type and production line for cars with major defects. To answer this question, we use the chi-square test for independence.

HYPOTHESIS

H_0 : Production line and car type are independent.

H_a : Production line and car type are not independent.

$\alpha = 0.05$

ASSUMPTIONS The sample was a random sample of cars with major defects.

The expected counts are all greater than or equal to 5, so we can proceed (see table below).

Expected Counts for Auto Errors

	Sedan	Wagon	Truck
Line A	14.1	9.5	10.4
Line B	16.9	11.5	12.6

TEST STATISTIC Since this is a χ^2 test for independence, the calculations would be

$$\chi^2 = \frac{(13-14.1)^2}{14.1} + \frac{(9-9.5)^2}{9.5} + \frac{(12-10.4)^2}{10.4} + \frac{(18-16.9)^2}{16.9} + \frac{(12-11.5)^2}{11.5} + \frac{(11-12.6)^2}{12.6} = 0.63$$

$df = 2$

$P = 0.73$

CONCLUSION Since 0.73 is not smaller than 0.05, we fail to reject the null hypothesis. Therefore, we do not have sufficient evidence of an association between car type and production line.

Notice, the steps in the chi-square hypothesis tests are the same as for all other tests: State hypotheses, identify the test by name or formula and check assumptions required for the test, calculate the value of the test statistic and the associated *P*-value, and write a conclusion in context that is linked to the *P*-value.

INTERPRETING AND COMMUNICATING THE RESULTS OF STATISTICAL ANALYSES

(*Statistics: Learning From Data*, 1st ed. pages 708–723)

This particular set of tests involved writing our conclusions in terms that match the setting of the individual test. Each chi-square test is performed in a different setting depending on the data we have to work with. It is important to state which of the three chi-square tests you are using when carrying out one of these tests. The goodness-of-fit, homogeneity, and independence tests all have different hypotheses. Be clear which test you are using.

CATEGORICAL DATA AND GOODNESS-OF-FIT TESTS: STUDENT OBJECTIVES FOR THE AP EXAM

- ▨ You will be able to determine degrees of freedom for chi-square tests of goodness-of-fit, homogeneity, and independence.
- ▨ You will be able to calculate expected counts for chi-square tests of goodness-of-fit, homogeneity, and independence.
- ▨ You will be able to carry out chi-square tests of goodness-of-fit, homogeneity, and independence.
- ▨ You will identify the correct hypothesis for each of the three chi-square scenarios.
- ▨ You will be able to interpret conclusions in context for chi-square tests of goodness-of-fit, homogeneity, and independence.

MULTIPLE-CHOICE QUESTIONS

1. Each car in a random sample of 200 cars sold in 2010 at a large car dealership was classified by color. Sixty of the cars were white, 80 were blue, 20 were silver, and 10 were red. What is the value of the chi-square statistic in a test of $H_0 : p_w = p_b = p_s = p_r = 0.25$?
   (A) 18
   (B) 32
   (C) 50
   (D) 70
   (E) 110

2. Which of the following is not true of the χ^2 probability distribution?
   (A) For small degrees of freedom, the distribution is right skewed.
   (B) The chi-square curves move farther to the right and spread out more as the number of degrees of freedom increases.
   (C) All of the area under a chi-square curve is associated with positive values.
   (D) The total area under the χ^2 curve is equal to one.
   (E) The mean of a chi-square distribution is 0.

3. There are four surgical methods currently being used to place medical implants in patients. After surgery, patients are monitored and their pain level is recorded as severe, moderate, or mild. What are the degrees of freedom for a test to determine if there is an association between surgical method and pain level?
   (A) 2
   (B) 3
   (C) 4
   (D) 6
   (E) 12

Questions 4–7 refer to the following set of data:

A group of AP Statistics students wanted to see if plain, peanut, and almond M&M's have the same color distribution. To test this, students took a random sample of each type of M&M and classified the candies in the sample by color. They plan to carry out a chi-square test to decide if there is evidence that the color distributions are not the same for the three types of M&M's. A total of 207 M&M's were classified.

	Red	**Blue**	**Yellow**	**Green**	**Orange**	**Brown**
Plain	20	18	15	10	14	12
Peanut	8	6	8	25	5	7
Almond	7	11	10	12	10	9

4. What is the expected count for green peanut M&M's?

(A) $\dfrac{25}{47}$

(B) $\dfrac{25 \times 47}{47}$

(C) $\dfrac{59 \times 47}{59}$

(D) $\dfrac{59 \times 47}{207}$

(E) $\dfrac{25 \times 47}{207}$

5. What are the correct degrees of freedom for the appropriate test?
(A) 206
(B) 18
(C) 10
(D) 7
(E) 4

6. What is an appropriate set of hypotheses?
(A) H_0: There is no association between the type of M&M's and color.
  H_a: There is an association between the type of M&M's and color.
(B) H_0: There is an association between the type of M&M's and color.
  H_a: There is no association between the type of M&M's and color.
(C) H_0: The color proportions are the same for all three types of M&M's.
  H_a: The color proportions are not the same for all three types of M&M's.
(D) H_0: The proportions of red M&M's are the same for all three types of M&M's.
  H_a: The proportions of red M&M's are not the same for all three types of M&M's.
(E) H_0: There is no difference in the type of M&M for green M&M's.
  H_a: There is a difference in the type of M&M for green M&Ms.

7. The value of the chi-square test statistic is 22.9. What conclusion would be reached in a test of the hypotheses of interest using a significance level of 0.05?
   (A) Fail to reject H_0 and conclude there is strong evidence of an association.
   (B) Fail to reject H_0 and conclude there is strong evidence that the color distributions are the same.
   (C) Fail to reject H_0 and conclude there is not strong evidence that the color distributions are the same.
   (D) Reject H_0 and conclude there is strong evidence of an association.
   (E) Reject H_0 and conclude there is strong evidence that the color distributions are not the same.

8. A cupcake store manager believes that half of the cupcakes sold are chocolate and that the other half is equally divided between vanilla and cherry. If the manager's belief is correct and 300 cupcakes are sold, how many would you expect to be cherry cupcakes?
   (A) 300
   (B) 200
   (C) 150
   (D) 100
   (E) 75

9. Each person in a random sample of patrons of a local mall was surveyed regarding a public smoking area outside one of the mall entrances. Each person was asked if they approved of the idea of a public smoking area in the mall. The resulting data is summarized in the table below. The mall management would like to know if there is a relationship between gender and approval of the smoking area. What would be an appropriate set of hypotheses?

Public Smokers

	Approve	Do Not Approve
Males	28	57
Females	39	31

   (A) H_0: Gender and approval are not independent.
       H_a: Gender and approval are independent.
   (B) H_0: Gender and approval are independent.
       H_a: Gender and approval are not independent.
   (C) H_0: Knowing a person does not approve of a public smoking area indicates their gender.
       H_a: Knowing a person does not approve of a public smoking area does not indicate their gender.
   (D) H_0: There is no difference in gender distribution based on approval.
       H_a: There is a difference in gender distribution based on approval.
   (E) H_0: There is an association between gender and approval.
       H_a: There is no association between gender and approval.

10. A local bagel shop makes six types of bagels and eight types of cream cheese toppings. Suppose that all customers order one type of bagel and one type of cream cheese. If customers are classified according to the type of bagel and type of cream cheese, how many degrees of freedom would be appropriate for a test to determine if there is an association between type of bagel and type of cream cheese purchased?
    (A) 7
    (B) 12
    (C) 14
    (D) 35
    (E) 48

11. In a recent opinion poll, likely voters were asked if they would continue to support health care reform if the cost per taxpayer citizen was increased by $800 per year. Political preference was also recorded. The P-value for a chi-square test of independence is 0.001. Which of the following is a correct interpretation of this result?

	Democrat	Republican	Independent	Libertarian
Support	125	87	99	48
Don't Support	75	113	101	52

    (A) Since $P = 0.001$, there is strong evidence that political preference and support are independent of each other. In other words, knowing their political preference gives no insight into their support of the increase.
    (B) Since $P = 0.001$, there is strong evidence that political preference and support are not independent of each other. In other words, knowing their political preference gives insight into their support of the increase.
    (C) Since $P = 0.001$, there is strong evidence that political preference and support are independent of each other. Since they are independent, there is no association between the two variables.
    (D) Since $P = 0.001$, there is insufficient evidence that political preference and support are independent of each other. Since they are not independent, there is no association between the two variables.
    (E) Since $P = 0.001$, there is insufficient evidence that political preference and support are not independent of each other. In this case, knowing their political preference would not help in knowing their support.

12. Two different cereal companies each make a bran cereal, a corn cereal, and a rice cereal. Each person in a random sample of 300 potential customers tasted all six cereals and selected their favorite. The choice was then classified according to type of cereal and the company that made the cereal. The data is shown in the table below. What type of test would be appropriate?

	Cereal 1	Cereal 2	Cereal 3
Company A	51	48	47
Company B	43	65	46

(A) Chi-square goodness-of-fit test
(B) Chi-square test of independence
(C) Chi-square test of homogeneity
(D) Either a chi-square test for independence or homogeneity is appropriate
(E) A chi-square test is not appropriate in this situation

13. Which is not true for a chi-square test of independence?
   I.   The test is based on bivariate categorical data.
   II. df is calculated using $k - 1$, where k = the number of categories.
   III. The test is an upper-tailed test.
   (A) I, II, and III
   (B) I and II
   (C) II and III
   (D) I only
   (E) II only

Questions 14–15 refer to the following set of data:

A movie theater recorded the type of snack item purchased and the type of movie the customer was attending for each person in a random sample of theater customers who made a snack bar purchase. The resulting data is given in the table below.

	Soda	Popcorn	Candy
Children's Movie	70	83	47
Action Movie	61	49	20

14. If the manager wanted to determine if there is an association between the type of movie and type of snack sold, which type test would be appropriate?
   (A) Chi-square goodness-of-fit
   (B) Chi-square test for homogeneity
   (C) Chi-square test for independence
   (D) Two-sample proportions test
   (E) Multiple sample proportions test

15. Using the data in the table above, the value of the chi-square statistic is 5.66. Which of the following is a correct statement about the P-value for this test?
(A) $P\text{-value} > 0.10$
(B) $0.05 < P\text{-value} < 0.10$
(C) $0.01 < P\text{-value} < 0.05$
(D) $0.001 < P\text{-value} < 0.01$
(E) $P\text{-value} < 0.001$

FREE-RESPONSE PROBLEMS

1. Regina is worried that the color of her new cardigan will attract the attention of killer bees in southern California where she is going to hike. To settle her nerves she looks at the American Killer Bee Association website. It shows that these bees are highly agitated by various colors. They have found that 75% of bees are agitated by green, 9% by blue, 6% by purple or pink, and the remaining 10% by other colors.
   (a) In a random sample of 200 killer bees, how many would you expect to be agitated by each color?
   (b) A recent study of 120 randomly selected people stung by killer bees last year found the individuals were wearing the colors shown in the table below. Do these data provide convincing evidence that the color distribution of colors worn by people stung by killer bees is different from the percentages given on the web site?

Color Worn By Individual Stung By Killer Bees

Green	Blue	Purple/Pink	Other
86	21	6	7

2. A restaurant offers both dine-in and take-out service. Customers can pay for their meal in cash, by credit card, or by debit card. The restaurant owner wonders if there is an association between the method of payment and the type of service. To investigate, a random sample is selected from the orders placed during the last year and the method of payment and the type of service is recorded for each of these orders. The data is summarized in the table below.

	Cash	**Credit**	**Debit**
Dine-in	34	122	32
Take-out	70	95	47

   (a) Should the restaurant owner carry out a test of homogeneity or a test of independence to answer his question?

   (b) Carry out a test to answer the question of interest to the owner. Use a significance level of 0.05 for your test.

Answers

MULTIPLE-CHOICE QUESTIONS

1. **D.** Since there are four different colors, each of the expected cell counts equal 50). The calculation would be found as follows:
    $$\chi^2 = \frac{(60-50)^2}{50} + \frac{(80-50)^2}{50} + \frac{(30-50)^2}{50} + \frac{(10-50)^2}{50} = 2 + 18 + 18 + 32 = 70$$
    (*Statistics: Learning From Data,* 1st ed. pages 698–716).

2. **E.** The χ^2 distribution is right skewed with all of its area associated with positive values. Therefore, the mean of a chi-square distribution can't be 0 (*Statistics: Learning From Data,* 1st ed. pages 698–716).

3. **D.** There are four surgical methods and three pain levels. $df = (r-1)(c-1) = (4-1)(3-1) = (3)(2) = 6$ (*Statistics: Learning From Data,* 1st ed. pages 698–716).

4. **D.** The expected counts in a two-way table are found using
    $$\frac{(\text{row total} \times \text{column total})}{\text{table total}} = 13.4.$$ (*Statistics: Learning From Data,* 1st ed. pages 698–716).

5. **C.** In a two-way table, degrees of freedom are calculated by $(r-1)(c-1)=(3-1)(6-1)=(2)(5)=10$ (*Statistics: Learning From Data,* 1st ed. pages 698–716).

6. **C.** This is a test of homogeneity. Three populations are being compared on the basis of a categorical variable (color) (*Statistics: Learning From Data,* 1st ed. page 711).

7. **E.** $\chi^2 = 22.9$ and $df = 10$, The table of χ^2 critical values gives a P-value between 0.01 and 0.02., which is less than 0.05. This means there is strong evidence to conclude that the color distributions are not all the same, so we reject the null hypotheses. The calculator χ^2-test gives a P-value of 0.011. (*Statistics: Learning From Data,* 1st ed. pages 698–716).

8. **E.** The expected count is calculated as (300–150)/2 = 75 (*Statistics: Learning From Data,* 1st ed. pages 698–716).

9. **B.** A test of independence between the variables would be used to answer the question of interest, because management wants to know if a relationship exists. (*Statistics: Learning From Data,* 1st ed. pages 715–716).

10. **D.** $(6-1)(8-1) = 5(7) = 35$ (*Statistics: Learning From Data,* 1st ed. pages 698–716).

11. **B.** The *P*-value is smaller than a significance level of 0.01, so the null hypothesis of independence would be rejected. This means there is strong evidence that an association exists, and that knowing a person's political party provides information about his or her support of this issue (*Statistics: Learning From Data,* 1st ed. pages 715–716).

12. **B.** Because a single sample is classified on the basis of two categorical variables, the appropriate test is a test for independence. The test is trying to determine whether or not the taste of the cereals is independent from the company that makes the cereals. (*Statistics: Learning From Data,* 1st ed. pages 715–716).

13. **E.** The degrees of freedom for a goodness-of-fit test are calculated using $k - 1$. However, for a chi-square test of independence, a two-way table is involved, so the degrees of freedom are calculated using $(r - 1)(c - 1)$ (*Statistics: Learning From Data,* 1st ed. pages 715–716).

14. **C.** This is a test of independence since we are trying to see if knowing the type of movie provides any information about the type of snack purchased or if these two items are independent. (*Statistics: Learning From Data,* 1st ed. pages 715–716).

15. **B.** With $df = 2$. the table of χ^2 critical values gives a *P*-value between 0.10 and 0.05. The calculator χ^2 -test gives a P-value of 0.06. (*Statistics: Learning From Data,* 1st ed. pages 698–716).

FREE-RESPONSE PROBLEMS

1. (a) The expected counts by color for 200 bees would be:
      Green: $200(0.75) = 150,$

      Blue: $200(0.09) = 18,$

      Purple/Pink: $200(0.06) = 12,$

      Other: $200(0.10) = 20.$
   (b) Hypotheses
      H_0 : The proportions of bee stings by color is as specified on the web site ($p_{green} = 0.75$, $p_{blue} = 0.09$, $p_{purple/pink} = 0.06$, $p_{other} = 0.10$)

      H_a : At least one of the color proportions is different from what is specified inthe null hypothesis

      $\alpha = 0.05$
      Test: Chi-square goodness-of-fit test
      Assumptions
      1. The problem states that the individuals in the sample were randomly selected.
      2. The expected counts are shown in the table below (note that the sample size is 150, not 200 as in part (a)). All expected counts are greater than 5, so the sample size is large enough.

Expected Counts for 150 Stings

Green	Blue	Purple/Pink	Other
(90.0)	(10.8)	(7.2)	(12.0)

Test Statistic

$\chi^2 = 12.09, \quad df = 3, \quad P = 0.007$

Conclusion

Since $0.007 < 0.05$, we reject the null hypotheses. There is strong evidence that the proportions of bee stings by killer bees are not all the same as reported on the web site. At least one of the proportions is not as stated (*Statistics: Learning From Data,* 1st ed. pages 698–716).

2. (a) This is a test of independence. There was only one sample and each individual in the sample was classified according to two categorical variables. The owner is trying to determine whether or not method of payment and type of service are independent.

 (b)

Hypotheses

H_0 : method of payment and order type are independent.

H_a : method of payment and order type are not independent.

$\alpha = 0.05$

Assumptions

(1) The sample was a random sample of orders.

(2) The expected counts are given in the table below. All expected counts are greater than 5, so the sample size is large enough.

Expected Counts

	Cash	Credit	Debit
Dine-in	$48.88	$101.99	$37.13
Take-out	$55.12	$115.01	$41.87

Test Statistic

$\chi^2 = \dfrac{(34 - 48.88)^2}{48.88} + \dfrac{(122 - 101.99)^2}{101.99} + \ldots = 17.291, \quad \text{with } df = 2$

Conclusion

We know from the table of χ^2 critical values that the P-value is less than 0.001. Therefore, we reject the null hypothesis. There is strong evidence that there is an association between method of payment and order type (*Statistics: Learning From Data,* 1st ed. pages 715–716).

12

INFERENCE FOR LINEAR REGRESSION AND CORRELATION

In Review Section 3, you learned how to describe and summarize bivariate data. You might wonder if the sample provides evidence that the two variables are associated in the population from which the sample was selected. In this section, you will learn inference techniques for bivariate numerical data.

OBJECTIVES

- Understand a simple regression model.
- Construct a confidence interval for the slope of the population regression line.
- Perform a test of significance for the slope of the population regression line.
- Interpret and communicate the results of the statistical analysis.

SIMPLE LINEAR REGRESSION MODEL

(Statistics: Learning From Data, 1st ed. pages 736–748)

Recall from Review Section 3 that we use one variable x (the explanatory variable) to help explain or shed light on the variability of another variable y (the response variable).

278

The simple linear regression model is the linear model that is assumed to describe the relationship between x and y in the population data. It is written as

$$y = \alpha + \beta x + e$$

The relationship between two variables is unlikely to result in a scatterplot with points that fall exactly on a straight line, so the error e accounts for the random variability that exists. Think of the residuals (Res = Actual – Predicted) as a sample from the population of errors. If a data point from the population lies above the regression line, e is positive and likewise, e is negative when a data point from the population lies below the regression line.

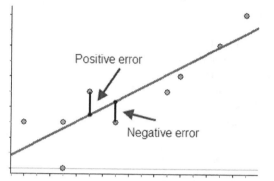

Our goal is to take a sample from a population and, using the data from our sample, be able to infer something about the population. In order to perform any inference procedures in the linear regression setting, we need to make four assumptions about the population and the associated errors.

The first assumption is that the mean of all the errors at any particular x value is equal to zero for each and every value of x.

Below is a scatterplot to illustrate this idea; notice that the mean of the errors is zero (remember zero is the point on the line since we are talking about errors which are directed distances above and below the line). We are assuming that the population appears like this for each and every value of x.

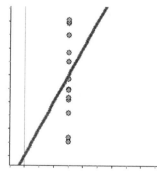

The second assumption we must make is that the spread of the distribution of e is the same for any particular value of x. This means that the standard deviation of e is the same for each and every value of x. The standard deviation of e is denoted σ_e.

Below is a scatterplot of what the population might look like for two specific x values; notice that the spread of the errors for each x is about the same. We are assuming that the population appears like this for each and every value of x.

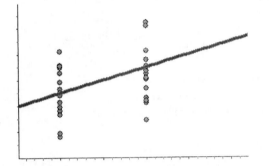

The third assumption we must make is that the distribution of e is normal for any particular value of x. Below is a scatterplot of what the population might look like for two specific x's; notice that the errors appear to be normally distributed about the regression line for each x. Assessing normality can be difficult. Remember, we are assuming that the population appears like this for each and every value of x.

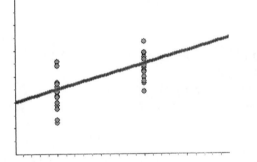

The fourth and final assumption is that each error that corresponds to an observation is independent of each and every error that corresponds to another observation.

Now, let's put all four conditions together. The observations must be independent and for every value of x the errors must be normally distributed with a mean of zero and the same standard deviation. Below is a picture illustrating all of these ideas together.

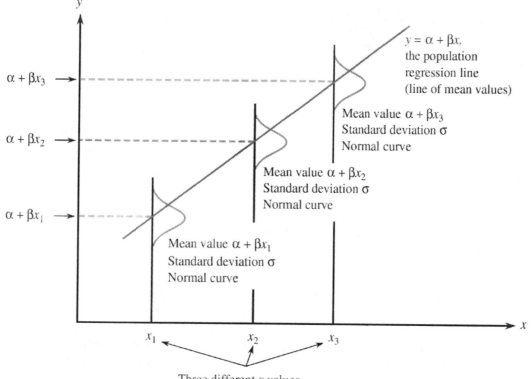

$y = \alpha + \beta x$, the population regression line (line of mean values)

Mean value $\alpha + \beta x_3$
Standard deviation σ
Normal curve

Mean value $\alpha + \beta x_2$
Standard deviation σ
Normal curve

Mean value $\alpha + \beta x_1$
Standard deviation σ
Normal curve

Three different x values

EXAMPLE The following data set gives the number of nights in a three-week time frame that students completed their homework and their quiz grade on the material taught during those three weeks. Is it reasonable to assume that the four assumptions of the linear regression model are true?

Number of Nights	11	12	8	12	7	6	11	7	10	6	10	11	7	9	8
Quiz Grade	81	88	76	88	66	70	80	75	85	68	76	86	71	77	75

First, we look at a scatterplot and residual plot for the data to confirm that a linear model is appropriate.

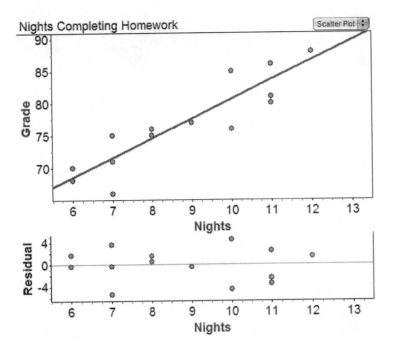

Based on the scatterplot and the residual plot, it appears that a linear model is appropriate. Now focus on the residual plot for assumptions 1–3.

Assumption 1: The mean of e at any given x value is equal to 0. This assumption is for each and every value of x. This assumption cannot be checked; the least squares regression line will always have residuals that have a mean of zero. We must assume that the mean of the errors is zero for the population. This assumption is reasonable when the pattern in the scatterplot is linear and there is no obvious pattern in the residual plot.

Assumption 2: σ_e is the same for each and every value of x. The scatter about the regression line does not appear to be much larger for some x values than for others, so this assumption seems reasonable.

Assumption 3: The distribution of e is normal for any particular value of x. For this assumption to be reasonable, points should tend to cluster near the line, with fewer points as you move away from the line. This appears to be the case here, so this assumption seems reasonable.

Assumption 4: The observations are independent. If no students worked together on the homework assignments or on the quiz, it seems reasonable to assume that students' homework and quiz grades are independent.

Since all four assumptions are reasonable, it makes sense to use a linear regression model.

Notice that all of the assumptions are about the population where the population is the collection of all students' number of nights of homework preparation and their corresponding quiz grade. Usually, we only have information from a sample taken from the desired population. Using a least squares regression line, we can estimate the population parameters from sample data.

The estimated regression line for the population is

$$\hat{y} = a + bx$$

Where a is a point estimate for α and b is a point estimate for β (which is usually the more important of the two variables).

Let x^* be a specific value of x. Then, $a + bx^*$ has two different meanings:

1. It is an estimate of the mean y for all observations when x is the $x = x^*$

2. It is the predicted y when we have substituted a specific x value.

EXAMPLE Refer to the previous example about number of nights a student completed their homework and their quiz grade. Estimate α and β for the population regression model.

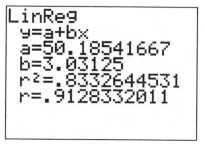

```
LinReg
 y=a+bx
 a=50.18541667
 b=3.03125
 r²=.8332644531
 r=.9128332011
```

The least squares regression line for our sample data is

$$\hat{y} = 50.185 + 3.031x$$

where x is the number of nights that a student completed his or her homework and $\hat{y}$ is the student's predicted quiz grade.

Therefore, our estimate for α is 50.185 points and our estimate for β is 3.031 points per night of homework completed.

SAMPLE PROBLEM 1 Data on outside temperature at 9 a.m. and number of students absent from school is given for 9 randomly selected days during the school year. Is a linear regression model reasonable? If so, estimate the equation of the population regression line. Find $\hat{y}$ when $x = 68$ and interpret this value in context.

Outside Temperature (°F)	60	62	63	65	67	71	72	74	75
Number of Students Absent	1	0	2	5	6	5	8	7	8

SOLUTION TO PROBLEM 1 First, look at a scatterplot and a residual plot for the data.

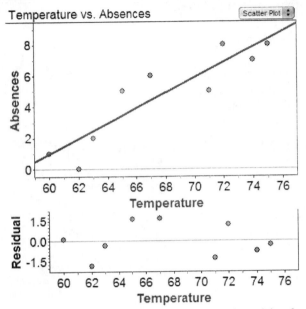

A linear model seems appropriate for this data. Now, we must check the four assumptions.

Assumption 1: The mean of e at each and every value of x is equal to zero. This assumption is reasonable when the pattern in the scatterplot is linear and there is no obvious pattern in the residual plot, which appears to be the case here.

Assumption 2: The standard deviation of the errors is the same for each and every value of x. The scatter about the regression line does not appear to be much larger for some x values than for others, so this assumption seems reasonable.

Assumption 3: The distribution of e is normal for any particular value of x. For this assumption to be reasonable, points should tend to cluster near the line, with fewer points as you move away from the line. This appears to be the case here, so this assumption seems reasonable.

Assumption 4: The observations are independent. Because the days were randomly selected, it is reasonable to assume that the observations are independent.

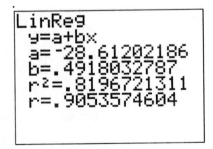

The least squares regression line for our sample data is

$$\hat{y} = -28.612 + 0.492x$$

where x is the outside temperature and $\hat{y}$ is the predicted number of students absent from class. This equation is the estimate of the population regression line.

When $x = 68$,

$$\hat{y} = -28.612 + 0.492(68) = -28.612 + 33.443 = 4.831$$

There are two ways to correctly interpret this value:

1. The average number of students absent on days when it is 68°F outside is 4.831.

2. Our model predicts that 4.831 students will be absent on an individual day where it is 68°F.

This example highlights what happens on most problems when we check the four assumptions—we did not have sufficient evidence to doubt the assumptions, so we assumed they were true and proceeded.

INFERENCES ABOUT THE SLOPE OF THE REGRESSION LINE

(*Statistics: Learning From Data,* 1st ed. pages 748–759)

Remember from Review Section 3 that the slope of the least squares regression line is the average change in y for a one-unit change in x. We now are concerned with β, the slope of the population regression line. It is interpreted the same way—only we are talking about the population, not a sample.

When the four assumptions for a linear regression model are met, the following is true for the slope of the sample regression line, b:

1. The mean value of b for all possible random samples is β. This means that b is an unbiased estimator of β.

2. The standard deviation of b for all possible random samples is

$$\sigma_b = \frac{\sigma}{\sqrt{S_{xx}}}.$$

3. The sample statistic b has a normal distribution.

Since the sample statistic b has a normal distribution, the statistic will have a standard normal distribution.

Just as before, this presents a problem. We rarely, if ever, know the value of σ. We use s_e to estimate σ and use a t interval (test) instead of a z interval (test). The distribution of $t = \dfrac{b - \beta}{\dfrac{s_e}{\sqrt{S_{xx}}}}$ has a t distribution with

$df = n - 2$.

EXAMPLE A teacher is interested in determining if there is a relationship between GPA and the number of minutes each day a student spends on social networking web sites. She selects a random sample of nine students. Use the data given below to construct an interval estimate the slope of the population regression line relating GPA and the number of minutes spent on social networking web sites.

Minutes Spent	20	60	15	30	0	30	0	2	30
Grade Point Average	3.70	3.41	3.10	3.90	3.80	3.67	3.80	3.85	3.83

We are interested in knowing about β (the true mean change in GPA per additional minutes spent on social networking web sites). The question asks for an estimate—this signals to us that we should find a confidence interval.

NAME OF INTERVAL: t interval for the slope of the population regression line

CONDITIONS: First, look at a scatterplot and a residual plot of the sample data (shown below). A linear model seems appropriate for this data.

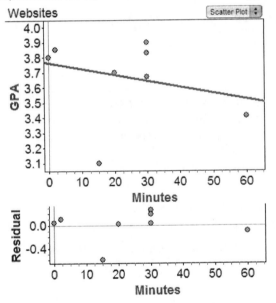

Students may have concerns for that point (14, 3.1) You may want to change the window for the y-axis if possible

Assumption 1: The mean of e for each and every value of x is equal to zero. This assumption is reasonable when the pattern in the scatterplot is linear and there is no obvious pattern in the residual plot, which appears to be the case here.

Assumption 2: The standard deviation of the errors is the same for each and every value of x. The scatter about the regression line does not appear to be much larger for some x values than for others, so this assumption seems reasonable.

Assumption 3: The distribution of e is normal for any particular value of x. For this assumption to be reasonable, points should tend to cluster near the line, with fewer points as you move away from the line. This appears to be the case here, so this assumption seems reasonable.

Assumption 4: Because the sample was a random sample, the observations are independent.

CALCULATIONS:

Model of Websites	Simple Regression ⬍
Response attribute (numeric): GPA	
Predictor attribute (numeric): Minutes	

```
Sample count:      9
Equation:          GPA = -0.00397999 Minutes + 3.7560
r:                 -0.299352
r-squared:         0.089612
Slope:             -0.00397999 +/- 0.0113378
SE Slope:          0.00479474
```

For a confidence level of 95% and df = 9 – 2 = 7, the appropriate t critical value is 2.365. From the given output, b = 0.00398 and $s_b = 0.00479$. The interval is then

estimate $\pm$ t^* (SE$_b$)

–0.00398 $\pm$ (2.365)(0.00479)

(–0.0153, 0.0074)

CONCLUSION We are 95% confident that the true mean change in GPA for each additional minute spent on social networking sites each day is between –0.0153 and 0.0074.

EXAMPLE Airlines are interested in knowing if there is a relationship between number of times people travel in a year and the weight of their luggage. A random sample of adult Americans who took at least one trip in the last year was used to produce the computer output below. Suppose it is reasonable to believe that the assumptions for inference are met. Does the sample provide evidence of a relationship between number of trips and luggage weight? Use the Minitab printout below.

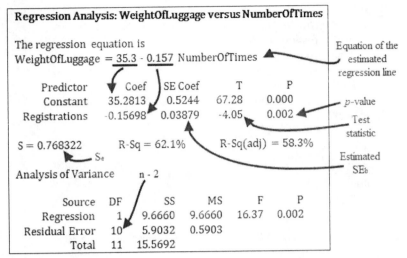

We are interested in knowing about β (the true mean change in baggage weight per additional trip). The question asked for a decision in the form of a yes or no answer—is there enough evidence to show that there is a relationship between number of times a person travels in a year and the weight of their luggage? To answer this question, we will perform a hypothesis test about the slope of the regression line.

NAME OF TEST Linear regression t test (two tailed)

HYPOTHESES

$$H_0 : \beta = 0$$
$$H_a : \beta \neq 0$$

β = the true mean change in baggage weight per additional trip
$\alpha = 0.05$

CONDITIONS The problem states that the conditions for inference are met.

CALCULATIONS From the given computer output, $t = -4.05$, P-value = 0.002, and df = 10.

CONCLUSION If the null hypothesis were true, we would get this result or one more extreme than this less than 0.2% of the time. Since this P-value is less than $\alpha = 0.05$, we reject the null hypothesis. We have sufficient evidence to conclude that there is a linear relationship between the number of times people travel and the weight of their luggage.

AP Tip

Be sure to know how to read a generic computer output. Know where you can find the statistics to answer the questions you might be asked. Some of the information on these outputs is extraneous and you will not need this information nor are you expected to know what it means. Be sure to practice finding the statistics you will need to know!

SAMPLE PROBLEM 2 Each year, most students take a test that measures their reading achievement. The score reports this in terms of the grade level at which the child is reading. The following scores are for a random sample of six children in a particular U.S. public school. Suppose it is reasonable to believe that the assumptions for inference are met. Is there convincing evidence that mean grade level score increases by more than 1 with each additional year of school?

Grade Level	1	3	4	7	9	12
Grade Level Equivalence	0.83	3.2	4.1	6.7	9.1	12.2

SOLUTION TO PROBLEM 2

Name of Test: t test (one tailed) for slope of regression line

Hypotheses:

$$H_0 : \beta = 1$$
$$H_a : \beta > 1$$

β = the true mean change in grade level score per additional year of school

$$\alpha = 0.05$$

Conditions: The problem states that the conditions for inference are met.

Calculations: Computer regression output is shown here, or a graphing calculator can be used.

```
The regression equation is
Grade Level Reading Score = - 0.068 + 1.01 Grade in School

Predictor            Coef   SE Coef      T      P
Constant          -0.0676   0.1714   -0.39  0.713
Grade in School   1.01488   0.02424   41.86  0.000


S = 0.222195    R-Sq = 99.8%    R-Sq(adj) = 99.7%
```

Remember that the hypothesized value of the slope is 1. From the computer output, we can compute the value of the test statistic and then find the associated P-value (the values of the test statistic and the P-value aren't the ones shown in the computer output; the default computer output is for a null hypothesis of $\beta = 0$.

$$t = \frac{b-1}{s_b} = \frac{.015-1}{0.024} = 0.625$$

$$df = 4$$

$$P - \text{value} = 0.283$$

CONCLUSION If it were true that we would see a result like this or one more extreme than this about 28.3% of the time. Since this p-value is larger than $\alpha = 0.05$, we fail to reject the null hypothesis. We do not have evidence that the mean increase in grade level score for each additional year of school is greater than 1.

INTERPRETATION OF RESULTS OF HYPOTHESIS TESTING

(Statistics: Learning From Data, 1st ed. pages 748–759)

Linear regression is often a useful way to summarize bivariate data. Researchers use linear regression hypothesis tests to make inferences about the way the two variables are related. You should ask yourself the following questions when you evaluate research that involves relating two variables.

1. Which variable is the response variable? Is it quantitative?

2. If the research uses a sample to make inferences about the population, is it reasonable to assume the conditions were met?

3. Does the model appear to be useful? Are the results of the hypothesis test given? What is the p-value of the test?

4. Has the linear model been used in an appropriate way? Has the research avoided extrapolation?

   Keep in mind the following limitations of linear regression inference:

   ▪ Small samples can show a weak linear pattern due to chance not due to a relationship in the population parameters.

   ▪ As with all inference, linear regression inference is not appropriate if the sample is not a random sample.

   ▪ As with all inference, linear regression inference procedures are not appropriate if the necessary conditions are not met.

INFERENCE FOR LINEAR REGRESSION AND CORRELATION: STUDENT OBJECTIVES FOR THE AP EXAM

   ▪ You should be able to find a point estimate of the slope of the population regression line.
   ▪ You should be able to construct a confidence interval to estimate the slope of the population regression line.
   ▪ You should be able to perform a hypothesis test for the slope of a population regression line.

MULTIPLE-CHOICE QUESTIONS

Questions 1–7 refer to the following information:

Data on x = number of powerboat registrations in Florida and y = number of manatee deaths in Florida for 31 randomly selected years was used to produce the following computer output. You can assume that all conditions needed for inference are met.

```
Regression Analysis: Manatees versus Registrations


    Predictor       Coef     SE Coef        T         P
     Constant     -40.641       5.840     -6.96     0.000
Registrations    0.125006    0.007867     15.89     0.000


S = 7.87985        R-Sq = 89.7%        R-Sq(adj) = 89.3%


Analysis of Variance


        Source   DF      SS      MS       F       P
    Regression    1   15677   15677  252.48   0.000
Residual Error   29    1801      62
         Total   30   17477
```

1. The estimate for the population regression line is
   (A) predicted manatees = –40.641 + 5.84(registrations)
   (B) predicted manatees = 0.125 + 0.007867(registrations)
   (C) predicted manatees = 5.840 + 0.007867(registrations)
   (D) predicted manatees = –40.641 + 0.125006(registrations)
   (E) predicted manatees = 7.87985 + 0.125006(registrations)

2. The standard error of the slope is shown to be 0.0079. Interpret this value in context.
   (A) The standard deviation of the number of powerboat registrations is 0.0079.
   (B) The standard deviation of the number of manatees killed is 0.0079.
   (C) If many samples were taken and the slope of each least squares regression line was recorded, the estimated standard deviation of these slopes was 0.0079.
   (D) If many samples were taken and the slopes of each least squares regression line were recorded, the difference in the estimated change in number of manatees killed per powerboat registration and the true change in number of manatees killed per powerboat registration is on average 0.0079.
   (E) The distance between the true change in number of manatees killed per powerboat registration and the sample change in number of manatees killed per powerboat registration is 0.0079.

3. Which of the following is a confidence interval for the mean change in number of manatee deaths associated with an increase of 1 powerboat registration?
   (A) $-40.641 \pm t^*(5.84)$
   (B) $0.125 \pm t^*(0.0079)$
   (C) $0.125 \pm t^*(15.89)$
   (D) $0.125 \pm t^*(7.88)$
   (E) $0.125 \pm t^*(5.84)$

4. The value of the t test statistic for testing the null hypothesis that the slope of the population regression line is equal to 0 is
   (A) 7.87985
   (B) -6.96
   (C) 15.89
   (D) 252.48
   (E) 89.3

5. What is the value of degrees of freedom that would be used in determining the P-value is a hypothesis test of $H_0 : \beta = 0$?
   (A) 1
   (B) 28
   (C) 29
   (D) 30
   (E) 32

6. In a test of $H_0 : \beta = 0$ versus $H_a : \beta \neq 0$ with a significance level of 0.05, the decision would be to
   (A) accept H_0
   (B) fail to reject H_a
   (C) fail to reject H_0
   (D) reject H_0
   (E) reject H_a

7. The estimated standard deviation of the residuals is
   (A) 5.84
   (B) 7.87985
   (C) 0.0079
   (D) 1801
   (E) 89.7

Questions 8–15 refer to the following information:

Each person in a random sample of adults was asked to estimate the average number of minutes of television watched per day and their yearly salary. The following data resulted.

Minutes of TV Watched	36	57	60	62	70	76	101
Annual Salary (in thousands)	89	45	30	50	55	67	30

This data was used to produce the following regression output.

Regression Analysis: Salary versus TV

```
The regression equation is
Salary = 94.4 - 0.638 TV


Predictor      Coef   SE Coef      T      P
Constant      94.40     25.60   3.69  0.014
TV          -0.6382    0.3736  -1.71  0.148


S = 18.2023    R-Sq = 36.9%    R-Sq(adj) = 24.2%
```

The researcher wishes to determine if these data provide evidence of a linear relationship between number of minutes spent watching television each day and annual salary. You can assume that all conditions required for inference are met.

8. The appropriate hypotheses are
    (A) $H_0 : \beta = 0$; $H_a : \beta \neq 0$
    (B) $H_0 : b = 0$; $H_a : b \neq 0$
    (C) $H_0 : \beta \neq 0$; $H_a : \beta = 0$
    (D) $H_0 : \beta = 0$; $H_a : \beta < 0$
    (E) $H_0 : \beta = 0$; $H_a : \beta > 0$

9. What is the value of the correlation coefficient for these data?
    (A) –0.369
    (B) –0.607
    (C) 0.148
    (D) 0.369
    (E) 0.607

10. What proportion of variability in salary can be explained by the linear relationship between TV watched and salary?
    (A) 0
    (B) 0.148
    (C) 0.369
    (D) 0.374
    (E) 18.202

11. The p-value for the test of $H_0 : \beta = 0$ versus $H_0 : \beta \neq 0$ is 0.148. A correct interpretation of this value in context is
    (A) The probability that we committed a Type I error is 0.148.
    (B) The probability that we committed a Type I or Type II error is 0.148.
    (C) If we were to take many samples of people and ask their TV viewing habits and their salary, 14.8% of the tests would yield different results.
    (D) If minutes watching TV and salary are not linearly related, we would expect to get samples like this one or more extreme 14.8% of the time just by chance.
    (E) If minutes watching TV and salary are associated, we would detect the relationship only 14.8% of the time.

12. Using a significance level of 0.05, the correct conclusion for the test of $H_0 : \beta = 0$ versus $H_0 : \beta \neq 0$ in context is
    (A) Since the p-value is greater than the significance level, we accept the null hypothesis that minutes watching TV and salary are linearly related.
    (B) Since the p-value is greater than the significance level, we reject the null hypothesis that minutes watching TV and salary are not linearly related.
    (C) Since the p-value is greater than the significance level, we fail to reject the null hypothesis that minutes watching TV and salary are not linearly related.
    (D) Since the p-value is greater than the significance level, we fail to reject the null hypothesis that minutes watching TV and salary are linearly related.
    (E) Since the p-value is greater than the significance level, we have evidence that minutes watching TV and salary are not linearly related.

13. Which of the following is a correct description of a Type II error in context?
    (A) We conclude that minutes watching TV and salary are not linearly related when, in fact, they are linearly related.
    (B) We conclude that minutes watching TV and salary are linearly related when, in fact, they are not linearly related.
    (C) We conclude that minutes watching TV and salary are not linearly related when, in fact, they are not linearly related.
    (D) We conclude that minutes watching TV and salary are linearly related when, in fact, they are linearly related.
    (E) A Type II error happens 2% of the time when we run similar tests on similar samples.

14. Which of the following is not a necessary condition for inference about the slope of the population regression line?
    (A) The observations must be independent of one another.
    (B) The residuals for every value of x must have a mean of zero.
    (C) The residuals must be normally distributed at each x value.
    (D) The residuals have the same standard deviation for each value of x.
    (E) The sample size must be large.

15. Which of the following is true of all significance tests for the slope of a population regression line?
    (A) The alternative hypothesis is that there is a linear relationship between x and y.
    (B) The degrees of freedom for this test is one less than the sample size.
    (C) We always assume the four necessary assumptions are true no matter what the residual plot displays.
    (D) If the slope of the sample regression line is small, then we know we will have a small p-value.
    (E) The null hypothesis assumes no linear relationship between x and y.

FREE-RESPONSE PROBLEMS

1. Each person in a random sample of 42 students at a large university was asked how much they paid per month for housing and how far from campus they lived (in miles). The resulting data were used to produce the following regression output. You can assume that any conditions needed for inference are met.

```
The regression equation is
Housing cost = 452 - 9.25 Distance

Predictor      Coef   SE Coef      T      P
Constant      452.1     178.1   2.54  0.039
Distance    -9.2472    0.2145 -43.11  0.000

S = 221.098   R-Sq = 99.6%   R-Sq(adj) = 99.6%
```

   (a) Estimate the slope of the population regression line using a 90% confidence interval.
   (b) Interpret this interval in context.
   (c) A recent news report stated that students pay a premium to live near campus. Evaluate this statement based on the interval from part (a).

2. Each person in a random sample of adult males with children was asked his height and the height of his first full grown child, resulting in the following data.

Height of Father (inches)	68	68	73	72	74	71	73	73	71	70
Height of Child (inches)	61	65	65	73	61	64	65	70	66	63

   Is there convincing evidence of a linear relationship between the height of fathers and the height of their child?

Answers

MULTIPLE-CHOICE QUESTIONS

1. **D.** On the computer output, the Coef value for Registrations is the slope, and the Coef value for Constant is the y-intercept (*Statistics: Learning From Data,* 1st ed. pages 205–206 and 752–756).

2. **C.** The standard error is associated with repeated samples (*Statistics: Learning From Data,* 1st ed. pages 752–756).

3. **B.** The mean change refers to the slope of the line (*Statistics: Learning From Data,* 1st ed. pages 749 and 752–756).

4. **C.** This test statistic is for testing the slope (*Statistics: Learning From Data,* 1st ed. pages 752–756).

5. **C.** df = n-2 for regression (*Statistics: Learning From Data,* 1st ed. pages 752–756).

6. **D.** Since the P-value = 0.000, we reject Ho (*Statistics: Learning From Data,* 1st ed. pages 752–756).

7. **B.** The standard deviation of the residuals appears as S in the computer output (*Statistics: Learning From Data,* 1st ed. pages 752–756).

8. **A.** Hypothesis test parameters. We have no reason to believe that any relationship that may exist is either positive or negative (*Statistics: Learning From Data,* 1st ed. pages 752–756).

9. **B.** . Since the slope is negative, r will also be negative (*Statistics: Learning From Data,* 1st ed. pages 185–187 and 752–756).

10. **C.** The coefficient of determination, r^2, is the proportion of variability that can be explained by the linear regression line (*Statistics: Learning From Data,* 1st ed. pages 217–220, 743 and 752–756).

11. **D.** This result refers to repeated sampling (*Statistics: Learning From Data,* 1st ed. pages 752–756).

12. **C.** We do not have sufficient evidence to reject the null hypotheses (*Statistics: Learning From Data,* 1st ed. pages 752–756).

13. **A.** We fail to reject Ho when Ho is false (*Statistics: Learning From Data,* 1st ed. pages 502–505).

14. **E.** The distribution of errors must be approximately normal for every *x* (*Statistics: Learning From Data,* 1st ed. pages 752–756).

15. E. We are testing Ho: β = 0 (*Statistics: Learning From Data,* 1st ed. pages 752–756).

FREE-RESPONSE QUESTIONS

1. (a) The question states that the conditions for inference are met, so it is OK to use a *t* confidence interval to estimate the slope. With *n* = 42, df = 40 and the *t* critical value for a 90% confidence interval is 1.68. The confidence interval is then

    $-9.2472 \pm (1.68)(0.2145)$

    -9.2472 ± 0.3604

    $(-9.6076, -8.8868)$

    (b) We can be confident that, on average, housing cost decreases by somewhere between $8.89 and $9.61 with each additional mile from campus.

    (c) Housing costs do tend to decrease as you move farther away from campus. A student who chose to live 5 miles from campus might expect to pay about $45 less per month than someone who lives adjacent to the campus, and a student who lives 10 miles from campus would expect to pay about $90 less per month.

    (*Statistics: Learning From Data,* 1st ed. pages 205–206 *and 749*).

2. **Name of Test:** *t* test for slope of regression line

    Hypotheses:

    $H_0 : \beta = 0$

    $H_a : \beta \neq 0$

    Conditions:

    First, look at a scatterplot and a residual plot of the sample data (shown below). A linear model seems appropriate for this data.

    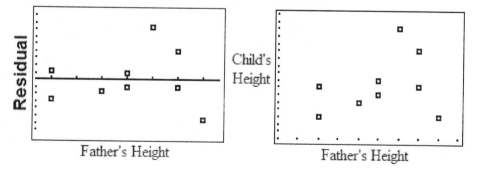

    Assumption 1: The mean of e for each and every value of *x* is equal to zero. This assumption is reasonable when the pattern in the scatterplot is linear and there is no obvious pattern in the residual plot, which appears to be the case here.

    Assumption 2: The standard deviation of the errors is the same for each and every value of *x*. The scatter about the regression line does not appear to be much larger for some *x* values than for others, so this assumption seems reasonable.

Assumption 3: The distribution of *e* is normal for any particular value of *x*. For this assumption to be reasonable, points should tend to cluster near the line, with fewer points as you move away from the line. This appears to be the case here, so this assumption seems reasonable.

Assumption 4: Because the sample was a random sample, the observations are independent.

Calculations:

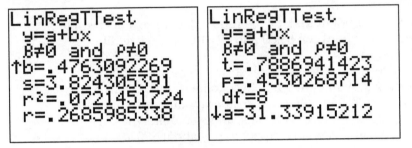

$T = 0.79$ with 8 degrees of freedom, *P*-value = 0.453

Conclusion:

Since the *P*-value of 0.453 is greater than any reasonable alpha level, we fail to reject the null hypothesis. There is not convincing evidence that the father's height and the child's height are linearly related.

(*Statistics: Learning From Data,* 1st ed. pages 752–756).

Part III

Practice Tests

PRACTICE TEST 1

This test will give you some indication of how you might score on the AP Statistics Exam. Of course, the exam changes every year, so it is never possible to predict a student's score with certainty. This test will also pinpoint your strengths and weaknesses on the key content areas covered by the exam.

AP STATISTICS EXAMINATION
Section I: Multiple-Choice Questions
Time: 90 minutes
40 Questions

Directions: Each of the following questions or incomplete statements is accompanied by five suggested answers or completions. Select the one that best answers the question or completes the statement.

1. In a *USA TODAY*/Gallup poll taken shortly after the game, 64% of respondents who described themselves as baseball fans said Major League Baseball should overturn an umpire's safe call that cost pitcher Armando Gallaraga a perfect game. The poll, based on 470 respondents, has a margin of error of 6%. Which of the following statements best describes what is meant by the 6% margin of error?
(A) About 6% of the baseball fans agreed with the umpire's call.
(B) About 6% of those polled were not actually baseball fans.
(C) The difference between the sample percentage and the population percentage is likely to be less than or equal to 6%.
(D) About 6% of the sample should not be included in the population.
(E) The difference between the percentage of people in favor of overturning the call in this survey and in a second similar survey would be less than 6%.

2. A university has 12,000 female students and 8,000 male students. Fifty percent of the women use the student gym and 40% of the men use the student gym. A simple random sample of 100 students is selected. What is the expected number of students in the sample who use the gym?
(A) 47
(B) 46
(C) 45
(D) 44
(E) 43

3. A manufacturer of batteries claims they have a shelf life of one year. The manufacturer contends that 99.5% of the batteries will function after sitting without use for one year. If the company's claim is true, what is the expected number of batteries in a random sample of 5,000 one-year-old batteries that will function?
(A) 4750
(B) 4,925
(C) 4,950
(D) 4,975
(E) 4,995

4. A fertilizer company is trying to convince corn farmers using a particular fertilizer that their new product CORNPLOSION will lead to an increase in yield over the currently used fertilizer. To determine which fertilizer results in a higher yield of corn, 10 one-acre fields are split into halves. The current fertilizer is used on one-half of each field and CORNPLOSION on the other half. The fertilizer that goes on the east half of the field is determined by coin flip. The yield of corn is measured in bushels per acre. The data are given below:

Field	Generic Fertilizer	CORNPLOSION
1	15	21
2	12	16
3	8	7
4	22	30
5	14	12
6	25	33
7	11	14
8	12	21
9	20	22
10	18	16

What is the number of degrees of freedom associated with the appropriate *t* test for testing to see if there is a difference between the mean corn yield for the two types of fertilizer?

(A) 9
(B) 10
(C) 13
(D) 18
(E) 20

5. A certain library has 10,000 books. Two thousand of these books are nonfiction books and the rest are fiction. In order to estimate the total number of pages in their nonfiction books, two plans are proposed.

Plan I:
(a) Sample 50 books at random
(b) Estimate the mean number of pages per book using a confidence interval.
(c) Multiply both ends of the interval by 10,000 to get an interval estimate of the total.

Plan II:
(a) Identify the 2,000 nonfiction books.
(b) Sample 50 nonfiction books at random.
(c) Estimate the mean number of pages for nonfiction books using a confidence interval.
(d) Multiply both ends of the interval by 2,000 to get an interval estimate of the total.

On the basis of the information given, which of the following is the better method for estimating the total number of pages in nonfiction books in the library?

(A) Choose plan I over plan II.
(B) Choose plan II over plan I.
(C) Choose either plan, since both are good and will produce equivalent results.
(D) Choose neither plan, since neither estimates the total number of nonfiction pages.
(E) The plans cannot be evaluated from the information given.

GO ON TO NEXT PAGE

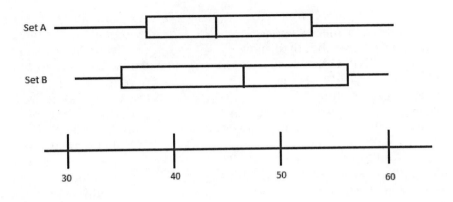

6. To check the effect of wet road conditions on automobile stopping distances, five economy-size cars from automaker A and five economy-size cars from automaker B are tested. The same driver is used to test all cars. The driver accelerates to 50 miles per hour and then brings the vehicle to a complete stop when it reaches a white line. The cars from automaker A are tested on a dry straightaway. The cars from automaker B are tested on the same track, but the track is hosed down with water before testing. The distance between the white line and where the car comes to a stop is measured for each car and the mean for the cars stopping on the dry road is compared to the mean for the cars stopping on the wet road. Is this a good experimental design?
(A) No, because the means are not proper statistics for comparison.
(B) No, because more than two brands should be used.
(C) No, because road conditions are confounded with brand.
(D) No, because the experiment should not be limited to economy-sized cars.
(E) Yes

7. The boxplots above summarize two data sets, A and B. Which of the following must be true?
   I. Set B has a higher median than set A.
   II. Set B contains more data than set A.
   III. The data in set B have a larger range than the data in set A.
(A) I only
(B) III only
(C) I and II only
(D) I and III only
(E) I, II, and III

8. Splatastic Paintball Range sells paintballs in bulk at wholesale prices, as well as individually at retail prices. Next year's sales will depend on market conditions, but company executives believe that the following probability distributions describe wholesale and retail sales for next year.

WHOLESALE SALES

Number Sold	50,000	100,000	150,000	200,000
Probability	0.2	0.3	0.4	0.1

RETAIL SALES

Number Sold	10,000	20,000	30,000
Probability	0.2	0.5	0.3

What is the expected profit for the next year if the profit from each paintball sold is $1 for wholesale and $2 for retail?
(A) $141,000
(B) $144,000
(C) $162,000
(D) $282,000
(E) $341,000

9. D'Andre and Claire are both going to a Chinese buffet. Based on their previous visits to the buffet, the probability distributions of the number of plates of food they will eat are given below.

Number of plates D'Andre will eat	1	2	3
Probability	0.1	0.2	0.7

Number of plates Claire will eat	1	2	3
Probability	0.4	0.5	0.1

Assuming D'Andre and Claire make their decisions independently, what is the probability that they will eat six plates of food combined?
(A) 0.04
(B) 0.07
(C) 0.10
(D) 0.70
(E) 0.80

10. A survey of 84 randomly selected military officers was conducted to determine whether or not they participated in high school athletics. The two-way table shows the numbers of officers by varsity status (varsity athlete, not varsity athlete) and branch of service (Navy, Air Force). Which of the following best describes the relationship between varsity status and branch of service?

	Varsity Athlete	Not Varsity Athlete	Total
Navy	32	16	48
Air Force	24	12	36
Total	56	28	84

(A) There appears to be an association, since more Navy officers than Air Force officers were varsity athletes.
(B) There appears to be an association, since twice as many officers were varsity athletes.
(C) There appears to be no association, since the proportion of officers that were varsity athletes was the same for both Navy and Air Force.
(D) There appears to be no association, since there are fewer Air Force officers who were not varsity athletes but more Navy officers who were varsity athletes.
(E) There is not enough information to decide whether or not there is evidence of an association.

GO ON TO NEXT PAGE

11. Which is the best estimate of the standard deviation of the normal distribution shown below?

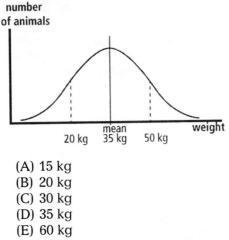

number of animals

mean
20 kg   35 kg   50 kg   weight

(A) 15 kg
(B) 20 kg
(C) 30 kg
(D) 35 kg
(E) 60 kg

12. A window manufacturer claims that its new seal-tight system will save homeowners in cold climates $100 a month on average during the winter. A consumer group is skeptical of this claim and thinks that the manufacturer may be overstating the savings. If μ represents the true mean savings for this new window system, which of the following gives the null and alternative hypotheses that the consumer group should test?
(A) H_0: $\mu \leq \$100$
     H_a: $\mu > \$100$
(B) H_0: $\mu = \$100$
     H_a: $\mu \neq \$100$
(C) H_0: $\mu < \$100$
     H_a: $\mu \geq \$100$
(D) H_0: $\mu = \$100$
     H_a: $\mu < \$100$
(E) H_0: $\mu = \$100$
     H_a: $\mu > \$100$

13. Stanley's boss told him that he did not deserve a raise and that Stanley was making an amount similar to other accountants with his experience. Stanley gathered some data and found that the z-score for his salary as compared to other accountants with the same level of experience is –1.25. Which of the following is the best interpretation of this z-score?
(A) Stanley makes 1.25% less than other accountants with his experience level.
(B) Stanley makes 1.25% more than other accountants with his experience level.
(C) Stanley makes $1.25 less per pay period than the average for accountants with his experience level.
(D) Stanley's salary is 1.25 standard deviations above the mean salary of other accountants with his experience level.
(E) Stanley's salary is 1.25 standard deviations below the mean salary of other accountants with his experience level.

14. Uncle Hugo's Electronics decided to see whether a local music CD section would be profitable at its 1,000 stores nationwide. It decided to set up local music displays at its 100 biggest locations. In this study, Uncle Hugo's marketing department found that sales of local artists were significantly greater in stores with the local music display. Which of the following statements explains why it would *not* be appropriate to recommend that all of the stores should add a local music section?
    I. The study only targeted Uncle Hugo's largest locations and there may not be as much good local music available in smaller locations.
    II. Only 100 stores had a local display while 900 stores did not have a local display.
    III. The sample consisted of more than 5% of the population.
    (A) I only
    (B) II only
    (C) III only
    (D) I and II only
    (E) I, II, and III

Questions 15–16 refer to the following information:

Every Wednesday, Burgerville has a "discount deck day." A customer who comes in with a child may choose to pick two cards from a standard deck of playing cards. If the two cards are the same number, then the child's junior burger is free with the purchase of a megaburger. If the two cards are the same suit, then the child's junior burger is half price with the purchase of a megaburger. The menu price for a junior burger is $2.00 and a megaburger is $4.00. Let X represent the amount paid by a customer with a child for a junior burger and a megaburger. The expected value of X is 5.65 and the standard deviation is 0.59.

15. If a customer with a child picks from the discount deck and purchases both a junior burger and a megaburger every Wednesday for 12 consecutive weeks, what is the total amount that the customer would expect to pay for all of the burgers?
    (A) $5.65
    (B) $56.50
    (C) $65.50
    (D) $67.80
    (E) $70.00

16. If a customer with a child picks from the discount deck every Wednesday and buys both a junior burger and a megaburger for 5 consecutive weeks, what is the standard deviation of the total amount paid for all of the burgers?
    (A) $0.26
    (B) $0.35
    (C) $1.32
    (D) $1.74
    (E) $2.95

17. Which of the following is not a property of all normal distributions?
    I. The distribution is symmetric.
    II. The distribution is centered at 0.
    III. The median and the mean are equal.
    (A) I only
    (B) II only
    (C) III only
    (D) I and II only
    (E) I, II, and III

GO ON TO NEXT PAGE

18. Before going to college, Chelsea wanted to see if she could sell her old car online. She selected a sample of the prices on 25 similar cars sold online in the past month. Later, Chelsea realized that the lowest-priced car in the sample was mistakenly recorded as half the price it actually sold for. However, after correcting the error, the corrected price was still less than or equal to any other car price in her sample. Which of the following sample statistics must have remained the same after the correction was made?
    (A) Mean
    (B) Median
    (C) Mode
    (D) Range
    (E) Variance

19. The back-to-back stem-and-leaf plot below gives the growth in millimeters of 23 onion bulbs planted by students in Mrs. Shelton's third grade class.

```
  After One Week |   | After Two Weeks
 _____|_|_____
                6 |0|
 9 8 8 7 6 6 6 5 4 4 2 1 1 0 |1| 7
        8 6 5 5 3 3 2 |2| 5 6 6 6 7 9
                    4 |3| 0 0 0 0 1 2 3 3 5 6 9
                      |4| 2 2 3 6 8
                      |5|
```

Which of the following statements is *not* justified by the data?
    (A) The plants shown grew on average less than 25 millimeters per week.
    (B) The mean size of the plants increased from week one to week two.
    (C) The spread between the shortest and tallest plant did not change much between the two weeks.
    (D) The size of every plant increased from week one to week two.
    (E) The median size of the plants increased from week one to week two.

20. A carnival game requires you to make three baskets in a row to win a prize. Andre is a basketball player who normally makes 80% of his shots, but because of the trick rim on the carnival game basket he can only make the shot 60% of the time. Assuming that successive shots are independent, what is the probability that Andre wins a prize if he decides to play the game once?
    (A) 0.2
    (B) 0.216
    (C) 0.4
    (D) 0.512
    (E) 0.6

21. Laws have been proposed in some border states that would allow police officers to ask motorists to prove that they are U.S. citizens during routine traffic stops. If it is determined that these motorists are not in the United States legally, then they will be deported. Thinking of this in terms of a hypothesis test with H_0 is that the motorist is in the United States legally and H_a is that the motorist is not in the United States legally. Which of the following is an example of a Type I error?
    (A) A motorist is in the country legally but is thought to not be in the country legally.
    (B) A motorist is not in the country legally but is thought to be in the country legally.
    (C) A motorist is not in the country legally but when stopped gets deported.
    (D) A motorist is not in the country legally but never gets stopped by the police.
    (E) A motorist is in the country legally and is not stopped by the police.

22. The number of people in a gym at noon on each of 60 randomly selected days produces a mean of 34.7 people and a standard deviation of 8.1 people. Which of the following is an approximate 90% confidence interval for the mean number of people in the gym at noon?
(A) (26.70, 42.70)
(B) (32.95, 36.45)
(C) (32.65, 36.75)
(D) (33.65, 35.75)
(E) (34.48, 34.92)

23. Mr. Branch and his son Dennis both played on their college basketball teams. Mr. Branch scored 12.5 points per game at a time when the average number of points scored by a player in college was 11 points a game and the standard deviation was 3.8 points. Dennis scored 15.8 points a game when the average number of points scored by a player in college was 14 points a game and the standard deviation was 6.1. Mr. Branch contends that while he's proud of his son, he was a better scorer compared to the players of his day than his son was compared to his contemporaries. Who is correct in saying that he did better when compared to his peers?
(A) Mr. Branch is correct.
(B) Dennis is correct.
(C) Neither is correct, they both did equally well.
(D) There is no basis for comparison, since they played in different eras for different teams.
(E) There is not enough information for comparison, because the number of players in each era is not known.

24. A die is rolled six times, and the number of times a 3 is rolled is counted. This procedure of six dice rolls is repeated 100 times and the results are summarized in a frequency distribution. Which of the frequency distributions below is most likely to contain the results from these 100 trials?

(A)

Number of 3s	Frequency
0	33
1	40
2	16
3	8
4	2
5	1
6	0

(B)

Number of 3s	Frequency
0	5
1	10
2	20
3	30
4	20
5	10
6	5

(C)

Number of 3s	Frequency
0	0
1	100
2	0
3	0
4	0
5	0
6	0

(D)

Number of 3s	Frequency
0	50
1	25
2	12
3	6
4	4
5	2
6	1

(E)

Number of 3s	Frequency
0	14
1	14
2	15
3	15
4	14
5	14
6	14

GO ON TO NEXT PAGE

25. An art gallery wants to conduct a survey to see what type of offerings people are most interested in. It wants to begin with a simple random sample of 50 people who have previously purchased art at the gallery. Which of the following methods will produce a simple random sample?
(A) Survey the 50 most recent purchasers.
(B) Survey the 50 purchasers who have spent the most money at the gallery.
(C) Use random numbers to choose 10 people from each of the following five groups: those who have purchased (1) watercolors, (2) sculptures, (3) canvas, (4) photographs, (5) mixed media.
(D) Hold a brunch for previous customers and put comment cards at 50 randomly assigned seats.
(E) Number a list of all people who have previously made a purchase. Use a table of random numbers to choose 50 people from this list and survey them.

26. There is a linear relationship between the amount of fat in a sandwich and the amount of calories in a sandwich. A least-squares line was fit using some data collected by a nutritionist, resulting in

$$\hat{y} = 217.3 + 35.2x$$

where x is the grams of fat in the sandwich and y is the estimated caloric content of the sandwich. What is the estimated increase in calories that corresponds to an increase of 8 grams of fat?
(A) 35.2
(B) 217.3
(C) 281.6
(D) 498.9
(E) 1738.4

27. In a test of the null hypothesis $H_0: \mu = 23$ against the alternative hypothesis $H_a: \mu > 23$, a random sample from a normal population with a known standard deviation produces a mean of 28.9. The z statistic for the test is 2.48 and the P-value is 0.0066. Based on these statistics, which of the following conclusions could be drawn?
(A) There is convincing evidence that $\mu > 23$.
(B) Due to random fluctuation, 49.34% of the time a sample produces a mean larger than 23.
(C) 0.66% of the time, rejecting the alternative hypothesis is an error.
(D) 0.66% of the time, the mean is above 23.
(E) 99.34% of the time, the mean is below 23.

28. The weights of watermelons are approximately normally distributed with a mean of 90 ounces and a standard deviation of 15 ounces. If a watermelon is in the 96th percentile for weight, then its weight, in ounces, is closest to
(A) 26
(B) 105
(C) 116
(D) 126
(E) 262

29. Ten golf carts were selected at random from a country club to be outfitted with an engine modification designed to increase their maximum rate of speed. The time needed to complete 18 holes was recorded for a group using the upgraded golf carts and a group using golf carts that had not been modified. The difference in means for the time needed to complete 18 holes was 24 minutes. With μ_1 denoting the mean time for unmodified carts and μ_2 denoting the mean time for modified carts, a 95% confidence interval estimate of the true difference in mean time, $\mu_1 - \mu_2$, is (16, 32). Which of the following statements is a correct decision based on this confidence interval?
(A) We can be confident that μ_1 is around 16 and μ_2 is around 32.
(B) The country club should not change to modified carts because we can be confident that μ_1 is less than μ_2.
(C) The country club should change to modified carts because we can be confident that μ_1 is greater than μ_2.
(D) The country club should not change to modified carts because we can be confident that there is no difference between μ_1 and μ_2.
(E) The country club should convert the modified carts back to the original engines because we are not confident that the modified carts are better.

30. The equation of the least-squares regression line for a set of data is $\hat{y} = 0.68 + 1.21x$. What is the residual for the point (3, 4)?
(A) –0.31
(B) –0.68
(C) –1.52
(D) –3.63
(E) –4.31

31. A random variable X has a normal distribution centered at 100. About 10% of the time, X takes on a value that is less than 80. What is the approximate value of the standard deviation of X?
(A) 10.0
(B) 12.2
(C) 15.6
(D) 20.0
(E) 40.0

32. Which of the following represents a distribution where you would expect that the mean is greater than the median?

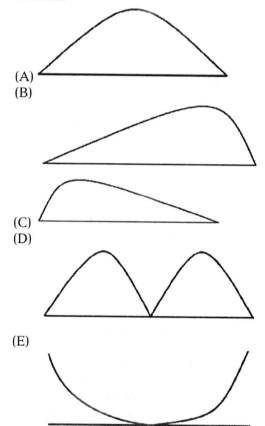

(A)
(B)

(C)
(D)

(E)

GO ON TO NEXT PAGE

33. The process of packaging candy corn yields bags with varying numbers of candy pieces. The candy manufacturer claims that, on average, bags of candy corn contain 350 pieces. The Consumer's Corner magazine tests a random sample of 40 bags. The mean number of pieces of candy corn for this sample is 346.8, while the standard deviation is 12.4. Which of the following describes the approximate P-value for a test to determine if there is evidence that the mean number of candies per bag is less than the candy maker claims?
(A) P-value < 0.001
(B) $0.001 < P$-value < 0.01
(C) $0.01 < P$-value < 0.05
(D) $0.05 < P$-value < 0.10
(E) P-value > 0.10

34. A survey was conducted to estimate the proportion of California workers who would rather live in a different state. In a random sample of 100 California workers, 28% indicated that they would rather live in another state. A 95% confidence interval for the proportion of all Californian workers who would rather live in another state is:
(A) (0.192, 0.368)
(B) (0.207, 0.353)
(C) (0.230, 0.330)
(D) (0.234, 0.326)
(E) (0.255, 0.305)

35. Sally realizes that all of her professors are going to give an exam of some kind in the half week that leads up to Thanksgiving break. If Sally takes four classes, and each class is equally likely to place its exam on either the Monday, Tuesday, or Wednesday before Thanksgiving break, what is the probability that Sally has all four of her exams on one day?
(A) 0
(B) 0.012
(C) 0.037
(D) 0.250
(E) 0.333

36. A study was conducted by psychologists who wanted to determine at what age children develop their first crush. They asked college freshman to select from one of four age categories, and the results are summarized below:

	Age 0–8	Age 9–12	Age 13–16	Age 16+
Male	2	11	35	10
Female	7	19	23	4

According to the two-way table, what percentage of males had their first crush before the age of 13?
(A) 3.4
(B) 18.9
(C) 22.4
(D) 34.5
(E) 49.1

37. Which of the following is least likely to be a potential confounding variable in a study of the effect of diet on weight loss in pounds?
(A) Gender
(B) Beginning weight
(C) Time spent exercising
(D) Name
(E) Time spent watching TV

38. The periodic table provides an atomic mass for each element, but these figures do not prove accurate for certain isotopes of some elements. Because isotopes are likely to contain a different mix of neutrons and electrons, the weights of some isotopes can be significantly more or less than the weight of the general element. Such isotopes occur much more rarely than the general element. If you were to plot the weights of the element and its isotopes, the isotopes could be clearly identified as falling far away from the accepted mean. The best word to describe the isotopes as they relate to such a plot in statistics would be
(A) outlier.
(B) residual.
(C) deviation.
(D) quartile.
(E) minimum.

39. The 23 members of the math club decide to hold the "random awards" in which they use a random number table to award 1st, 2nd, and 3rd prizes. Each member is given a two-digit number from 01 to 23, then the prizes are drawn in reverse order with 3rd prize being the first member who appears in the random number table below when digits are read two at a time:

92646 90110 79365 04891 39174 39823

Which numbered members of the math club won the three prizes?
(A) 11, 04, 17
(B) 9, 2, 6
(C) 01, 10, 17
(D) 01, 11, 10
(E) 23, 17, 10

40. Sales of jelly (X) and sales of peanut butter (Y) at a local store are recorded weekly. The correlation between X and Y is 0.9. Which is the best explanation of what this correlation means?
(A) 90% of customers either buy peanut butter or jelly.
(B) 90% of the time when a customer buys peanut butter, they also buy the jelly.
(C) 90% of the variability in peanut butter sales can be explained by a linear relationship between jelly sales and peanut butter sales.
(D) 81% of the variability in peanut butter sales can be explained by a linear relationship between jelly sales and peanut butter sales.
(E) 90% of the time when the price of peanut butter goes up, so does the price of jelly.

STOP

END OF SECTION 1
IF YOU FINISH BEFORE TIME IS CALLED, YOU MAY CHECK YOUR WORK ON THIS SECTION. DO NOT GO ON TO SECTION II UNTIL YOU ARE TOLD TO DO SO.

Section II: Free-Response Questions
Part A
Questions 1–5
Spend about 65 minutes on this part of the exam
Percent of Section II grade—75

Directions: Show all your work. Clearly indicate the methods you use as you will be graded on the correctness of the methods as well as the accuracy of your results and explanation.

FREE-RESPONSE PROBLEMS

1. The table below provides the cumulative proportions for times in the Men's 200-meter freestyle Olympic swimming race. Cumulative proportions for the 2004 and 2008 Olympics are shown below.

Cumulative Proportions for Men's 200-Meter Freestyle Olympic Entry Times

Time	2004	2008	Time	2004	2008
1:43	0.125	0	1:48	0.875	0.625
1:44	0.125	0	1:49	1	0.750
1:45	0.25	0.125	1:50	1	0.875
1:46	0.5	0.25	1:51	1	1
1:47	0.875	0.375	1:52	1	1

The cumulative proportions are graphed below:

(a) Use the cumulative proportions graph to approximate the median finishing time for each year. Show supporting work on the graph.

(b) Use the cumulative proportion graph to approximate the interquartile range for each year.

(c) Using the results from parts (a) and (b), write a sentence or two for a newspaper column comparing the two distributions of times.

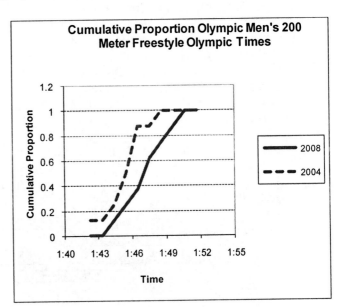

2. A cereal manufacturer wants to investigate if having a display at the end of an aisle in a grocery store would increase sales of a new cereal compared to having the new cereal shelved with the rest of the cereals that the store carries. Sixteen grocery stores have agreed to participate in the study, which will last two weeks. For each store, the cereal can be placed on a display at the end of an aisle or in with the rest of the cereals. In any particular store, different locations can be used for week 1 and for week 2 if desired, but the cereal can only be in one of the two locations in any given week.

    The marketing manager proposes that 8 of the 16 stores be selected at random. These 8 stores will display the cereal at the end of an aisle and the other 8 stores will display the cereal with the rest of the cereals that the store sells. At the end of the two-week period, mean sales for stores with aisle displays will be compared to mean sales for stores that did not have aisle displays.
    (a) Describe an alternate way to carry out the study that would improve on the study design proposed by the marketing manager.
    (b) Given the design you proposed in part (a) and assuming that any conditions for inference are met, what hypothesis test would be appropriate to determine if the data from the study supports the statement that mean sales are higher when an end-of-aisle display is used?

3. Mrs. Murdock gives her statistics students a mock exam before the real AP exam. In her class, all students are required to take both the mock exam and the AP Statistics exam. Ten percent of the students who pass the AP exam did not pass the mock exam and 7% of the students who do not pass the AP test passed the mock exam. Overall, 58% of her students pass the AP test.
    (a) What is the probability that a student selected at random from Mrs. Murdock's students passed the mock exam? Show your work.
    (b) What is the probability that a student selected at random from Mrs. Murdock's students who passed the mock exam also passes the AP exam?

4. Each person in a random sample of 375 parents of children who attend a large urban school with over 6000 students was interviewed at the start of a new school year. Of 375 parents, 280 said that they thought the principal was doing an excellent job. However, when she implemented a stricter grading policy a month into the school year, it was thought that the proportion of parents who thought the principal was doing an excellent job decreased. To investigate, a new random sample of 523 parents was selected, and 358 thought the principal was doing an excellent job. The data from these two samples will be used to test $H_0 : p_b - p_a = 0$ versus

    $H_a : p_b - p_a > 0$, where p_b is the proportion of all parents who thought the principal was doing an excellent job before the change and p_a is the proportion of all parents who thought the principal was doing an excellent job after the change.
    (a) Describe Type I and Type II errors in the context of this problem.
    (b) Do these data support the conclusion that there was a decrease in the proportion of parents who thought the principal was doing an excellent job after the new grading policy was implemented? Give appropriate statistical evidence to support your answer.

5. The percentage of high school teachers who have taken advanced course work after receiving a teaching credential was recorded for both science teachers and mathematics teachers for 12 randomly selected school districts in California. The resulting data is given in the table below.

School District	Percentage Math Teachers	Percentage Science Teachers
1	41	36
2	38	34
3	44	37
4	30	32
5	21	30
6	24	32
7	33	36
8	40	38
9	48	40
10	41	37
11	35	35
12	30	36

With μ_M denoting the mean percentage of math teachers with advanced course work for school districts in California and μ_S denoting the mean percentage of science teachers with advanced course work for school districts in California, a researcher used a two-sample t test to test $H_0 : \mu_M - \mu_S = 0$ versus $H_a : \mu_M - \mu_S > 0$. This resulted in a test statistic of $t = 0.07$ and a P-value of 0.474. Based on this test, the researcher concluded that there was no evidence that the mean percentage of teachers who have taken advanced course work for school districts in California is greater for math teachers than for science teachers.

(a) Explain why the two-sample t test is not an appropriate choice for testing these hypotheses.

(b) Carry out an appropriate test using a significance level of 0.05.

Section II
Part B
Question 6
Spend about 25 minutes on this part of the exam
Percent of Section II grade—25

Directions: Show all your work. Clearly indicate the methods you use as you will be graded on the correctness of the methods as well as the accuracy of your results and explanation.

6. A high school counselor at a large public high school is interested in developing a model to predict SAT math scores. He takes a random sample of 50 seniors at the school who took the SAT exam and recorded SAT math score, the number of math courses completed at the high school and the number of science courses completed at the high school for each of the students in the sample. Computer output resulting from fitting a linear regression model with

y = SAT math score (an integer between 200 and 800)
and
x = number of math courses completed is given below.

Regression Analysis: SAT math score versus math courses

```
The regression equation is
SAT math score = 505 + 47.4 math courses


Predictor       Coef  SE Coef     T      P
Constant      505.14    13.54  37.32  0.000
math courses  47.436     3.595  13.20  0.000


S = 29.6836    R-Sq = 78.4%    R-Sq(adj) = 77.9%
```

(a) Construct and interpret a 90% confidence interval for the slope of the population regression line. You may assume that the conditions for inference are met.
(b) Use the least squares regression line to predict the SAT math score of a student who has taken four math courses in high school.
(c) Interpret the value of r^2 in context.
(d) A multiple regression model is one that incorporates more than one predictor variable. The regression output below resulted from fitting a multiple regression model that incorporates both number of math courses and number of science courses as predictors. Use the regression equation to predict the SAT math score of a student who has taken four math courses and three science courses. How does this prediction compare to the prediction in part (b)?

GO ON TO NEXT PAGE

Regression Analysis: SAT Math Score versus Math Courses, Science Courses

```
The regression equation is
SAT math score = 515 + 47.7 math courses - 3.16 science courses


Predictor          Coef   SE Coef      T      P
Constant         514.70     18.00  28.59  0.000
math courses     47.682     3.621  13.17  0.000
science courses  -3.164     3.909  -0.81  0.422


S = 29.7907    R-Sq = 78.7%    R-Sq(adj) = 77.8%
```

(e) The values of s_e, R^2, and the standard deviations of the regression coefficients are interpreted in the same way as they are interpreted in linear regression. Do you think that using the multiple regression model that includes both number of math courses completed and number of science courses completed would result in predictions that are much more accurate than those produced by the linear regression model that only uses number of math courses completed? Explain your reasoning.

ANSWERS TO PRACTICE TEST 1

MULTIPLE-CHOICE ANSWERS

Using the table below, score your test.

Determine how many questions you answered correctly and how many you answered incorrectly. You will find explanations of the answers on the following pages.

1. C	9. B	17. B	25. E	33. D
2. B	10. C	18. B	26. C	34. A
3. D	11. A	19. D	27. A	35. C
4. A	12. D	20. B	28. C	36. C
5. B	13. E	21. A	29. C	37. D
6. C	14. A	22. B	30. A	38. A
7. A	15. D	23. A	31. C	39. C
8. C	16. C	24. A	32. C	40. D

ANSWERS AND EXPLANATIONS

MULTIPLE-CHOICE ANSWERS

1. **C.** The margin of error reflects sampling variability and describes how results from a sample might differ from the underlying population. A 6% margin of error indicates that we expect the population percentage to be between 58% and 70% (*Statistics: Learning from Data,* 1st ed. pages 443–444).

2. **B.** The first step is figuring out what portion of your sample is likely to be women and which portion is likely to be men. Since the campus has a 12,000/8,000 ratio, the ratio of our sample should be 60/40. Of the 60 women, 50% would likely use the gym, which is 30 students. Of the 40 men, 40% would likely use the gym, which is 16 students. 30 students plus 16 students is a total of 46 from the sample that we expect to use the gym (*Statistics: Learning from Data,* 1st ed. page 356–360).

3. **D.** Multiply the 5,000 batteries by the 0.995 functional percentage and the result is 4,975 batteries (*Statistics: Learning from Data,* 1st ed. pages 356–360).

4. **A.** Because the samples are paired, the degrees of freedom is equal to the number of pairs minus 1. In this case, df = 10 – 1 = 9 (*Statistics: Learning from Data,* 1st ed. pages 580–581).

5. **B.** Plan I does not differentiate between fiction and nonfiction books. Plan II is an effective use of sampling and outlines a way to accurately estimate the total number of nonfiction pages in the books library. Our population of interest is the nonfiction books so

our sample should come from the non fiction books only (*Statistics: Learning from Data,* 1st ed. pages 24–41).

6. **C.** Road conditions are confounded with brand. The stated purpose is to check the effect of road conditions on stopping distances. To correct for this, we would need to include both wet and dry conditions for brand A and brand B, or we would need to focus on only brand A or brand B but not both (*Statistics: Learning from Data,* 1st ed. page 25).

7. **A.** The median is represented by the center line in the box. Set B's median is located further to the right than set A. It is unclear from a box plot how many data points are being represented, so you cannot definitively say which set contains more data. The range of the data is represented by the endpoints of the whiskers, and in this case these endpoints are farther apart in set A than in set B (*Statistics: Learning from Data,* 1st ed. pages 151–155).

8. **C.** In order to find the expected profit, you need to multiply the profit per item by the number of items sold and the probability of selling that number, then total it for all possible outcomes. On wholesale sales the company is expected to make $1(50,000(0.2) + 100,000(0.3) + 150,000(0.4) + 200,000(0.1)) = (10,000 + 30,000 + 60,000 + 20,000) = $120,000. On retail sales the company is expected to make $2(10,000(0.2) + 20,000(0.5) + 30,000(0.3)) = 2(2,000 + 10,000 + 9,000) = $42,000. The total is $120,000 + $42,000 = $162,000 (*Statistics: Learning from Data,* 1st ed. pages 352–363).

9. **B.** The only way for the two of them to eat six plates combined is for each of them to eat three. Since both events have to happen and are independent as stated, the probabilities are multiplied. 0.7 multiplied by 0.1 gives you 0.07 as a final probability (*Statistics: Learning from Data,* 1st ed. pages 285–287).

10. **C.** There would be an association only if one of the two branches had a higher proportion of varsity athletes than the other. The proportion of varsity athletes is 32/48 = 2/3 for Navy and 24/36 = 2/3 for Air Force. Because the proportion is the same for both Navy and Air Force, there does not appear to be an association (*Statistics: Learning from Data,* 1st ed. pages 290–302).

11. **A.** For a normal distribution, the standard deviation is the distance from the mean to the inflection point on the normal curve shown by the dotted line. This distance is 15 kg (*Statistics: Learning from Data,* 1st ed. pages 87 and 382).

12. **D.** The null hypothesis is that the windows will save $100. The consumer group is not concerned with the case where the windows save more than $100, only the case where the windows save less than $100, which goes against the company's claim (*Statistics: Learning from Data,* 1st ed. pages 496–497).

13. **E.** The *z*-score describes how far away a data point is from the mean in terms of the standard deviation. A *z*-score of –1.25 tells us that Stanley's salary is below the mean because of the negative sign. The numerical value 1.25 tells us how many standard deviations his salary is away from the mean. So Stanley's salary is 1.25 standard deviations below the mean (*Statistics: Learning from Data,* 1st ed. pages 375–380).

14. **A.** The issue of concern is the way in which the sample was selected, not the sample size. The fact that the 100 stores that had local displays were not chosen at random introduces bias (*Statistics: Learning from Data,* 1st ed. pages 19–20).

15. **D.** To calculate the expected cost over 12 weeks, you take the expected cost for one week (5.65) and multiply it by the number of weeks (12). $5.65 \times 12 = \$67.80$ (*Statistics: Learning from Data,* 1st ed. pages 356–360).

16. **C.** We need to find the square root of the sum of the variances, not the sum of the standard deviations. With X_1 = amount paid in week 1, X_2 = amount paid in week 2, and so on, the total amount paid is $T = X_1 + X_2 + \ldots X_5$. The variance of T is $\text{Var}(X_1) + \text{Var}(X_2) + \ldots + \text{Var}(X_5) = 5(0.59^2) = 1.74$. The standard deviation of T is the square root of 1.74, which is 1.32 (*Statistics: Learning from Data,* 1st ed. pages 352–353).

17. **B.** All normal distributions are symmetric and have mean = median. While some normal distributions are centered at 0, this is not the case for *all* normal distributions (*Statistics: Learning from Data,* 1st ed. pages 349–350).

18. **B.** The median is the only number that must remain unchanged because the correction was made in a place that could not be the center of the data. The mean will change because you have added no additional cars, but the total of all bid prices has changed. The mode may or may not change depending on whether there is another car in the sample that sold for the corrected price. The range must change because the data point that was corrected was an endpoint of the data. The variance also must change because the distance from that data point to the mean has changed as have all other data points' distances to the new mean. The median, which is resistant, is the only statistic listed that could not possibly change with the given information (*Statistics: Learning from Data,* 1st ed. pages 135–136 and 143).

19. **D.** We cannot be sure that every individual plant grew from week one to week two. Since we do not know how the data points are connected, it is possible that the plant that was 26 millimeters after one week was still 26 millimeters after two weeks. Without further information about the data, we cannot be sure that each plant increased in size (*Statistics: Learning from Data,* 1st ed. pages 24 and 416).

20. **B.** To win a prize, Andre would have to make all three shots. If his probability of making each shot is 0.6 and the shots are independent, then the probability of making all three is (0.6 × 0.6 × 0.6) = 0.216 (*Statistics: Learning from Data,* 1st ed. pages 285–287).

21. **A.** A type I error occurs if we reject the null hypothesis when it is true. If a motorist is in the country legally but is thought to not be in the country legally, this would be equivalent to rejecting the null hypothesis when the null hypothesis is actually true (*Statistics: Learning from Data,* 1st ed. pages 502–505).

22. **B.** Because the population standard deviation is not known, a t interval with 59 degrees of freedom would be used. With a 90% confidence interval the *t*-value is 1.67. The given standard deviation of the sample is 8.1 with a sample size of 60. So the confidence interval is $34.7 \pm \dfrac{8.1}{\sqrt{60}}$ (*Statistics: Learning from Data,* 1st ed. pages 578–591).

23. **A.** Mr. Branch has a *z*-score of (12.5 – 11)/3.8 = 0.395. Dennis has a *z*-score of (15.8 – 14)/6.1 = 0.295. Because Mr. Branch's *z*-score is farther above the mean than Dennis's *z*-score, Mr. Branch did better when compared to the players of his era (*Statistics: Learning from Data,* 1st ed. page 375).

24. **A.** The frequency distribution for A correctly describes the behavior of this variable. This distribution of six rolls is binomial with n=6, p=1/6 and x (the number of successes) going from 0 to 6. P(0)=0.33, p(1)=0.40, p(2)=0.20, p(3)=0.05, p(4)=0.01, p(5)=0.0006 and p(6)=0.00002. The proper frequency table should peak at the most common number of 3s rolled over the course of six rolls and taper off from there, in a manner similar to these probabilities (*Statistics: Learning from Data,* 1st ed. pages 363–367).

25. **E.** This is the only method which results in a simple random sample, one in which every person who has recently purchased art has an equal chance of being chosen. Method A only reaches the most recent clients, not the whole list. Method B may only reach the wealthiest clients. Method C is not a simple random sample because it stratifies by type of art. Method D would only sample those willing to attend the brunch (*Statistics: Learning from Data,* 1st ed. pages 13–14).

26. **C.** The slope of the line is the approximate increase associated with an increase of 1 gram of fat. For 8 grams, the approximate increase would be 8 × 35.2 = 281.6 (*Statistics: Learning from Data,* 1st ed. pages 198 and 201).

27. **A.** This *z* statistic is large and the *P*-value is very small (< 0.01_) at 0.0066. For this reason, we are able to reject the null hypothesis and conclude that $\mu > 23$ (*Statistics: Learning from Data,* 1st ed. pages 591–604).

28. **C.** If the watermelon is in the 96th percentile, then for the normal distribution this occurs at 1.75 standard deviations above the mean. The mean of 90 ounces plus the standard deviation (15) times 1.75 gives us $(90 + (15 \times 1.75)) = (90 + 26) = 116$ ounces (*Statistics: Learning from Data,* 1st ed. pages 375–380).

29. **C.** Because both endpoints of the interval are positive, we can be confident that the mean time for the unmodified carts is greater than the mean time for the modified carts by somewhere between 16 and 32 minutes (*Statistics: Learning from Data,* 1st ed. pages 649–651).

30. **A.** The residual is calculated by first substituting the x value into the regression equation and then finding the difference between the observed y and the predicted y. $\hat{y} = 0.68 + 1.21(3) = 4.31$, residual $= y - \hat{y} = 4 - 4.31 = -0.31$ (*Statistics: Learning from Data,* 1st ed. pages 205–206).

31. **C.** For a standard normal distribution, about 10% of the area is below –1.28. This means that 80 must be 1.28 standard deviations below the mean. Solving $1.28\sigma = 20$ gives $\sigma = 15.625$ (*Statistics: Learning from Data,* 1st ed. page 375).

32. **C.** When the histogram is skewed right, the outlying values in the upper tail pull the mean up. So the mean will lie above the median (*Statistics: Learning from Data,* 1st ed. pages 83–84).

33. **D.** Because the value of the population standard deviation is unknown, this would be a t test. The test statistic is $t = \dfrac{346.8 - 350}{\dfrac{12.4}{\sqrt{40}}} = -1.63$. The P-value is the area to the left of -1.63 under the t curve with df = 39. This area is 0.055 (*Statistics: Learning from Data,* 1st ed. pages 593–595).

34. **A.** To find the confidence interval you need to use the equation $p \pm 1.96\sqrt{\dfrac{\hat{p}\left(1 - \hat{p}\right)}{n}} = 0.28 \pm 1.96\sqrt{\dfrac{(0.28)(0.72)}{100}} = 0.28 \pm 0.089$. The 95% confidence interval is then (0.191, 0.369) (*Statistics: Learning from Data,* 1st ed. pages 467–480).

35. **C.** The chance of all four exams occurring on one particular day is $(1/3)^4 = 0.0123$. When that is multiplied by the three possible days on which it could occur, you get $0.0123 \times 3 = 0.037$ (*Statistics: Learning from Data,* 1st ed. pages 285–287).

36. **C.** First find the total number of males who participated in the survey: $2 + 11 + 35 + 10 = 58$. Then find the total number who had their first crush before the age of 13: $2 + 11 = 13$. To find the percentage, take 13/58 and multiply by 100 to get 22.4% (*Statistics: Learning from Data,* 1st ed. pages 715–716).

37. **D.** While the rest of the variables could be related to weight loss, name is not related to weight loss (*Statistics: Learning from Data*, 1st ed. pages 13–14).

38. **A.** A data point that lies far away from the accepted mean is called an outlier (*Statistics: Learning from Data*, 1st ed. page 73 and 153–156).

39. **C.** As you read across and cross off 2 numbers at a time, the first number under 23 comes after crossing out 92, 64, and 69. The next two numbers are 01, 10. Both count as below 23. Then after crossing out 79, 36, 50, 48, 91, and 39, you come to 17 for the top prize in the random awards (*Statistics: Learning from Data*, 1st ed. page 320).

40. **D.** The square of the correlation coefficient, r^2, is a measure of the proportion of variability in Y that can be explained by a linear relationship between X and Y (*Statistics: Learning from Data*, 1st ed. pages 217–220).

SECTION II: FREE-RESPONSE ANSWERS

1. (a) The median time will be the time that corresponds to a cumulative proportion of 0.50. Half of the times will be smaller than this value and half larger. By finding 0.50 on the cumulative proportion axis of the graph, moving over to the plot for 2004, and then looking down to find the corresponding time on the time axis, you can determine that the median finishing time for 2004 is approximately 1:45. The median finishing time for 2008 is approximately 1:47. You should show this process on the graph.

   (b) The quartiles can be found in a way that is similar to the way you found the median by finding the times that correspond to cumulative proportions of 0.25 and 0.75. For 2004, the quartiles are approximately 1:44 and 1:46, and the IQR is about 2 seconds. For 2008, the quartiles are approximately 1:45 and 1:48, and the IQR is about 3 seconds.

   (c) The 200-meter freestyle times were slower in 2004 than in 2008. However, the variability in times was a bit larger in 2008 than in 2004.

   (*Statistics: Learning from Data*, 1st ed. pages 149–160).

Scoring Question 1:

   Each part of this problem can be scored as essentially correct (E), partially correct (P), or incorrect (I).

   Part (a) is scored
      E if correct medians are given for both years and supporting work is indicated on the graph.
      P if correct medians are given for both years but no supporting work is shown on the graph.

Part (b) is scored
  E if both IQRs are correctly computed
  P if quartiles for both years are correctly computed but the IQR is reported as an interval rather than a single number.

Part (c) is scored
  E if the response comments on (1) times tend to be faster in 2004 than 2008 and (2) variability tends to be greater in 2008 than 2004.
  P if the response comments on only one of the two things required for an E.

Question 1 is scored a
  4 if all three parts are E
  3 if two parts are E and the one is P
  2 if two parts are E and no parts are P, or one part is E and two parts are P, or if three parts are P
  1 if two parts are P and one part is I or if one part is E and no parts are P
  0 if one or no parts are P

2. (a) Because some stores may have a higher volume of sales overall than other stores and this may be related to how much cereal is sold, it would be a good idea to try both locations in each store. For each store, the cereal will be displayed at the end of an aisle for one of the two weeks of the study and with the rest of the cereals for the other week. For each store, flip a coin to determine which of the two weeks will use the end-of-aisle display—if the coin lands heads, assign the end-of-aisle display to week 1 and if the coin lands tails, assign the end-of-aisle display to week 2. Record sales at each store for each of the two weeks.
    (b) The design described in part (a) will result in paired data (paired by store). A paired t test would be used.

(*Statistics: Learning from Data,* 1st ed. pages 9, 10, 33, and 483).

Scoring Question 2:

Each part of this problem can be scored as essentially correct (E), partially correct (P), or incorrect (I).

Part (a) is scored
  E if the proposed design uses both locations at each store and uses random assignment to determine which of the two weeks will have the end-of-aisle display.
  P if the proposed design uses both locations at each store but does not explicitly state that random assignment will be used to determine which of the two weeks will have the end-of-aisle display.

Part (b) is scored
    E if a paired *t* test is recommended
    P if the response fails to recognize the pairing and
        recommends a two-sample *t* test
    I if another test is recommended

Question 2 is scored a
    4 if both parts are E
    3 if one part is E and the other is P
    2 if one part is E and one part is I or if both parts are P
    1 if one part is P and one part is I
    0 if both parts are I

3. A good place to start is to organize the probability information
given using a tree diagram. The diagram is not required, but it
makes the required computations easier.

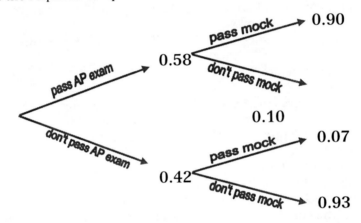

(a) Let M = event that the selected student passed the mock
exam and A = event that the selected student passes the AP
exam.
Then
P(M)=(0.58)(0.90)+(0.42)(0.07)=0.5514

(b) $P(A \mid M) = \dfrac{(0.58)(0.9)}{0.5514} = 0.9467$

(*Statistics: Learning from Data,* 1st ed. pages 290–302).

Scoring Question 3:

**Each part of this problem can be scored as essentially
correct (E), partially correct (P), or incorrect (I).**

Part (a) is scored
    E if the a correct probability is computed with appropriate
        supporting work
    P if a correct probability is computed but appropriate
        supporting work is not given.

Part (b) is scored
    E if the a correct probability is computed with appropriate
        supporting work
    P if a correct probability is computed but appropriate
        supporting work is not given.

Question 2 is scored a
    4 if both parts are E
    3 if one part is E and the other is P
    2 if one part is E and one part is I or if both parts are P
    1 if one part is P and one part is I
    0 if both parts are I

4. (a) Type I error: Rejecting the null hypothesis when it's true. In
       this context, this is believing that the proportion who think
       the principal is doing an excellent job has decreased when
       in fact it has not.
       Type II error: Failing to reject the null hypothesis when it
       should be rejected. In this context this is believing that the
       proportion who think the principal is doing an excellent job
       has not decreased when it actually has decreased.
   (b) The hypotheses were given in the question.

   Test: two-sample z test for difference in proportions.

   Assumptions:
   The two samples are independent random samples. We
   also need to check to make sure that the sample sizes are
   large enough.

   $$\hat{p}_b = \frac{280}{375} = 0.75 \qquad\qquad \hat{p}_a = \frac{358}{523} = 0.68$$

   $n_a \hat{p}_a = 375(0.75) > 10 \qquad\qquad n_b \hat{p}_b = 523(0.68) > 10$

   $n_a(1 - \hat{p}_a) = 375(0.25) > 10 \qquad\qquad n_b(1 - \hat{p}_b) = 523(0.32) > 10$

   Because $n_a \hat{p}_a$, $n_b \hat{p}_b$, $n_a(1 - \hat{p}_a)$, and $n_b(1 - \hat{p}_b)$ are all greater
   than 10, the sample sizes are large enough.

   Calculations: $z = 2.025$, P-value $= 0.0214$

   Conclusion:
   Since the P-value is less than 0.05, we will reject the null
   hypothesis. There is strong evidence that the proportion
   who think the principal is doing an excellent job decreased
   after the new policy was implemented.

   (*Statistics: Learning from Data,* 1st ed. pages 507–511).

Scoring Question 4:

**Part (a) is scored as essentially correct (E), partially correct
(P), or incorrect (I). There are three parts to the hypothesis
test in part (b): test and assumptions, calculations, and
conclusion (The hypotheses were given in the question, so**

this part of a hypothesis test is not scored). **Each of these parts can be scored as essentially correct (E), partially correct (P), or incorrect (I).**

Part (a) is scored as
  E if a Type I error and a type II error are correctly described in context.
  P if Type I and Type II errors are correctly defined, but context is missing, or if only one of the two errors is correctly described..

For Part (b)
Test and assumptions are scored as
  E if (1) the correct test is identified either by name or by formula and (2) both assumptions for the test are adequately addressed.
  P if the response includes only one of the two components required for an E.

Calculations are scored as
  E if correct values of the test statistic and P-value are given. If the calculator command is used, the formula must be shown and the statistics in the formula must be correctly identified.
  P if either the value of the test statistic or the P-value is omitted or incorrect.
Note: Calculations should be consistent with the hypothesized values given in the null hypothesis of part (a).

Conclusion is scored as
  E if a correct conclusion is given in context and the conclusion is linked to the P-value.
  P if a correct decision to reject or fail to reject the null hypothesis is given but the conclusion is not in context or the conclusion is not linked to the P-value.

Each part scored as essentially correct counts as 1 and each part scored as partially correct counts as ½ to compute an overall score. Then Question 4 is scored a
  4 if the overall score is 3 or 3.5
  3 if the overall score is 2.5
  2 if the overall score is 2 or 1.5
  1 if the overall score is 1
  0 if the overall score is 0 or 0.5

5. (a) The two samples are paired by school district. A two-sample t test is only appropriate for independent samples.
   (b) Test and assumptions:
   Because the samples are paired, we will use the paired t test.

Assumptions:

The school districts were selected at random.

The number of pairs is not large, so we need to be willing to assume that the population distribution of differences is approximately normal. The differences are

5   4   7   −2   −9   −8   −3   2   8   4   0   −6

A boxplot of these differences is given here:

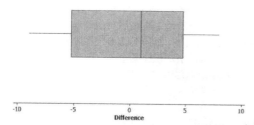

The boxplot is approximately symmetric and there are no outliers, so it is reasonable to assume that the difference distribution is approximately normal.

Calculations:

$t = 0.10$, P-value $= 0.461$, df $= 11$

Conclusion:

Because the P-value is greater than 0.05, we fail to reject the null hypothesis. There is not convincing evidence that the mean percentage of teachers with advanced course work for school districts in California is greater for math teachers than for science teachers.

(*Statistics: Learning from Data,* 1st ed. pages 625–626).

Scoring Question 5:

Part (a) is scored as essentially correct (E), partially correct (P), or incorrect (I). There are three parts to the hypothesis test in part (b): test and assumptions, calculations, and conclusion (The hypotheses were given in the question, so this part of a hypothesis test is not scored). Each of these parts can be scored as essentially correct (E), partially correct (P), or incorrect (I).

Part (a) is scored as

E if the response states that the two-sample t test is only appropriate for independent samples and notes that these samples are paired.

P if the response makes a credible attempt to assess the assumptions for a two-sample t test, but fails to note that the two samples are paired.

For Part (b)

Test and assumptions are scored as
  E if (1) the correct test is identified either by name or by formula and (2) both assumptions for the test are adequately addressed.
  P if the response includes only one of the two components required for an E.

Calculations are scored as
  E if correct values of the test statistic and P-value are given. If the calculator command is used, the formula must be shown and the statistics in the formula must be correctly identified.
  P if either the value of the test statistic or the P-value is omitted or incorrect.

Note: Calculations should be consistent with the hypothesized values given in the null hypothesis of part (a).

Conclusion is scored as
  E if a correct conclusion is given in context and the conclusion is linked to the P-value.
  P if a correct decision to reject or fail to reject the null hypothesis is given but the conclusion is not in context or the conclusion is not linked to the P-value.

Each part scored as essentially correct counts as 1 and each part scored as partially correct counts as ½ to compute an overall score. Then Question 5 is scored a
  4 if the overall score is 3 or 3.5
  3 if the overall score is 2.5
  2 if the overall score is 2 or 1.5
  1 if the overall score is 1
  0 if the overall score is 0 or 0.5

6. (a) We will use a t confidence interval for the slope. With a sample size of 50, df = $50 - 2 = 48$. The appropriate t critical value is 1.68. Using values from the computer output, the confidence interval is then

$$47.436 \pm (1.68)(3.595) = 47.436 \pm 6.040 = (41.396, 53.476)$$

Based on the sample data, we expect, on average, the SAT math score will increase by somewhere between 41 and 53 points for each additional math course taken.

   (b) $\hat{y} = 505.14 + 47.436(4) = 694.884$. Because SAT math scores must be integers, we would round this to 695. We predict an SAT math score of 695 for a student who has completed 4 math courses.

   (c) From the computer output, $r^2 = 0.784$. This means that about 78% of the variability in SAT math scores can be attributed to an approximate linear relationship between the number of math courses completed and SAT score.

(d) $\hat{y}$ =514.70+47.682(4)-3.164(3)=695.936. Because SAT math scores must be integers, we would round this to 696. We would predict an SAT math score of 696 for a student who has completed four math courses and three science courses. This prediction is only one point higher than the prediction in part (b).

(e) In the linear model, the value of the standard deviation s_e provides information about expected accuracy of predictions based on the regression line. Since we are told that s_e is interpreted in a similar way for multiple regression, we would want to compare the values of s_e for the two models. For the linear model, $s_e = 29.6836$ and for the multiple regression model $s_e = 29.7907$. Since these two standard deviations are nearly equal, we would not expect predictions based on the multiple regression model to be more accurate than those based on the linear model. The values of the two coefficients of determination are also very similar.

(*Statistics: Learning from Data,* 1st ed. pages 219–225, 229–231 and 752–756).

Scoring Question 6:

Parts (b) and (c) are scored together. Parts (a) (b+c), (d) and (e) are each scored as essentially correct (E), partially correct (P), or incorrect (I).

Part (a) is scored
  E if the confidence interval is correctly computed and interpreted in context
  P if an incorrect *t* critical value or standard deviation is used in the calculation of the confidence interval or if the interpretation is missing or not in context.

Part (b + c) is scored
  E if a correct prediction is made in Part (b) and a correct interpretation is given in context in Part (c)
  P if the calculation of the prediction in Part (a) is not correct or if the interpretation in Part (b) is missing, incorrect or not in context

Part (d) is scored
  E if a correct predicted value is computed and a comparison to the prediction from Part (b) is made.
  P if the prediction is incorrectly computed and this prediction is compared to the prediction in Part (b) OR if the prediction is correctly computed but no comparison is made.

Part (e) is scored

  E if the response says that the predictions based on the multiple regression model will not be much more accurate than those based on the linear model and justifies this by appealing either to the values of s_e for both models or to the values of r^2 for both models.

  P if the response says that the predictions based on the multiple regression model will not be much more accurate than those based on the linear model but the explanation is weak or is not linked to the values of r^2 or s_e.

Each of the four parts (a, b + c, d, and e) scored as essentially correct counts as 1 and each part scored as partially correct counts as ½ to compute an overall score. Then Question 6 is scored a

  1 if the overall score is 3 or 3.5

  3 if the overall score is 2.5

  2 if the overall score is 2 or 1.5

  1 if the overall score is 1

  0 if the overall score is 0 or 0.5

CALCULATING YOUR SCORE

SECTION I: MULTIPLE-CHOICE QUESTIONS

[_____] × 1.25 = _____
Number Correct   Weighted Section I Score
(out of 40)   (do not round)

SECTION II: FREE-RESPONSE PROBLEMS

Question 1 _____ × (1.875) = _____
(out of 4)   (Do not round)

Question 2 _____ × (1.875) = _____
(out of 4)   (Do not round)

Question 3 _____ × (1.875) = _____
(out of 4)   (Do not round)

Question 4 _____ × (1.875) = _____
(out of 4)   (Do not round)

Question 5 _____ × (1.875) = _____
(out of 4)   (Do not round)

Question 6 _____ × (3.1255) = _____
(out of 4)   (Do not round)

Sum = _____
Weighted Section II Score
(Do not round)

COMPOSITE SCORE

_____ + _____ = _____
Weighted   Weighted   Composite Score
Section I Score Section II Score (Round to nearest
                                   whole number)

Composite Score Range	Approximate AP Grade
70–100	5
57–69	4
44–56	3
33–43	2
0–32	1

Practice Test 2

This test will give you some indication of how you might score on the AP Statistics Exam. Of course, the exam changes every year, so it is never possible to predict a student's score with certainty. This test will also pinpoint strengths and weaknesses on the key content areas covered by the exam.

AP STATISTICS EXAMINATION
Section I: Multiple-Choice Questions
Time: 90 minutes
40 Questions

Directions: Each of the following questions or incomplete statements is accompanied by five suggested answers or completions. Select the one that best answers the question or completes the statement.

1. John thinks that whenever he takes the late bus home after school, he is hungrier than when he takes the regular bus home. He decides to record whether or not he goes back for seconds at the dinner table each night and cross references those data with which bus he took home from school. He found that he had seconds at dinner eight of the nine recorded days in which he took the late bus as compared to having seconds at dinner only seven of the 21 times he took the early bus. John turned in a paper to his statistics teacher that stated that the late bus was causing his hunger. When John's paper was returned to him, the professor had written, "Were the days you took the late bus the same days you played after-school sports?" What potential problem is his professor is trying to point out?

(A) The possibility of a confounding variable
(B) The low power of a test with small sample sizes
(C) The lack of blinding
(D) The unequal sample sizes
(E) The lack of a control group

2. Four statistics are being considered for estimating a population characteristic. The statistics are as follows. Which of these statistics would be the best one to use?

Statistic 1: unbiased with standard deviation 0.03

Statistic 2: biased with standard deviation 0.03

Statistic 3: biased with standard deviation 0.5

Statistic 4: unbiased with standard deviation 1.3

(A) Statistic 1
(B) Statistic 2
(C) Statistic 3
(D) Statistic 4
(E) Either Statistic 1 or 4 would be a good choice.

GO ON TO NEXT PAGE

3. When determining which students will represent their high school in a local competition, the principal selected two students at random from a list of all freshmen at the school, two students at random from all sophomores, two students at random from all juniors, and two students at random from all seniors. This is an example of
   (A) Simple random sampling
   (B) Cluster sampling
   (C) Systematic sampling
   (D) Stratified sampling
   (E) Convenience sampling

4. A security watch group insists that the least secure area of the United States is its ports. To justify this claim, the group surveyed randomly selected members of the dockworkers' unions to ask whether or not they thought a terrorist shipment would be intercepted using current inspection procedures. Of the 457 dockworkers surveyed, only 114 felt confident that a shipment would be intercepted. Which of the following represents a 95% confidence interval for the true proportion of dockworkers who think that a terrorist shipment would be intercepted?
   (A) (0.17, 0.33)
   (B) (0.25, 0.25)
   (C) (0.19, 0.31)
   (D) (0.23, 0.27)
   (E) (0.21, 0.29)

5. It is sometimes stated that the American judicial system is built on the idea that it is better to let 100 guilty men go free than to lock up one innocent man. If it is assumed from the saying "presumed innocent" that the null hypothesis of a criminal trial is that the accused is not guilty, which type of error is represented in the following two situations?

   Situation 1: A guilty man is found not guilty.

   Situation 2: An innocent man is found guilty.

   (A) Both represent Type I error.
   (B) Both represent Type II error.
   (C) Situation 1 represents Type I error, Situation 2 represents Type II error.
   (D) Situation 1 represents Type II error, Situation 2 represents Type I error.
   (E) Both represent both types of error.

6. Herman has applied for three jobs. He believes that the probability that he gets a job offer is 0.20 for each of the three jobs and that whether or not he gets an offer for one of the jobs is independent of whether or not he gets an offer for either of the other two jobs. What is the probability that Herman gets at least one job offer?
   (A) 0.008
   (B) 0.488
   (C) 0.512
   (D) 0.600
   (E) 0.800

7. A researcher focusing on Native American studies wants to determine the current attitude of Native Americans toward the United States government. The researcher creates a survey of five statements which respondents can mark with a number 1 to 5 indicating how strongly they agree or disagree with each statement. Some of the sampling methods described below would introduce the potential for bias. Which of the following sampling methods would produce potential <u>selection</u> bias?
(A) The researcher posts his survey to a popular Native American web site and allows any Native American to take the survey.
(B) Federal government employees administer the survey on reservations while wearing badges that are fully visible to the respondents.
(C) All members from the three largest tribes still existing in America are given the survey.
(D) The researcher accesses a list of all of the Native Americans living in the United States from the U.S. Census department and randomly selects 100 people from that list to take the survey.
(E) The survey is given to all Americans.

8. While attending his 10-year high school reunion, Chester ran into his old friend Gary who immediately began talking about the new job he landed in New York City. Even though Gary and Chester both became accountants, Gary makes more money. Gary said his salary was $100,000 and Chester felt embarrassed by his salary of $80,000. Chester's wife, a statistician, tells Chester that he is doing better compared to the people in his town than Gary is compared to the people in New York. She did some research and found that the average salary for an accountant in New York was $80,000 and the standard deviation was $8,000 while the average salary for an accountant in Chester's home town was $60,000 and the standard deviation was $5,000. What are the associated z-scores for Chester and Gary?
(A) Chester's $z = 1.25$, Gary's $z = 4$
(B) Chester's $z = 2.5$, Gary's $z = 4$
(C) Chester's $z = 4$, Gary's $z = 4$
(D) Chester's $z = 1.25$, Gary's $z = 2.5$
(E) Chester's $z = 4$, Gary's $z = 2.5$

9. Thomas has $1,000 that he wants to invest in a mutual fund. After some research, he has narrowed down his choices to one of three mutual funds. He plans to hold on to the fund for five years, and his research gives him the following probabilities and payouts based on his $1,000 investment five years later.

Payout:	500	1000	2000
Fund X	0.5	0	0.5
Fund Y	0.33	0.33	0.34
Fund Z	0.25	0.5	0.25

Order the funds from highest expected value to lowest expected value.
(A) Fund X, Fund Y, Fund Z
(B) Fund Y, Fund Z, Fund X
(C) Fund Z, Fund X, Fund Y
(D) Fund Y, Fund X, Fund Z
(E) Fund X, Fund Z, Fund Y

10. The distribution of the weights of fish in a lake is approximately normal and is centered at 2 pounds. It is known that 15% of the fish in the lake are larger than 3 pounds. What is the standard deviation of the weights of fish in the lake?
(A) 0.15
(B) 0.67
(C) 0.85
(D) 0.96
(E) 1.00

GO ON TO NEXT PAGE

Questions 11–12 refer to the following information:

The Commonwealth of Virginia wants to determine how many more people would use the Chesapeake Bay Bridge-Tunnel if the price were lowered from $12 to $10 with a valid Virginia ID. At first it was suggested that the Department of Motor Vehicles ask a random sample of people in the state if they had used the bridge in the past and how likely they would be to use it in the future should the rate remain the same and should the rate decrease. But objections were raised that a majority of Virginians do not live close enough to the bridge to be likely to use the bridge regardless of the price. It was then suggested that the survey be conducted at the tollbooth to the bridge. Again, there were objections. This time, the complaint was that you would not reach those who had decided not to use the bridge. Finally, someone suggested that the 10 counties closest to the bridge be identified and the survey be conducted by selecting a random sample of drivers in these counties. Everyone agreed that this was the preferable method to use in order to gauge increased traffic that would be generated by a reduction in rate.

11. For the final sampling method used, what is the population, and what is the sample?
    (A) Sample: All drivers in Virginia, Population: All drivers who consider using the bridge
    (B) Sample: All drivers who use the Chesapeake Bay Bridge, Population: All drivers in Virginia
    (C) Sample: Selected drivers in the counties closest to the bridge, Population: All drivers in the counties closest to the bridge
    (D) Sample: Selected drivers in the counties closest to the bridge, Population: All drivers in Virginia
    (E) Sample: All drivers who do not currently use the bridge, Population: All drivers in Virginia

12. In switching from the first sampling method where people were randomly chosen from all drivers in Virginia to the final sampling method, the state researchers were attempting to address the problem of people who live far from the bridge not using it. In this circumstance, what is the best description of how proximity to the bridge impacts this study?
    (A) It creates nonresponse bias
    (B) It is a variable that could influence the response
    (C) It defines the sampling frame
    (D) It creates outliers
    (E) It is used for clustering

13. An apple producer owns a large orchard that is pesticide free. The producer contends that 98.2% of the apples will be worm free after ripening on the tree. If the company's claim is true, what is the expected number of worm-infested apples in a random sample of 25,000 apples from the pesticide free orchard?
(A) 24,550
(B) 24,500
(C) 24,450
(D) 550
(E) 450

14. To illustrate the danger of using a regression equation to make predictions that are outside the range of the data, an AP Statistics teacher used data on x = year and y = median house price for years from 1997 to 2007 to fit a least-squares regression line. During this time period, the median price increased from each year to the next. The line was then used to predict median house price for the years 2008, 2009, 2010 and 2011. During this time period, housing prices declined significantly. Which of the following best describes how the predicted median prices compare to the actual median prices for 2008, 2009, 2010 and 2011?
(A) Some of the predictions would be greater than the actual median price and some would be less than the actual median price.
(B) The predictions would tend to be less than the actual values and would be closer to the actual value in 2011 than 2008.
(C) The predictions would tend to be less than the actual values and would be closer to the actual value in 2008 than 2011.

(D) The predictions would tend to be greater than the actual values and would be closer to the actual value in 2011 than 2008.
(E) The predictions would tend to be greater than the actual values and would be closer to the actual value in 2008 than 2011.

15. Environmental scientists enlisted to help with a beach erosion problem in a coastal community studied the accumulation of sand along the shoreline. They took measurements from a sample of randomly selected sand dunes, recording the distance from the top of the dune to the highest point of water at high tide. One month later they came back to the same locations and measured again. The table below summarizes the measurements at each of the two times.

	n	$\bar{x}$	S
June measurement	135	14.2	1.8
July measurement	135	13.8	2.1
Difference (June–July)	135	0.4	0.2

Using the null hypothesis that there is no difference between mean distance in June and mean distance in July, what is the value of the t-statistic associated with a test to determine if the mean distance from dune to ocean was smaller in July than it was in June?
(A) $t = 0.10$
(B) $t = 1.68$
(C) $t = 2.00$
(D) $t = 17.74$
(E) $t = 23.24$

16. In a population of cigarette smokers, 26% quit smoking in 2005. Of those who quit, 41% have not had a single cigarette in the five years after quitting. If a person is selected at random from the original population of cigarette smokers, what is the probability that the selected person quit smoking in 2005 and did not have a single cigarette in the 5 years after quitting?
   (A) 0.88
   (B) 0.83
   (C) 0.43
   (D) 0.17
   (E) 0.11

17. Researchers at a prestigious medical school were asked how they could be sure that their own biases toward the drug they were studying were not influencing their observations of patient behavior. The researchers responded that their study had been structured in a way that neither the patients nor the people measuring the response knew who was being given the drug and who was being given a placebo. What term describes this aspect of the study?
   (A) Replication
   (B) Direct control
   (C) Blocking
   (D) Single blind
   (E) Double blind

18. A study was done to estimate the amount of time spent exercising by residents of a nursing home. Investigators recorded the number of minutes per week spent exercising by each of 26 randomly selected residents. The sample had a mean of 42 minutes and a standard deviation of 15 minutes. Assuming exercise times have a distribution that is approximately normal, which of the following is a 95% confidence interval for the mean time spent exercising per week by the nursing home residents?
   (A) (40, 44)

(B) (36, 48)
(C) (30, 54)
(D) (39, 45)
(E) (27, 57)

19. Wheelchairs have braking mechanisms that minimize speeds as they go down hills. In order to test how long it takes different wheelchairs to come to a complete stop after going down a hill, a consumer group tests 20 different wheelchair models on a ramp at their facility. The testers roll down the ramp then apply the brake when they reach a white line at the base of the ramp and the distance from the white line to where the wheelchair came to a complete stop is measured and recorded. A relative frequency histogram of the stopping times is shown below.

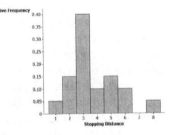

What proportion of wheelchairs tested were able to stop in 3.5 feet or less?
(A) 0.05
(B) 0.15
(C) 0.40
(D) 0.60
(E) 0.70

20. The air conditioner in Bob's apartment is broken. This is the third time it has malfunctioned since Bob moved in 6 months ago. After speaking with his neighbors, Bob realized that the air conditioning is a consistent problem throughout his apartment complex. Bob wants to collect data to determine if there is evidence that the mean time between malfunctions is less than 90 days. He plans to test a null hypothesis that says the mean time between malfunctions is equal to 90 days against the alternative hypothesis that says that the mean time between malfunctions is less than 90. Which of the following represents the conclusion that there insufficient evidence to conclude that the mean time between malfunctions is less than 90 days?
(A) Reject H_0
(B) Fail to reject H_0
(C) Reject H_a
(D) Fail to reject H_a
(E) Accept H_a

21. For a particular population, the sampling distribution of the sample mean for random samples of size 9 has a mean of 350 and a standard deviation of 63. What are the mean and standard deviation of the population?
(A) $\mu = 117, \sigma = 21$
(B) $\mu = 350, \sigma = 21$
(C) $\mu = 350, \sigma = 63$
(D) $\mu = 350, \sigma = 189$
(E) $\mu = 3150, \sigma = 567$

22. In trying to describe the weight of dinosaurs to her fifth-grade class, Mrs. Crofton decides to put dinosaur weights in terms the children can understand. She researches weights of Tryannosaurs and finds out that the weights were approximately normally distributed with a mean of 7.5 tons and a standard deviation of 0.5 tons. Which of the following is a correct statement about Tryannosaur weights?
(A) About 70% of the Tryannosaurs were between 4.45 and 4.95 tons.
(B) About that 70% of the Tryannosaurs were between 6.5 and 8.5 tons.
(C) About 70% of the Tryannosaurs were between 7.0 and 7.5 tons.
(D) About 70% of the Tryannosaurs were between 7.0 and 8.0 tons.
(E) About 70% of the Tryannosaurs were between 7.5 and 8.0 tons.

23. Below is a relative frequency distribution for the number of cars per household from a study of 50 families from suburban Boston.

Number of Cars	Frequency	Relative Frequency
0	8	0.16
1	17	0.34
2	15	0.30
3	7	0.14
4 or more	3	0.06

According to the table, what percentage of families in this study had more than two cars?
(A) 6
(B) 14
(C) 20
(D) 30
(E) 80

GO ON TO NEXT PAGE

24. Gym of the Jungle is a fitness center focusing on unique class offerings beyond the customary classes offered at rival gyms. The schedule of classes and their registration numbers are given in the chart below:

	1 pm	6 pm	9 pm	Total
Rope Swinging	14	32	12	58
Stump Jump	18	23	21	62
Saunastenics	12	17	15	44
Total	44	72	48	164

What is the probability that a randomly selected person registered to take a class at 6 PM is in Saunastenics?
(A) 0.104
(B) 0.170
(C) 0.236
(D) 0.386
(E) 0.439

25. Meat Market sandwich shop offers five different kinds of bread, four different kinds of meat, and seven different sauces on their sandwiches. Each sandwich contains one bread, one sauce, and two different meats. How many different sandwiches are possible?
(A) 16
(B) 19
(C) 140
(D) 420
(E) 560

26. A church choir wants to attract more members between the ages of 16 and 30. The current choir members believe that early Sunday morning practices may be a problem for this group, so they decided to take a simple random sample of 30 church members in the targeted age group to determine a convenient time to hold practices. Each person in the sample was sent a survey. Ten people responded to the survey, and half of the responses identified Friday nights as their best practice time. If the choir used this information to make a decision to change the practice time to Friday nights, which type of bias might be impacting their decision?
(A) Selection bias
(B) Measurement or response bias
(C) Nonresponse bias
(D) Sampling frame bias
(E) Replacement bias

27. A video game manufacturer noticed that most of its competitors set the retail price of their new games at $60 and then wait a number of months before lowering the prices. The marketing department feels this might not be the best price for maximum profit. They are considering a price of $50 for their next new game. The company knows that most new purchases are made by people who have previously purchased a game from the company. With the $60 price for new games, about 50% of those who have made a previous purchase also buy a new game when it is released with a price of $60. The company believes that a price of $50 would result in a higher profit if more than 5% of those who have made a previous purchase would buy the new game with a $50 price but not with a $60 price. The company plans to survey a large random sample selected from those who have made a previous purchase to see what proportion would purchase with a $50 price but not with a $60 price. What hypotheses should the company test?

(A) $H_0 : \mu = 50 \quad H_a : \mu = 60$

(B) $H_0 : p = 0.50 \quad H_a : p > 0.50$

(C) $H_0 : p = 0.05 \quad H_a : p > 0.05$

(D) $H_0 : p = 50 \quad H_a : p = 60$

(E) $H_0 : \mu = 50 \quad H_a : \mu > 50$

28. As part of an adult education statistics class, the professor asked the students in each of her two sections to report their ages. She then constructed a back-to-back stem-and-leaf display to compare the two sections. This plot is reproduced below: Section 1
Section 2

```
          9|1|7 9
9 9 7 6 2 2 1|2|3 5 8 9
    8 6 3 3 1|3|0 0 0 2 3 3 7 8
        6 5 2|4|2 7 7
        9 3|5|9
```

If the two sections were combined into a single data set, what is the mode of this data set?
(A) 0
(B) 9
(C) 29
(D) 30
(E) 33

29. A game is played in which two players (player A and player B) each toss a fair coin. If both tosses result in heads or both tosses result in tails, the players will toss again. If the tosses result in one head and one tail, the player whose toss was the head is the winner. What is the probability that each player must toss exactly three times before a winner is declared and that player A is the winner?
(A) 1/4
(B) 1/2
(C) $(1/2)^3$
(D) $(1/2)^2(1/4)$
(E) $(1/4)^3$

GO ON TO NEXT PAGE

30. Below is one possibility for a standard BINGO card.

B	I	N	G	O
3	28	32	56	66
15	18	45	48	67
12	22	FREE	51	70
8	20	41	53	73
9	17	38	47	68

In the game of BINGO, there are 75 balls numbered 1–75 placed in a cylinder and selected at random. What is the probability that the first ball selected matches one of the numbers on the bingo card above?

(A) 0.66
(B) 0.50
(C) 0.33
(D) 0.32
(E) 0.07

31. Which of the scatterplots below represents a negative linear relationship between two variables?

(A)

(B)

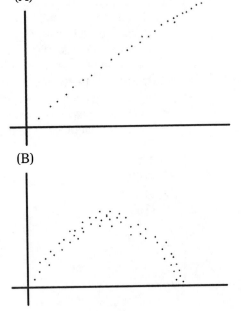

(C)

(D)

(E)

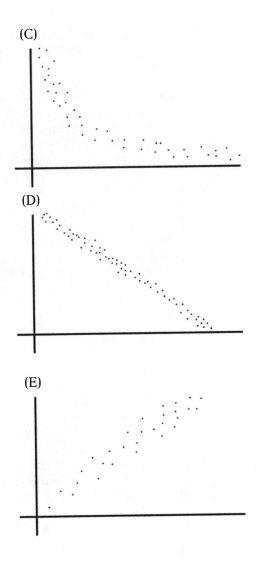

32. After being randomly selected for a prize at a concert, you are told that you can choose between two options. Option one is that you take a $25 gift card. Option two is that you draw a slip of paper from a hat that has two slips that say zero and one that says 100. If you get one of the zeros, you get nothing. If you get the 100, then you get a $100 gift card. Which of the following best compares the expected value and standard deviation of the amount won for the two options?

(A) Option one has a larger expected value and larger standard deviation than option two.

(B) Option one has a larger expected value and smaller standard deviation than option two.

(C) Option one has a smaller expected value and smaller standard deviation than option two.

(D) Option one has a smaller expected value and larger standard deviation than option two.

(E) Both options have equal expected values and standard deviations.

33. A shoe company wants to determine the effect of factory layoffs on production. After compiling data, they determine that a regression line that relates number of employees to factory shoe output is a good fit. The equation of the least-squares regression line is $\hat{y} = 3{,}000 + 140x$, where $\hat{y}$ is the predicted number of shoes produced in a day and x is the number of employees working at the factory. If the company is considering cutting 10 employees, how much does the regression line predict that shoe output would drop?

(A) 140 shoes
(B) 700 shoes
(C) 1,400 shoes
(D) 3,000 shoes
(E) 30,000 shoes

34. A vaccine has been developed that is designed to reduce the chance of catching a cold during the winter months. In an experiment to evaluate the effectiveness of this vaccine, 500 volunteers were randomly assigned to one of two groups. Those in one group received the vaccine and those in the other group received an injection with no active ingredients. The no active ingredient injection is an example of

(A) blinding.
(B) a confounding variable.
(C) a blocking factor.
(D) a volunteer treatment.
(E) a placebo.

35. Suppose that the number of dandelions in a one-acre grassland plot is approximately normally distributed with a mean of 750 and a standard deviation of 80. Approximately what percentage of one-acre grassland plots have between 590 and 910 dandelions?

(A) 50%
(B) 68%
(C) 75%
(D) 95%
(E) 99.7%

36. Danny's friend Joe claims that he is the best coin flipper in the history of coins. When Danny asks what he means, Joe says he can get heads flipping a coin 75% of the time. To prove his point, Joe flips a coin four times and gets three heads. Danny explains that this is not that unusual with only four tosses, but that if Joe were to continue to toss the coin a large number of times, the percentage of heads would get close to 50%. What statistical principle is Danny describing?

(A) Central Limit Theorem
(B) Law of Large Numbers
(C) Law of Total Probability
(D) Law of Small Numbers
(E) Empirical Rule

GO ON TO NEXT PAGE

37. At Studytown International Airport, customers have their choice of three airlines for both international and domestic travel. The table below lists the number of travelers on each airline on July 1, 2010.

Airline	International	Domestic	Total
Skyway	114	361	475
Avian Air	45	212	257
Patriot Air	0	845	845
Total	159	1418	1577

What is the probability that a randomly selected passenger traveling domestically from Studytown on the first of July would have flown on Patriot Air?
(A) 0.254
(B) 0.404
(C) 0.536
(D) 0.596
(E) 0.899

38. A printer company believes that its printers use less ink than a competitor's printer. To support this claim, the company plans to test 50 printers made by the company and 50 printers made by the competitor. Each printer will print the same page over and over and the number of pages printed before running out of ink will be recorded. The resulting data will be used to carry out a hypothesis test to determine if there is evidence that the mean number of pages printed before running out of ink differs for the two brands of printers. They plan to use a significance level of 0.05. Which of the following would increase the power of the test?
I. Use 25 printers of each brand instead of 50.
II. Use 100 printers of each brand instead of 50.
III. Use a significance level of 0.01 instead of 0.05.
IV. Use a significance level of 0.10 instead of 0.05

(A) II only
(B) I and III only
(C) I and IV only
(D) II and III only
(E) II and IV only

39. After showing up late for work again, Lafayette is called into his boss's office. The boss presents him with a cumulative relative frequency distribution of Lafayette's arrival times. In each case it is based on the time that the system clocked him in. The system records the time rounded to the next 15-minute interval, so if Lafayette clocked in at 8:51 it would record his arrival as 9:00. The chart is displayed below:

Time of Day	Cumulative Relative Frequency
8:45	0.04
9:00	0.15
9:15	0.27
9:30	0.61
9:45	0.83
10:00	1

According to the chart, during what time interval is Lafayette most likely to clock into work?
(A) 8:46 to 9:00
(B) 9:01 to 9:15
(C) 9:16 to 9:30
(D) 9:31 to 9:45
(E) 9:46 to 10:00

40. Suppose that in 2010, the proportion of all cars for sale in California that were hybrids was 0.08. Suppose that a new government program provided incentives for production of hybrids. In 2011, a random sample of cars for sale in California included 32 hybrids. In trying to establish whether or not the government program was effective in increasing the availability of hybrid vehicles, statisticians are asked to use the sample to determine if this program has led to a statistically significant increase in the proportion of hybrids. Assuming that the conditions for inference are met, what is the z test statistic for this test?

(A) 0.11
(B) 0.27
(C) 0.68
(D) 1.70
(E) 5.10

STOP

END OF SECTION 1
IF YOU FINISH BEFORE TIME IS CALLED, YOU MAY CHECK YOUR WORK ON THIS SECTION. DO NOT GO ON TO SECTION II UNTIL YOU ARE TOLD TO DO SO.

Section II: Free-Response Questions
Part A
Questions 1–5
Spend about 65 minutes on this part of the exam
Percent of Section II grade—75

Directions: Show all your work. Clearly indicate the methods you use as you will be graded on the correctness of the methods as well as the accuracy of your results and explanation.

FREE-RESPONSE PROBLEMS

1. Rolling Hills Country Club wants to hold a charity golf tournament to benefit a cancer foundation. Thirty-two players will be divided into eight teams of four that will then compete to win prizes. The tournament director would like to create the teams of four in a way that doesn't tend to favor any one team.

   (a) One possibility for creating teams is to let people that usually play together during the year play together on the same team for the tournament. Is this a good idea? Explain why or why not.

   (b) Two other possibilities for creating teams are

   Method 1 (random assignment): Assign the 32 golfers to the eight teams at random.

   Method 2: Order the golfers by ability using their average score for the past 12 months. Create four groups of golfers, with the first group consisting of the golfers ranked 1 – 8 (the eight best golfers), the second group consisting of the golfers ranked 9 – 16, the third group consisting of the golfers ranked 17 – 24, and the fourth group consisting of the golfers ranked 25 – 32 (the eight worst golfers). Form the one team of four by selecting one golfer at random from each of the four groups. Form the second team by selecting one golfer at random from the remaining golfers in each group, and so on.

   Which of these two methods is more likely to create teams that are comparable in terms of ability? Explain your reasoning.

   (c) Describe how you would use a table of random numbers or a random number generator to create the teams using the method you selected in part (b).

2. A spinner at a carnival is divided into 20 segments of equal size, as shown below.

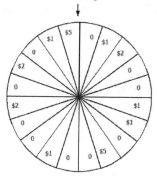

Ten of the segments are labeled "0", five are labeled "$1", three are labeled "$2", and two are labeled "$5". If the spinner is fair, each of the 20 segments is equally likely to be the one that comes to rest at the top of the wheel on any given spin. In order to decide if there is evidence that the spinner is not fair, an investigator records the outcomes of 500 spins, resulting in the data summarized in the following table:

Outcome	0	$1	$2	$5
Frequency	280	140	55	25

Do these data provide convincing evidence that the spinner is not fair?

3. In order to assess the effect of boxing on hearing loss, researchers conducted a study using a large number of retired boxers over age 50 and a large number of adults over 50 who had never boxed. Five distinct tones used in hearing tests were played and the number of these tones that a person was able to hear was recorded. These data were used to obtain the approximate probability distribution for X = number of tones heard for boxers and for non-boxers.

Retired Boxers

Tones heard	0	1	2	3	4	5
Probability	0.10	0.20	0.20	0.25	0.15	0.10

Non-boxers

Tones heard	0	1	2	3	4	5
Probability	0.05	0.10	0.10	0.25	0.20	0.30

(a) Based on these probability distributions, what is the expected value of X for retired boxers over 50?

(b) Suppose that a randomly selected retired boxer identifies two tones and a randomly selected non-boxer identifies four. The difference (retired boxer – non-boxer) in the scores is –2. List all combinations of possible scores for a retired boxer and a non-boxer that will produce a difference of –2.

(c) Find the probability that the difference in the scores for a randomly selected boxer and a randomly selected non-boxer (retired boxer – boxer) is –2.

GO ON TO NEXT PAGE

(d) Based on this hearing test, 30% of retired boxers over 50 and 15% of non-boxers over 50 were classified as having severe hearing loss. How many of the five tones would someone need to hear in order to be classified as <u>not</u> having severe hearing loss?

4. The television game show "Jeopardy" has contestants compete for prize money by answering bits of trivia in the form of a question. After many years on TV with the same dollar values for each question, "Jeopardy" producers decided to double the dollar value of each question. Below are comparative boxplots of the winning contestant's prize money before and after the change in dollar values:

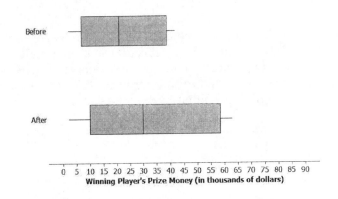

(a) In a few sentences, describe the similarities and differences in the distributions of amount won before and after the change.
(b) Did doubling the dollar value of questions tend to result in greater amounts won? Provide a justification based on the given boxplots.
(c) If doubling the dollar value of the questions tended to produce prize amounts that were twice as large as before the dollar values were doubled, what could you have expected the "after" boxplot to like? Answer by sketching in an "after" boxplot in the figure below.

5. After Mrs. Simmons' middle school students learned that Native Americans used to wake themselves for battle by drinking a lot of water the night before their battle, they wanted to conduct an experiment to see if this really works. Eighty students who volunteered to be a part of the experiment and who received parental approval to participate were the subjects in the experiment. Subjects were assigned at random to one of two treatment groups. Those assigned to the "water" treatment drank 24 ounces of water between 7:30 and 8:00 p.m. Those assigned to the "no water" treatment group did not drink any water after 7:30 p.m. All subjects were instructed to go to bed at 9:00 p.m. and then to record the time when they first woke up. The next day, each subject computed the number of minutes of sleep before waking. The resulting data is summarized below.

Group	n	mean	standard deviation
Water	40	289	56
No water	40	406	53

(a) Let μ_W denote the mean sleep time (in minutes) for the water treatment and μ_N denote the mean sleep time for the no water treatment. What hypotheses should the students test if they want to determine if there is evidence that sleep time for the water treatment is shorter than for the no water treatment?

(b) Carry out an appropriate test for the hypotheses in part (a).

GO ON TO NEXT PAGE

Section II
Part B
Question 6
Spend about 25 minutes on this part of the exam
Percent of Section II grade—25

Directions: Show all your work. Clearly indicate the methods you use as you will be graded on the correctness of the methods as well as the accuracy of your results and explanation.

6. A film executive believes that the days of the epic film are returning and that it is becoming more profitable to create longer movies. To investigate this theory, he recorded the length (in minutes) and the amount of money the movie made (domestic gross, in millions of dollars) for the 12 most profitable movies made between 2001 and 2005 and for the 12 most profitable movies made between 2006 and 2010. The resulting data are shown in the tables below:

2001–2005

Movie	Year	Domestic Gross (millions of dollars)	Length (minutes)
Shrek 2	2004	436	93
Spider-Man	2002	403	121
Star Wars Episode 3: Revenge of the Sith	2005	380	140
The Lord of the Rings: The Return of the King	2003	377	201
Spider-Man 2	2004	373	127
Passion of the Christ	2004	370	127
The Lord of the Rings: The Two Towers	2002	340	179
Finding Nemo	2003	339	100
Harry Potter and the Sorcerer's Stone	2001	317	152
The Lord of the Rings: The Fellowship of the Ring	2001	313	178
Star Wars Episode 2: Attack of the Clones	2002	310	142
The Pirates of the Caribbean: The Curse of the Black Pearl	2003	305	143

2006–2010

Movie	Year	Domestic Gross (millions of dollars)	Length (minutes)
Avatar	2009	754	162
The Dark Knight	2008	533	152
Pirates of the Caribbean: Dead Man's Chest	2006	423	151
Toy Story 3	2010	405	103
Transformers: Revenge of the Fallen	2009	402	150
Spider-Man 3	2007	336	139
Alice in Wonderland	2010	334	108
Shrek the Third	2007	320	93
Transformers	2007	318	144
Iron Man	2008	318	126
Indiana Jones and the Kingdom of the Crystal Skull	2008	317	122
Iron Man 2	2010	312	124

(a) Construct two dotplots, one for 2001–2005 and one for 2006–2010, that would allow you to compare the length distributions of the top 12 movies in each time period.

(b) Based only on the dotplots in part (a), write a few sentences comparing the two movie length distributions. Do the dotplots support the statement that top 12 movies tended to be longer for the time period 2006–2010 than for the time period 2001–2005?

(c) Below are a scatterplot and linear regression output for each of the two time periods separately and for the data set that consists of all 24 films from 2001 to 2010 combined. Does there appear to be a relationship between domestic gross and length? Do you think the relationship between domestic gross and length is the same in each of the two time periods? Support your answer by providing an explanation based on the scatterplots and computer output.

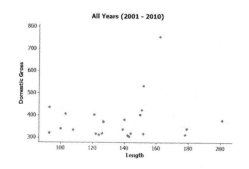

```
The regression equation is
Domestic Gross = 297 + 0.583 Length


Predictor      Coef    SE Coef     T       P
Constant      296.8      103.4   2.87   0.009
Length       0.5834     0.7428   0.79   0.441


S = 97.7172    R-Sq = 2.7%    R-Sq(adj) = 0.0%
```

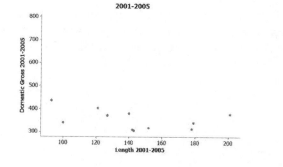

```
The regression equation is
Domestic Gross 2001-2005 = 431 - 0.535 Length 2001-2005


Predictor              Coef   SE Coef     T      P
Constant             431.17     53.76   8.02  0.000
Length 2001-2005    -0.5350    0.3703  -1.44  0.179


S = 39.3706   R-Sq = 17.3%   R-Sq(adj) = 9.0%
```

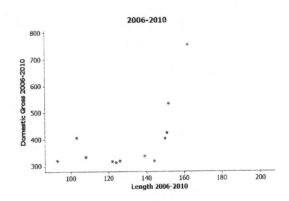

```
The regression equation is
Domestic Gross 2006-2010 = - 69 + 3.56 Length 2006-2010


Predictor              Coef   SE Coef     T      P
Constant             -69.0     199.1   -0.35  0.736
Length 2006-2010     3.558     1.499    2.37  0.039


S = 109.203   R-Sq = 36.0%   R-Sq(adj) = 29.6%
```

(d) Would you recommend using any of the three regression models in part (c) to predict the domestic gross for a new movie coming out this year that was 215 minutes long? If so, which model and why? If not, explain why not.

ANSWERS TO PRACTICE TEST 2

MULTIPLE-CHOICE ANSWERS

Using the table below, score your test.

Determine how many questions you answered correctly and how many you answered incorrectly. You will find explanations of the answers on the following pages.

1. A	9. A	17. E	25. D	33. C
2. A	10. D	18. B	26. C	34. E
3. D	11. C	19. D	27. C	35. D
4. E	12. B	20. B	28. E	36. B
5. D	13. E	21. D	29. D	37. D
6. B	14. E	22. D	30. D	38. E
7. C	15. E	23. C	31. D	39. C
8. E	16. E	24. C	32. C	40. D

ANSWERS AND EXPLANATIONS

MULTIPLE-CHOICE ANSWERS

1. **A.** Because John did not randomly assign days to the early bus and late bus, there is a potential for a confounding variable. While he might conclude that there is an association, he cannot conclude that the lateness was the cause of his hunger; the additional energy expended when he was engaged in sports after school may have contributed to his hunger (*Statistics: Learning From Data*, 1st ed. pages 25 and 29).

2. **A.** Unbiased statistics are generally preferred. When you have a choice between unbiased statistics, the best statistic is the one with the smallest standard deviation (*Statistics: Learning From Data*, 1st ed. pages 19 and 20).

3. **D.** Because the population of students was divided into four subgroups and a random sample was selected from each subgroup, this is an example of stratified sampling. The strata are the four classes (*Statistics: Learning From Data*, 1st ed. page 14).

4. **E.** In order to construct the 95% confidence interval, you must first determine the sample proportion. 114/457 = .25. Using the large sample confidence interval method with a z critical value of 1.96 for a 95% confidence level, you get

$$0.25 \pm 1.96\sqrt{\frac{0.25(1-0.25)}{457}} = (0.21, 0.29)$$ (*Statistics: Learning From Data*, 1st ed. pages 467–480).

5. **D.** Type I error occurs when you reject the null hypothesis, even though the null hypothesis is actually true. If the null hypothesis is that a man is not guilty, then in this case we reject that hypothesis and find him guilty even though he was actually not guilty. Situation 2 represents this Type I error. Type II error occurs when you fail to reject the null hypothesis even though it is false. If the null hypothesis is that a man is not guilty, failing to reject this hypothesis means that you continue to believe he's not guilty. If you have made a Type II error, then you continue to believe the man is not guilty even though he is actually guilty, the situation described in situation 1 (*Statistics: Learning From Data*, 1st ed. pages 502–505).

6. **B.** In order to determine the probability that Herman gets at least one job offer, the best option is to first look at the probability that he does not get any offers and then subtract that from 1. For each of his new applications, there is a probability of 0.80 that he does not get the job. Taking into account that he applied for 3 jobs and that job offers are independent, the probability that he does not get an offer from any of the three is (0.8 × 0.8 × 0.8) = 0.512. Subtracting this from 1 gives the probability of at least one offer of 1 – 0.512 = 0.488 (*Statistics: Learning From Data*, 1st ed. pages 285–287).

7. **C.** By limiting the survey to only the three largest tribes, you systematically exclude all the remaining tribes. This is selection bias. Answer A is an example of nonresponse bias. Answer B is an example of measurement or response bias. Answer D would be a viable sampling method. Answer E includes respondents who are not in the desired population (*Statistics: Learning From Data*, 1st ed. pages 18–20).

8. **E.** To calculate the z-scores, we use the formula $z = (x - \mu)/\sigma$. For Chester this score is $z = (80,000 - 60,000)/5,000 = 4$. For Gary this score is $z = (100,000 - 80,000)/8,000 = 2.5$. Chester is doing better (*Statistics: Learning From Data*, 1st ed. pages 375–385).

9. **A.** To find expected value, you multiply payout by probability and then add. For Fund X the calculation is 0.5(500) + 0.5(2,000) = 1,250. For Fund Y the calculation is 0.33(500) + 0.33(1,000) + 0.34(2,000) = 1175. For Fund Z the calculation is 0.25(500) + 0.50(1,000) + 0.25(2,000) = 1125. The order from highest to lowest expected value is therefore X, Y, Z (*Statistics: Learning From Data*, 1st ed. pages 356–360).

10. **D.** For a standard normal distribution, about 15% of the area is to the right of 1.04. This means that 3 pounds must be 1.04 standard deviations above the mean. Solving $1.04=(3-2)/ \sigma$ or $1.04\sigma = 1$ gives $\sigma = 0.96$ (*Statistics: Learning From Data*, 1st ed. pages 376–380).

11. **C.** The sample was selected from all drivers in the counties closest to the bridge, so this is the population. The sample is those drivers

in these counties that are actually selected for the survey (*Statistics: Learning From Data,* 1st ed. pages 13–14).

12. **B.** Proximity to the bridge could influence the response since those far from the bridge will probably not use the bridge no matter what the cost is (*Statistics: Learning From Data,* 1st ed. pages 18–21).

13. **E.** In order to find the number of worm-infested apples, you first take the percentage of worm-free apples and multiply it by the number of apples in the sample. 25,000(0.982) = 24,550. Subtract this number from the total apples and you get 25,000 – 24,550 = 450 (*Statistics: Learning From Data,* 1st ed. pages 356–360).

14. **E.** Because prices were increasing from 1997 to 2007, the slope of the regression line will be positive. This means the predicted value will increase each year. Because prices actually declined from 2008 to 2011, the predictions would be too large and the prediction would be farther from the actual value in 2011 than in 2008 (*Statistics: Learning From Data,* 1st ed. pages 205–206).

15. **E.** Because the samples are paired, the paired t test would be used. The value of the test statistic is $\dfrac{\bar{x}_d - 0}{\frac{s_d}{\sqrt{n}}} = \dfrac{0.4}{\frac{0.2}{\sqrt{135}}} = 23.24$ (*Statistics: Learning From Data,* 1st ed. pages 633–634).

16. **E.** This question asks for P(Q and N), where Q is the event "quits smoking" and N is the event "no cigarette for five years." This probability can be computed as P(Q and N) = P(Q)P(N|Q) = (0.41)(0.26) = 0.11. (*Statistics: Learning From Data,* 1st ed. pages 285–287 and 290–302).

17. **E.** Because neither the people measuring the response nor the patients know who is in the drug group and who is in the placebo group, the study is double blind (*Statistics: Learning From Data,* 1st ed. page 35).

18. **B.** To calculate the t confidence interval, you use the t critical value of 2.06 for a 95% confidence interval with 25 degrees of freedom. Using the sample mean of 42 minutes, you get a confidence interval of $42 \pm (2.06(15/\sqrt{26})) = (42 - 6, 42 + 6) = (36, 48)$ (*Statistics: Learning From Data,* 1st ed. pages 582, 644, and 645).

19. **D.** To find the proportion of wheelchairs that stop within 3.5 feet, add the relative frequencies corresponding to stopping distances in the first three bars of the histogram to get 0.05 + 0.15 + 0.40 = 0.60 (*Statistics: Learning From Data,* 1st ed. page 62).

20. **B.** If statistical analysis does not provide evidence that the air conditioners are lasting less than 90 days before needing to be serviced, then the null hypothesis has not been rejected. We

therefore fail to reject the null hypothesis (*Statistics: Learning From Data,* 1st ed. pages 591–604).

21. **D.** The population mean is the same as the mean of the sampling distribution. The standard deviation of the sampling distribution is $\frac{\sigma}{\sqrt{n}} = 63$. So $\sigma = (63)(\sqrt{9}) = 189$ (*Statistics: Learning From Data,* 1st ed. pages 568–578).

22. **D.** Because dinosaur weights are approximately normally distributed, we can use the Empirical Rule. This says that approximately 68% (which is about 70%) of the weights will be within 1 standard deviation of the mean weight (*Statistics: Learning From Data,* 1st ed. page 161).

23. **C.** The families that have more than two cars are the ones that have been listed as either having three cars (0.14) or four or more cars (0.06). Adding these numbers gives 0.14 + 0.06 = 0.20. Converting this decimal to a percentage gives an answer of 20% (*Statistics: Learning From Data,* 1st ed. pages 568–578).

24. **C.** There are 72 total people registered for 6 PM classes. Of those 72, 17 are in Saunastenics. This fraction is 17/72 = 0.236, which is a conditional probability (*Statistics: Learning From Data,* 1st ed. pages 290–306).

25. **D.** The total number of possible sandwiches can be found by multiplying $5 \times 7 \times 4 \times 3 = 420$ (*Statistics: Learning From Data,* 1st ed. pages 273).

26. **C.** Because the choir only received 10 responses to the 30 surveys that were sent out, the sample may differ from the corresponding population. Data were not obtained from all the individuals in the simple random sample, thus this is an example of nonresponse bias (*Statistics: Learning From Data,* 1st ed. page 19).

27. **C.** With p denoting the proportion of previous purchasers who would buy the new game with a price of $50 but not with a price of $60, the question of interest is whether $p > 0.05$ (*Statistics: Learning From Data,* 1st ed. pages 507–511).

28. **E.** The mode is the most frequently occurring number. In this case, the number 33 occurs twice in each class for a total of four times between the two classes (*Statistics: Learning From Data,* 1st ed. page 86).

29. **D.** For this outcome to occur, we must have two tosses where both players toss heads or both players toss tails followed by a toss where player A tosses a head and player B tosses a tail. The probability of HH or TT is ½ and the probability of HT is ¼, so the probability of interest is $(1/2)^2(1/4)$ since the events are independent (*Statistics: Learning From Data,* 1st ed. pages 285–287).

30. **D.** In order to calculate the probability of one of the numbers being selected, you take the total number of numbers on the card (24 because you do not count the free space) and divide it by the total number of balls that could be selected (75). 24/75 = 0.32 (*Statistics: Learning From Data,* 1st ed. pages 274–283).

31. **D.** This scatterplot slopes down and to the right, and the pattern is linear. Answer choice C also shows a negative relationship, but it is curved rather than linear (*Statistics: Learning From Data,* 1st ed. pages 96–98).

32. **C.** There is only one possible outcome under option one, so expected value for option one is 25. For option two it is 0(2/3) + 100(1/3) = 33.3. Option one has a smaller expected value. The standard deviation of option one is equal to zero because the amount won is always 25. Option two has a standard deviation that is larger than 0 because there will be variability in the amount won. Option one has a smaller standard deviation (*Statistics: Learning From Data,* 1st ed. pages 356–360).

33. **C.** By cutting 10 employees, you would be making a change in x of −10. Multiplying this −10 by the slope (140), which is the average number of shoes produced in a day by each employee, we get a change in output of (140)(−10) = −1,400. The regression line predicts that shoe production would fall by 1,400 shoes (*Statistics: Learning From Data,* 1st ed. pages 211–213).

34. **E.** Those in the no vaccine group still received an injection so the volunteers would not know who was receiving the drug and who was not. This is an example of a placebo treatment (*Statistics: Learning From Data,* 1st ed. pages 24 and 28).

35. **D.** With a mean of 750, 590, and 910 are each 160 away from the mean. That is two times the standard deviation of 80. According to the empirical rule, approximately 95% of the observations are within two standard deviations of the mean (*Statistics: Learning From Data,* 1st ed. page 161).

36. **B.** The law of large numbers states that as the number of repetitions of a chance experiment increases, the chance that the relative frequency of occurrence for an event will differ from the true probability of the event by more than any small number approaches zero (*Statistics: Learning From Data,* 1st ed. page 271).

37. **D.** To find the probability from this table, you need to find the conditional probability P(patriot|domestic). Look first at the total of the domestic category, and the table entry is 1,418. Then look for how many of those domestic travelers flew on Patriot Air (845). 845/1,418 = 0.596 (*Statistics: Learning From Data,* 1st ed. pages 290–306).

38. **E.** Two ways to increase the power of a test are to increase the significance level of the test and to increase the sample size (*Statistics: Learning From Data,* 1st ed. pages 527–530).

39. **C.** To find a relative frequency from a cumulative relative frequency distribution, you subtract the cumulative probability of the option above from the option below. For 9:00, this number is 0.11, for 9:15 it is 0.12, for 9:30 it is 0.34, for 9:45 it is 0.22, and for 10:00 it is 0.17. Of those numbers, 0.34 is the largest and thus the most likely time for Lafayette to clock in is between 9:16 and 9:30 (*Statistics: Learning From Data,* 1st ed. pages 62, 87, and 88).

40. **D.** The sample proportion is $\hat{p} = \dfrac{32}{300} = 0.1067$. The z test statistic is

$$z = \frac{0.1067 - 0.08}{\sqrt{\dfrac{(0.08)(1-0.08)}{300}}} = \frac{0.0267}{0.0157} = 1.70 \; (\textit{Statistics: Learning From Data,}$$

1st ed. pages 507–511).

SECTION II: FREE-RESPONSE ANSWERS

1. (a) This is not a good idea. People who usually play golf together may be similar in ability and so with this method of creating teams, one team might have all very good golfers and other teams might not have any very good golfers.
   (b) The second method is more likely to create comparable teams because each team will consist of a golfer with a high ranking, a golfer with a low ranking, and two golfers ranked near the middle. With random assignment, it is possible that one team might contain two or more highly ranked golfers.
   (c) Number the golfers from 1 to 32 based on their ranking. Generate a random number between 1 and 8 and assign the corresponding player to Team 1. Generate a random number between 9 and 16 and assign the corresponding player to Team 1. Generate a random number between 17 and 24 and assign the corresponding player to Team 1. Generate a random number between 25 and 32 and assign the corresponding player to Team 1. Team 1 is now complete. Next, form Team 2 by following the same procedure. If any of the random numbers generated correspond to a golfer who has already been assigned to a team, continue to generate random numbers in the appropriate range until a number corresponding to an unassigned golfer is obtained. Continue this process until all teams have been formed.

(*Statistics: Learning From Data,* 1st ed. pages 26–37).

Scoring Question 1:

Each part of this problem can be scored as essentially correct (E), partially correct (P), or incorrect (I).

Part (a) is scored

    E if the response states that this is not a good idea and gives a plausible explanation for why this method might not create comparable teams.

    P if the response gives a plausible explanation for why this method might not create comparable teams, but fails to answer the question about whether or not this is a good idea.

Part (b) is scored

    E if the student chooses Method 2 and provides a clear explanation of why this method is more likely to result in comparable teams.

    P if the student chooses Method 2 but the explanation is weak or poorly communicated.

Part (c) is scored

    E if a correct method is clearly described.

        Note: If Method 1 is chosen in part (b), the description in part (c) should be consistent with that selection. A correct method for random assignment that is clearly described would then be scored as E for part (c).

    P if a correct method is implied, but the description is not complete or is not clearly communicated.

Question 1 is scored a

    4 if all three parts are E

    3 if two parts are E and the one is P

    2 if two parts are E and no parts are P, or one part is E and two parts are P, or if three parts are P

    1 if two parts are P and one part is I or if one part is E and no parts are P

    0 if one or no parts are P

2. This question can be answered using a chi-square goodness-of-fit test. In the long run, if the spinner is fair, it should land on a segment labeled 0 about 10/20 of the time, a segment labeled $1 about 5/20 of the time, a segment labeled $5 about 3/20 of the time and a segment labeled $5 about 2/20 of the time.

Hypotheses:

$H_0 : p_1 = 0.50, \ p_2 = 0.25, \ p_3 = 0.15, \ p_4 = 0.10$

$H_a :$ at least one of the p_i's is different from the value specified in H_0

where

p_1 = the true proportion of the time that the spinner lands on a 0

p_2 = the true proportion of the time that the spinner lands on a $1

p_3 = the true proportion of the time that the spinner lands on a $2

p_4 = the true proportion of the time that the spinner lands on a $5

Test statistic and assumptions:

The test is a chi-square goodness of fit, with test statistic

$$X^2 = \sum \frac{(\text{observed} - \text{expected})^2}{\text{expected}}$$

Check assumptions

(1) Spins are independent, so it is reasonable to regard the 500 observations as if it were a random sample of spins of this spinner.

(2) Large sample. The expected counts are 500(0.50) = 250, 500(0.25) = 125, 500(0.15) = 75 and 500(0.10) = 50. All expected counts are greater than 5, so the sample size is large enough.

Calculations:

$$X^2 = \frac{(280 - 250)^2}{250} + \frac{(140 - 125)^2}{125} + \frac{(55 - 75)^2}{75} + \frac{(25 - 50)^2}{50} = 23.23$$

$df = 4 - 1 = 3$

$P - \text{value} = 0.000$

Conclusion

Because the P-value is so small, e.g., less than 0.001, we reject the null hypothesis and conclude that there is convincing evidence that the spinner is not fair.

(*Statistics: Learning From Data,* 1st ed. pages 698–716).

Scoring Question 2:

There are four parts to this hypothesis test: hypotheses, test and assumptions, calculations, and conclusion. Each of these parts can be scored as essentially correct (E), partially correct (P), or incorrect (I).

Hypotheses are scored as
E if a correct pair of hypotheses is given and the parameters used in the null hypothesis are defined.
P if hypotheses are correct but parameters are not defined OR if the form of the hypotheses is correct but hypothesized proportions other than 0.50, 0.25, 0.15, and 0.10 are specified in the null hypothesis.

Test and assumptions are scored as
E if (1) the correct test is identified either by name or by formula and (2) both assumptions for the test are adequately addressed. P if the response includes only one of the two components required for an E.

Calculations are scored as
E if correct values of the test statistic and P-value are given. A calculator may be used to perform the actual calculations.

P if either the value of the test statistic or the *P*-value is omitted or incorrect.

Note: Calculations should be consistent with the hypothesized values given in the null hypothesis of part (a).

Conclusion is scored as
E if a correct conclusion is given in context and the conclusion is linked to the *P*-value.
P if a correct decision to reject or fail to reject the null hypothesis is given but the conclusion is not in context or the conclusion is not linked to the *P*-value.

Each part scored as essentially correct counts as 1 and each part scored as partially correct counts as ½ to compute an overall score. Then Question 2 is scored a
1 if the overall score is 3 or 3.5
3 if the overall score is 2.5
2 if the overall score is 2 or 1.5
1 if the overall score is 1
0 if the overall score is 0 or 0.5

3. (a) The expected number of tones heard for boxers over 50 is
$E(X) = 0(0.10) + 1(0.20) + 2(0.20) + 3(0.25) + 4(0.15) + 5(0.10) = 2.45$ tones.

(b) Combinations of scores that result in a difference (boxer – non-boxer) of –2 are

Boxer	Non-boxer
3	5
2	4
1	3
0	2

(c) To find the probability that the difference between a randomly selected boxer and a randomly selected non-boxer is –2, we need to compute the probability of observing each of the pairs listed in part (b) and then add these probabilities. Since the selection of the boxer and the selection of the non-boxer are independent,
$P(3,5) = P(\text{boxer} = 3 \text{ and non-boxer} = 5) = (0.25)(0.30) = 0.075$.
Similarly, $P(2,4) = (0.20)(0.25) = 0.50$, $P(1, 3) = (0.20)(0.25) = 0.05$, and $P(0, 2) = (0.10)(0.10) = 0.010$. This gives
$P(\text{difference} = -2) = 0.75 + 0.050 + 0.050 + 0.010 = 0.185$

(d) For boxers, 30% heard only 0 or 1 tone. For non-boxers, 15% heard only 0 or 1 tone. This means that those with severe hearing loss heard only 0 or 1 tone. It follows that someone who heard two or more tones was not classified as having severe hearing loss.

(*Statistics: Learning From Data*, 1st ed. pages 285–287 and 356–360).

Scoring Question 3:

Parts (a) and (b) are scored together. Parts (a+b), (c) and (d) are each scored as essentially correct (E), partially correct (P), or incorrect (I).

Part (a + b) is scored
    E if (1) the expected value is correctly computed in Part (a) and (2) all four pairs of scores that result in a difference of –2 are specified in Part (b)
    P if the response includes only 1 of the components required for an E

Part (c) is scored
    E if a correct probability is given with appropriate supporting work
    P if the probability is incorrect because some but not all of the four possible combinations of scores that result in a difference of –2 were considered
    I if a probability is given but there is no supporting work

Part (d) is scored
    E if the response identifies 2 or more tones (or equivalently more than 1 tone or 2, 3, 4, or 5 tones) as the description of someone who does not have severe hearing loss.
    P if the response only describes sever hearing loss as 0 or 1 tones heard, but does not explicitly describe NOT severe hearing loss.

Question 3 is scored a
    4 if all three parts are E
    3 if two parts are E and the one is P
    2 if two parts are E and no parts are P, or one part is E and two parts are P, or if three parts are P
    1 if two parts are P and one part is I or if one part is E and no parts are P
    0 if one or no parts are P

4. (a) Both distributions appear to be approximately symmetric. The minimum values and lower quartiles are similar for both distributions. The most noticeable difference is in variability, with prize amounts are more variable after the change than before the change. The median prize amount is greater after the change, and there were larger prize amounts after the change.
   (b) Overall, prize amounts tended to be greater after the dollar amounts were doubled. From the boxplot, we can see that the lower quartile, the median and the upper quartile prize amounts are all greater after the change than before the change.
   (c) A boxplot should be consistent with the following: approximate minimum = 4; approximate lower quartile = 14; approximate median = 42; approximate upper quartile = 78; approximate maximum = 90.

(*Statistics: Learning From Data,* 1st ed. pages 151–155).

Scoring Question 4:

Each part of this problem can be scored as essentially correct (E), partially correct (P), or incorrect (I).

Part (a) is scored
    E if the two distributions are compared on the basis of all three of shape, center, and spread.
    P if the distributions are compared on only 2 of shape, center, and spread
    I if shape, center and spread are listed for both distributions, but no comparative statements are made.

Part (b) is scored
    E if the response says that doubling did tend to result in greater amounts won and provides an appropriate justification based on the given boxplots.
    P if the response says that doubling did tend to result in greater amounts won but the justification is weak or is not tied to characteristics of the boxplot.

Part (c) is scored
    E if a box plot is drawn that is consistent with doubling. The drawn box plot should be drawn so that the approximate values used to construct the plot are: Minimum ≈ 4, lower quartile ≈ 14, median ≈ 42, upper quartile ≈ 78, and maximum ≈ 90.
    P if a boxplot is drawn that has similar variability to the boxplot in part (a) but is shifted to the right so that the median of the after boxplot is approximately twice the median of the before boxplot.

Question 4 is scored a
    4 if all three parts are E
    3 if two parts are E and the one is P
    2 if two parts are E and no parts are P, or one part is E and two parts are P, or if three parts are P
    1 if two parts are P and one part is I or if one part is E and no parts are P
    0 if one or no parts are P

5. (a)

$$H_0 : \mu_W - \mu_N = 0 \quad (or \ \mu_W = \mu_N)$$
$$H_a : \mu_W - \mu_N < 0 \quad (or \ \mu_W < \mu_N)$$

    (b)

Test and Assumptions:
Test: two-sample t test

Assumptions:
(1) Random assignment: The question states that subjects were assigned to the treatment groups at random.

(2) Large samples or sleep time distributions approximately normal: both sample sizes were 40, which is larger than 30. The sample sizes are large enough to proceed.

Calculations:

$$t = \frac{\bar{x}_W - \bar{x}_N - 0}{\sqrt{\dfrac{s_W^2}{n_W} + \dfrac{s_N^2}{n_N}}} = \frac{(289 - 406)}{\sqrt{\dfrac{(56)^2}{40} + \dfrac{(53)^2}{40}}} = \frac{117}{12.19} = 9.598$$

$df = 77$

$P-value \approx 0$

Note: df=40-1=39 may also be used.

Conclusions:
Because the P-value is so small, e.g., less than 0.001, we reject the null hypothesis. There is convincing evidence that the mean sleep time for the water treatment is less than the mean sleep time for the no water treatment.

(*Statistics: Learning From Data,* 1st ed. pages 618–633).

Scoring Question 5:

There are four parts to this hypothesis test: hypotheses (part (a)), test and assumptions, calculations, and conclusion. Each of these parts can be scored as essentially correct (E), partially correct (P), or incorrect (I).

Hypotheses are scored as
   E if both hypotheses are correctly specified.
   Note: A null hypothesis of $\mu_W - \mu_N \geq 0$ is acceptable, as long as the equal case is included.
   P if only one of the hypotheses is correctly specified.

Test and assumptions are scored as
   E if (1) the correct test is identified either by name or by formula and (2) both assumptions for the test are adequately addressed.
   P if the response includes only one of the two components required for an E.

Calculations are scored as
   E if correct values of the test statistic and *P*-value are given. If the calculator command is used, the formula must be shown and the statistics in the formula must be correctly identified.
   P if either the value of the test statistic or the *P*-value is omitted or incorrect.

Conclusion is scored as

   E if a correct conclusion is given in context and the conclusion
      is linked to the *P*-value.
   P if a correct decision to reject or fail to reject the null
      hypothesis (based on the *P*-value in the response) is given
      but the conclusion is not in context or the conclusion is not
      linked to the *P*-value.

Each part scored as essentially correct counts as 1 and each part scored as
   partially correct counts as ½ to compute an overall score. Then
   Question 5 is scored a
      1 if the overall score is 3 or 3.5
      3 if the overall score is 2.5
      2 if the overall score is 2 or 1.5
      1 if the overall score is 1
      0 if the overall score is 0 or 0.5

6. (a)

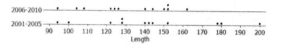

   (b) There is more variability in movie length for the 2001 – 2005
       movies than for the 2006 – 2010 movies. There were a few
       unusually long movies in the 2001–2005 data, and so the mean
       length was probably greater for this group of movies than the
       movies in the 2006–2010 data set. The data do not support the
       statement that the top 12 movies tended to be longer for the 2006–
       2010 time period.
   (c) For the data set that consists of all 24 films, there does not appear
       to be any strong relationship between domestic gross and length.
       Although there is an outlier in the scatterplot (Avatar, with a
       domestic gross of 754 million) there is no apparent pattern and the
       r^2 value is only 0.027, indicating that very little of the variability in
       domestic gross is explained by a linear relationship length.
           Looking at the two time periods separately, there is no
       evidence of a linear relationship for the 2001–2005 time period, and
       the r^2 value for this time period is only 0.173. For the 2006–2010
       time period, the scatterplot does show an interesting pattern, with
       the most profitable movies having the greatest lengths. r^2 is higher
       here at 0.360. However, a nonlinear equation would be a better
       choice for describing the relationship between domestic gross and
       length.
           So, it appears that the relationship between domestic gross
       and length is not the same for the most profitable movies in these
       two time periods.
   (d) No, it would not be a good idea to use any of the three regression
       equations to make a prediction. There are two key reasons why
       this is not a good idea. First, the data sets consisted only of the 12
       most profitable movies in each time period and are not
       representative of all movies made in that time period. Second,
       using any of these regression equations to make a prediction for a
       movie made outside the time period 2001–2010 is not a good idea
       because there is no reason to think that the relationship between

domestic gross and length would be the same for a different time period.

(*Statistics: Learning From Data, 1st ed. pages 206 and 229*).

Scoring Question 6:

Each part of this problem can be scored as essentially correct (E), partially correct (P), or incorrect (I).

Part (a) is scored
> E if correct dotplots are provided, the dotplots are labeled by time period, a scale is indicated, and the same scale is used for both dotplots.
> P if correct dotplots are given with scales indicated, but the dotplots are not labeled by time period and/or different scales are used for the two dotplots.

Part (b) is scored
> E if the two distributions are compared on the basis of center and spread, and the response states that the dotplots do not support the statement that top 12 movies tended to be longer in the 2006–2010 time period.
> P if the distributions are compared on only 1 (center or spread) OR if the distributions are compared on both center and spread but the response does not explicitly address the question of support for the statement that top 12 movies tended to be longer in the 2006–2010 time period.

Part (c) is scored
> E if the response (1) indicates that the relationship is not the same for the two time periods, (2) an appropriate justification based on the scatterplots and regression output is provided, and (3) the response makes note of the nonlinear pattern in the 2006–2010 scatterplot.
> Note: Because the movies are not a random sample from some population of movies, inference about the slope is not meaningful. Regression is merely being used here as a descriptive too. If the response appeals to the *t* test for slope, this can be overlooked as long as this is not the only justification provided.
> P if only 2 of the 3 components required for an E are included in the response.

Part (d) is scored
> E if the response states that it is not a good idea to use any of the regression equations to make the prediction and gives a valid reason to support this statement.
> P if E if the response states that it is not a good idea to use any of the regression equations to make the prediction but does not give a valid reason to support this statement.

Each part scored as essentially correct counts as 1 and each part scored as partially correct counts as ½ to compute an overall score. Then
Question 6 is scored a

    1 if the overall score is 3 or 3.5
    3 if the overall score is 2.5
    2 if the overall score is 2 or 1.5
    1 if the overall score is 1
    0 if the overall score is 0 or 0.5

CALCULATING YOUR SCORE

SECTION I: MULTIPLE-CHOICE QUESTIONS

[_____] × 1.25 = _____
Number Correct   Weighted Section I Score
(out of 40)   (do not round)

SECTION II: FREE-RESPONSE PROBLEMS

Question 1 _____ × (1.875) = _____
(out of 4)   (Do not round)

Question 2 _____ × (1.875) = _____
(out of 4)   (Do not round)

Question 3 _____ × (1.875) = _____
(out of 4)   (Do not round)

Question 4 _____ × (1.875) = _____
(out of 4)   (Do not round)

Question 5 _____ × (1.875) = _____
(out of 4)   (Do not round)

Question 6 _____ × (3.1255) = _____
(out of 4)   (Do not round)

Sum = _____
Weighted Section II Score
(Do not round)

COMPOSITE SCORE

_____ + _____ = _____
Weighted   Weighted   Composite Score
Section I Score   Section II Score (Round to nearest
                                    whole number)

Composite Score Range	Approximate AP Grade
70–100	5
57–69	4
44–56	3
33–43	2
0–32	1

Part IV

Equations and Constants

DESCRIPTIVE STATISTICS

$$\overline{x} = \frac{\sum x_i}{n}$$

$$s_x = \sqrt{\frac{1}{n-1}\sum(x_i - \overline{x})^2}$$

$$s_p = \sqrt{\frac{(n_1 - 1)s_1^2 + (n_2 - 1)s_2^2}{(n_1 - 1) + (n_2 - 1)}}$$

$$\widehat{y} = b_0 + b_1 x$$

$$b_1 = \frac{\sum(x_i - \overline{x})(y_i - \overline{y})}{\sum(x_i - \overline{x})^2}$$

$$b_0 = \overline{y} - b_1 \overline{x}$$

$$r = \frac{1}{n-1}\sum\left(\frac{x_i - \overline{x}}{s_x}\right)\left(\frac{y_i - \overline{y}}{s_y}\right)$$

$$b_1 = r\frac{s_y}{s_x}$$

$$s_{b_1} = \frac{\sqrt{\dfrac{\sum(y_i - \widehat{y}_i)^2}{n-2}}}{\sqrt{\sum(x_i - \overline{x})^2}}$$

PROBABILITY

$$P(A \cup B) = P(A) + P(B) - P(A \cap B)$$

$$P(A \mid B) = \frac{P(A \cap B)}{P(B)}$$

$$E(X) = \mu_x = \sum x_i p_i$$

$$Var(X) = \sigma_x^2 = \sum(x_i - \mu_i)^2 p_i$$

If X has a binomial distribution with parameters n and p, then:

$$P(X = k) = \binom{n}{k}p^k(1-p)^{n-k}$$

$$\mu_x = np$$

$$\sigma_x = \sqrt{np(1-p)}$$

$$\mu_{\widehat{p}} = p$$

$$\sigma_{\widehat{p}} = \sqrt{\frac{p(1-p)}{n}}$$

If $\overline{x}$ is the mean of a random sample of size n from an infinite population with μ and standard deviation σ, then:

$$\mu_{\overline{x}} = \mu$$

$$\sigma_{\overline{x}} = \frac{\sigma}{\sqrt{n}}$$

INFERENTIAL STATISTICS

Standardized test statistic: $\dfrac{\text{statistic} - \text{parameter}}{\text{standard deviation of statistic}}$

Confidence interval: statistic $\pm$ (critical value) $\cdot$ (standard deviation of statistic)

Single-Sample

Statistic	Standard Deviation of Statistic
Sample Mean	$\dfrac{\sigma}{\sqrt{n}}$
Sample Proportion	$\sqrt{\dfrac{p(1-p)}{n}}$

Two-Sample

Statistic	Standard Deviation of Statistic
Difference of sample means	$\sqrt{\dfrac{\sigma_1^2}{n_1} + \dfrac{\sigma_2^2}{n_2}}$ Special case when $\sigma_1 = \sigma_2$ $\sigma\sqrt{\dfrac{1}{n_1} + \dfrac{1}{n_2}}$
Difference of sample proportions	$\sqrt{\dfrac{p_1(1-p_1)}{n_1} + \dfrac{p_2(1-p_2)}{n_2}}$ Special case when $p_1 = p_2$ $\sqrt{p(1-p)}\sqrt{\dfrac{1}{n_1} + \dfrac{1}{n_2}}$

Chi-square test statistic $= \displaystyle\sum \dfrac{(\text{observed} - \text{expected})^2}{\text{expected}}$

Table A: Standard normal probabilities

Table entry for *z* is the probability lying below z.

Probability

z

z	.00	.01	.02	.03	.04	.05	.06	.07	.08	.09
-3.4	.0003	.0003	.0003	.0003	.0003	.0003	.0003	.0003	.0003	.0002
-3.3	.0005	.0005	.0005	.0004	.0004	.0004	.0004	.0004	.0004	.0003
-3.2	.0007	.0007	.0006	.0006	.0006	.0006	.0006	.0005	.0005	.0005
-3.1	.0010	.0009	.0009	.0009	.0008	.0008	.0008	.0008	.0007	.0007
-3.0	.0013	.0013	.0013	.0012	.0012	.0011	.0011	.0011	.0010	.0010
-2.9	.0019	.0018	.0018	.0017	.0016	.0016	.0015	.0015	.0014	.0014
-2.8	.0026	.0025	.0024	.0023	.0023	.0022	.0021	.0021	.0020	.0019
-2.7	.0035	.0034	.0033	.0032	.0031	.0030	.0029	.0028	.0027	.0026
-2.6	.0047	.0045	.0044	.0043	.0041	.0040	.0039	.0038	.0037	.0036
-2.5	.0062	.0060	.0059	.0057	.0055	.0054	.0052	.0051	.0049	.0048
-2.4	.0082	.0080	.0078	.0075	.0073	.0071	.0069	.0068	.0066	.0064
-2.3	.0107	.0104	.0102	.0099	.0096	.0094	.0091	.0089	.0087	.0084
-2.2	.0139	.0136	.0132	.0129	.0125	.0122	.0119	.0116	.0113	.0110
-2.1	.0179	.0174	.0170	.0166	.0162	.0158	.0154	.0150	.0146	.0143
-2.0	.0228	.0222	.0217	.0212	.0207	.0202	.0197	.0192	.0188	.0183
-1.9	.0287	.0281	.0274	.0268	.0262	.0256	.0250	.0244	.0239	.0233
-1.8	.0359	.0351	.0344	.0336	.0329	.0322	.0314	.0307	.0301	.0294
-1.7	.0446	.0436	.0427	.0418	.0409	.0401	.0392	.0384	.0375	.0367
-1.6	.0548	.0537	.0526	.0516	.0505	.0495	.0485	.0475	.0465	.0455
-1.5	.0668	.0655	.0643	.0630	.0618	.0606	.0594	.0582	.0571	.0559
-1.4	.0808	.0793	.0778	.0764	.0749	.0735	.0721	.0708	.0694	.0681
-1.3	.0968	.0951	.0934	.0918	.0901	.0885	.0869	.0853	.0838	.0823
-1.2	.1151	.1131	.1112	.1093	.1075	.1056	.1038	.1020	.1003	.0985
-1.1	.1357	.1335	.1314	.1292	.1271	.1251	.1230	.1210	.1190	.1170
-1.0	.1587	.1562	.1539	.1515	.1492	.1469	.1446	.1423	.1401	.1379
-0.9	.1841	.1814	.1788	.1762	.1736	.1711	.1685	.1660	.1635	.1611
-0.8	.2119	.2090	.2061	.2033	.2005	.1977	.1949	.1922	.1894	.1867
-0.7	.2420	.2389	.2358	.2327	.2296	.2266	.2236	.2206	.2177	.2148
-0.6	.2743	.2709	.2676	.2643	.2611	.2578	.2546	.2514	.2483	.2451
-0.5	.3085	.3050	.3015	.2981	.2946	.2912	.2877	.2843	.2810	.2776
-0.4	.3446	.3409	.3372	.3336	.3300	.3264	.3228	.3192	.3156	.3121
-0.3	.3821	.3783	.3745	.3707	.3669	.3632	.3594	.3557	.3520	.3483
-0.2	.4207	.4168	.4129	.4090	.4052	.4013	.3974	.3936	.3897	.3859
-0.1	.4602	.4562	.4522	.4483	.4443	.4404	.4364	.4325	.4286	.4247
-0.0	.5000	.4960	.4920	.4880	.4840	.4801	.4761	.4721	.4681	.4641

Table A: Standard normal probabilities *(Continued)*

Table entry for z is the probability lying below z.

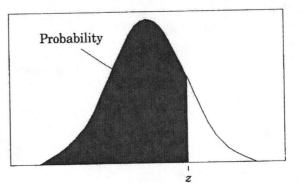

Probability

z	.00	.01	.02	.03	.04	.05	.06	.07	.08	.09
0.0	.5000	.5040	.5080	.5120	.5160	.5199	.5239	.5279	.5319	.5359
0.1	.5398	.5438	.5478	.5517	.5557	.5596	.5636	.5675	.5714	.5753
0.2	.5793	.5832	.5871	.5910	.5948	.5987	.6026	.6064	.6103	.6141
0.3	.6179	.6217	.6255	.6293	.6331	.6368	.6406	.6443	.6480	.6517
0.4	.6554	.6591	.6628	.6664	.6700	.6736	.6772	.6808	.6844	.6879
0.5	.6915	.6950	.6985	.7019	.7054	.7088	.7123	.7157	.7190	.7224
0.6	.7257	.7291	.7324	.7357	.7389	.7422	.7454	.7486	.7517	.7549
0.7	.7580	.7611	.7642	.7673	.7704	.7734	.7764	.7794	.7823	.7852
0.8	.7881	.7910	.7939	.7967	.7995	.8023	.8051	.8078	.8106	.8133
0.9	.8159	.8186	.8212	.8238	.8264	.8289	.8315	.8340	.8365	.8389
1.0	.8413	.8438	.8461	.8485	.8508	.8531	.8554	.8577	.8599	.8621
1.1	.8643	.8665	.8686	.8708	.8729	.8749	.8770	.8790	.8810	.8830
1.2	.8849	.8869	.8888	.8907	.8925	.8944	.8962	.8980	.8997	.9015
1.3	.9032	.9049	.9066	.9082	.9099	.9115	.9131	.9147	.9162	.9177
1.4	.9192	.9207	.9222	.9236	.9251	.9265	.9279	.9292	.9306	.9319
1.5	.9332	.9345	.9357	.9370	.9382	.9394	.9406	.9418	.9429	.9441
1.6	.9452	.9463	.9474	.9484	.9495	.9505	.9515	.9525	.9535	.9545
1.7	.9554	.9564	.9573	.9582	.9591	.9599	.9608	.9616	.9625	.9633
1.8	.9641	.9649	.9656	.9664	.9671	.9678	.9686	.9693	.9699	.9706
1.9	.9713	.9719	.9726	.9732	.9738	.9744	.9750	.9756	.9761	.9767
2.0	.9772	.9778	.9783	.9788	.9793	.9798	.9803	.9808	.9812	.9817
2.1	.9821	.9826	.9830	.9834	.9838	.9842	.9846	.9850	.9854	.9857
2.2	.9861	.9864	.9868	.9871	.9875	.9878	.9881	.9884	.9887	.9890
2.3	.9893	.9896	.9898	.9901	.9904	.9906	.9909	.9911	.9913	.9916
2.4	.9918	.9920	.9922	.9925	.9927	.9929	.9931	.9932	.9934	.9936
2.5	.9938	.9940	.9941	.9943	.9945	.9946	.9948	.9949	.9951	.9952
2.6	.9953	.9955	.9956	.9957	.9959	.9960	.9961	.9962	.9963	.9964
2.7	.9965	.9966	.9967	.9968	.9969	.9970	.9971	.9972	.9973	.9974
2.8	.9974	.9975	.9976	.9977	.9977	.9978	.9979	.9979	.9980	.9981
2.9	.9981	.9982	.9982	.9983	.9984	.9984	.9985	.9985	.9986	.9986
3.0	.9987	.9987	.9987	.9988	.9988	.9989	.9989	.9989	.9990	.9990
3.1	.9990	.9991	.9991	.9991	.9992	.9992	.9992	.9992	.9993	.9993
3.2	.9993	.9993	.9994	.9994	.9994	.9994	.9994	.9995	.9995	.9995
3.3	.9995	.9995	.9995	.9996	.9996	.9996	.9996	.9996	.9996	.9997
3.4	.9997	.9997	.9997	.9997	.9997	.9997	.9997	.9997	.9997	.9998

Table B: *t* distribution critical values

Table entry for p and C is the point t^* with probability p lying above it and probability C lying between $-t^*$ and t^*.

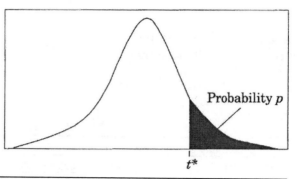

Probability p

t^*

df	Tail probability p											
	.25	.20	.15	.10	.05	.025	.02	.01	.005	.0025	.001	.0005
1	1.000	1.376	1.963	3.078	6.314	12.71	15.89	31.82	63.66	127.3	318.3	636.6
2	.816	1.061	1.386	1.886	2.920	4.303	4.849	6.965	9.925	14.09	22.33	31.60
3	.765	.978	1.250	1.638	2.353	3.182	3.482	4.541	5.841	7.453	10.21	12.92
4	.741	.941	1.190	1.533	2.132	2.776	2.999	3.747	4.604	5.598	7.173	8.610
5	.727	.920	1.156	1.476	2.015	2.571	2.757	3.365	4.032	4.773	5.893	6.869
6	.718	.906	1.134	1.440	1.943	2.447	2.612	3.143	3.707	4.317	5.208	5.959
7	.711	.896	1.119	1.415	1.895	2.365	2.517	2.998	3.499	4.029	4.785	5.408
8	.706	.889	1.108	1.397	1.860	2.306	2.449	2.896	3.355	3.833	4.501	5.041
9	.703	.883	1.100	1.383	1.833	2.262	2.398	2.821	3.250	3.690	4.297	4.781
10	.700	.879	1.093	1.372	1.812	2.228	2.359	2.764	3.169	3.581	4.144	4.587
11	.697	.876	1.088	1.363	1.796	2.201	2.328	2.718	3.106	3.497	4.025	4.437
12	.695	.873	1.083	1.356	1.782	2.179	2.303	2.681	3.055	3.428	3.930	4.318
13	.694	.870	1.079	1.350	1.771	2.160	2.282	2.650	3.012	3.372	3.852	4.221
14	.692	.868	1.076	1.345	1.761	2.145	2.264	2.624	2.977	3.326	3.787	4.140
15	.691	.866	1.074	1.341	1.753	2.131	2.249	2.602	2.947	3.286	3.733	4.073
16	.690	.865	1.071	1.337	1.746	2.120	2.235	2.583	2.921	3.252	3.686	4.015
17	.689	.863	1.069	1.333	1.740	2.110	2.224	2.567	2.898	3.222	3.646	3.965
18	.688	.862	1.067	1.330	1.734	2.101	2.214	2.552	2.878	3.197	3.611	3.922
19	.688	.861	1.066	1.328	1.729	2.093	2.205	2.539	2.861	3.174	3.579	3.883
20	.687	.860	1.064	1.325	1.725	2.086	2.197	2.528	2.845	3.153	3.552	3.850
21	.686	.859	1.063	1.323	1.721	2.080	2.189	2.518	2.831	3.135	3.527	3.819
22	.686	.858	1.061	1.321	1.717	2.074	2.183	2.508	2.819	3.119	3.505	3.792
23	.685	.858	1.060	1.319	1.714	2.069	2.177	2.500	2.807	3.104	3.485	3.768
24	.685	.857	1.059	1.318	1.711	2.064	2.172	2.492	2.797	3.091	3.467	3.745
25	.684	.856	1.058	1.316	1.708	2.060	2.167	2.485	2.787	3.078	3.450	3.725
26	.684	.856	1.058	1.315	1.706	2.056	2.162	2.479	2.779	3.067	3.435	3.707
27	.684	.855	1.057	1.314	1.703	2.052	2.158	2.473	2.771	3.057	3.421	3.690
28	.683	.855	1.056	1.313	1.701	2.048	2.154	2.467	2.763	3.047	3.408	3.674
29	.683	.854	1.055	1.311	1.699	2.045	2.150	2.462	2.756	3.038	3.396	3.659
30	.683	.854	1.055	1.310	1.697	2.042	2.147	2.457	2.750	3.030	3.385	3.646
40	.681	.851	1.050	1.303	1.684	2.021	2.123	2.423	2.704	2.971	3.307	3.551
50	.679	.849	1.047	1.299	1.676	2.009	2.109	2.403	2.678	2.937	3.261	3.496
60	.679	.848	1.045	1.296	1.671	2.000	2.099	2.390	2.660	2.915	3.232	3.460
80	.678	.846	1.043	1.292	1.664	1.990	2.088	2.374	2.639	2.887	3.195	3.416
100	.677	.845	1.042	1.290	1.660	1.984	2.081	2.364	2.626	2.871	3.174	3.390
1000	.675	.842	1.037	1.282	1.646	1.962	2.056	2.330	2.581	2.813	3.098	3.300
<x>	.674	.841	1.036	1.282	1.645	1.960	2.054	2.326	2.576	2.807	3.091	3.291
	50%	60%	70%	80%	90%	95%	96%	98%	99%	99.5%	99.8%	99.9%

Confidence level C

Table C χ^2 critical values

Table entry for p is the point (χ^2) with probability p lying above it

Probability p

df	Tail probability p										
	.25	.20	.15	.10	.05	.025	.02	.01	.005	.0025	.001
1	1.32	1.64	2.07	2.71	3.84	5.02	5.41	6.63	7.88	9.14	10.83
2	2.77	3.22	3.79	4.61	5.99	7.38	7.82	9.21	10.60	11.98	13.82
3	4.11	4.64	5.32	6.25	7.81	9.35	9.84	11.34	12.84	14.32	16.27
4	5.39	5.99	6.74	7.78	9.49	11.14	11.67	13.28	14.86	16.42	18.47
5	6.63	7.29	8.12	9.24	11.07	12.83	13.39	15.09	16.75	18.39	20.51
6	7.84	8.56	9.45	10.64	12.59	14.45	15.03	16.81	18.55	20.25	22.46
7	9.04	9.80	10.75	12.02	14.07	16.01	16.62	18.48	20.28	22.04	24.32
8	10.22	11.03	12.03	13.36	15.51	17.53	18.17	20.09	21.95	23.77	26.12
9	11.39	12.24	13.29	14.68	16.92	19.02	19.68	21.67	23.59	25.46	27.88
10	12.55	13.44	14.53	15.99	18.31	20.48	21.16	23.21	25.19	27.11	29.59
11	13.70	14.63	15.77	17.28	19.68	21.92	22.62	24.72	26.76	28.73	31.26
12	14.85	15.81	16.99	18.55	21.03	23.34	24.05	26.22	28.30	30.32	32.91
13	15.98	16.98	18.20	19.81	22.36	24.74	25.47	27.69	29.82	31.88	34.53
14	17.12	18.15	19.41	21.06	23.68	26.12	26.87	29.14	31.32	33.43	36.12
15	18.25	19.31	20.60	22.31	25.00	27.49	28.26	30.58	32.80	34.95	37.70
16	19.37	20.47	21.79	23.54	26.30	28.85	29.63	32.00	34.27	36.46	39.25
17	20.49	21.61	22.98	24.77	27.59	30.19	31.00	33.41	35.72	37.95	40.79
18	21.60	22.76	24.16	25.99	28.87	31.53	32.35	34.81	37.16	39.42	42.31
19	22.72	23.90	25.33	27.20	30.14	32.85	33.69	36.19	38.58	40.88	43.82
20	23.83	25.04	26.50	28.41	31.41	34.17	35.02	37.57	40.00	42.34	45.31
21	24.93	26.17	27.66	29.62	32.67	35.48	36.34	38.93	41.40	43.78	46.80
22	26.04	27.30	28.82	30.81	33.92	36.78	37.66	40.29	42.80	45.20	48.27
23	27.14	28.43	29.98	32.01	35.17	38.08	38.97	41.64	44.18	46.62	49.73
24	28.24	29.55	31.13	33.20	36.42	39.36	40.27	42.98	45.56	48.03	51.18
25	29.34	30.68	32.28	34.38	37.65	40.65	41.57	44.31	46.93	49.44	52.62
26	30.43	31.79	33.43	35.56	38.89	41.92	42.86	45.64	48.29	50.83	54.05
27	31.53	32.91	34.57	36.74	40.11	43.19	44.14	46.96	49.64	52.22	55.48
28	32.62	34.03	35.71	37.92	41.34	44.46	45.42	48.28	50.99	53.59	56.89
29	33.71	35.14	36.85	39.09	42.56	45.72	46.69	49.59	52.34	54.97	58.30
30	34.80	36.25	37.99	40.26	43.77	46.98	47.96	50.89	53.67	56.33	59.70
40	45.62	47.27	49.24	51.81	55.76	59.34	60.44	63.69	66.77	69.70	73.40
50	56.33	58.16	60.35	63.17	67.50	71.42	72.61	76.15	79.49	82.66	86.66
60	66.98	68.97	71.34	74.40	79.08	83.30	84.58	88.38	91.95	95.34	99.61
80	88.13	90.41	93.11	96.58	101.9	106.6	108.1	112.3	116.3	120.1	124.8
100	109.1	111.7	114.7	118.5	124.3	129.6	131.1	135.8	140.2	144.3	149.4